Techniques and Applications of UAV-Based Photogrammetric 3D Mapping

Techniques and Applications of UAV-Based Photogrammetric 3D Mapping

Editors

Wanshou Jiang
San Jiang
Xiongwu Xiao

MDPI • Basel • Beijing • Wuhan • Barcelona • Belgrade • Manchester • Tokyo • Cluj • Tianjin

Editors

Wanshou Jiang
State Key Laboratory of
Information Engineering in
Surveying, Mapping and
Remote Sensing,
Wuhan University,
Wuhan, China

San Jiang
School of Computer Science,
China University
of Geosciences,
Wuhan 430074, China

Xiongwu Xiao
State Key Laboratory of
Information Engineering in
Surveying, Mapping and
Remote Sensing,
Wuhan University,
Wuhan 430072, China

Editorial Office
MDPI
St. Alban-Anlage 66
4052 Basel, Switzerland

This is a reprint of articles from the Special Issue published online in the open access journal *Remote Sensing* (ISSN 2072-4292) (available at: https://www.mdpi.com/journal/remotesensing/special_issues/UAV_Image).

For citation purposes, cite each article independently as indicated on the article page online and as indicated below:

LastName, A.A.; LastName, B.B.; LastName, C.C. Article Title. *Journal Name* **Year**, *Volume Number*, Page Range.

ISBN 978-3-0365-5067-1 (Hbk)
ISBN 978-3-0365-5068-8 (PDF)

Cover image courtesy of San Jiang.

Contents

About the Editors vii

Preface to "Techniques and Applications of UAV-Based Photogrammetric 3D Mapping" . . . ix

Wanshou Jiang, San Jiang and Xiongwu Xiao
Editorial on Special Issue "Techniques and Applications of UAV-Based Photogrammetric 3D
Mapping"
Reprinted from: *Remote Sens.* **2022**, *14*, 3804, doi:10.3390/rs14153804 . . . . **1**

Guobiao Yao, Alper Yilmaz, Fei Meng and Li Zhang
Review of Wide-Baseline Stereo Image Matching Based on Deep Learning
Reprinted from: *Remote Sens.* **2021**, *13*, 3247, doi:10.3390/rs13163247 . . . . **5**

Wei Huang, San Jiang and Wanshou Jiang
Camera Self-Calibration with GNSS Constrained Bundle Adjustment for Weakly Structured
Long Corridor UAV Images
Reprinted from: *Remote Sens.* **2021**, *13*, 4222, doi:10.3390/rs13214222 . . . . **27**

Teng Xiao, Qingsong Yan, Weile Ma and Fei Deng
Progressive Structure from Motion by Iteratively Prioritizing and Refining Match Pairs
Reprinted from: *Remote Sens.* **2021**, *13*, 2340, doi:10.3390/rs13122340 . . . . **51**

Wei Huang, San Jiang, Sheng He and Wanshou Jiang
Accelerated Multi-View Stereo for 3D Reconstruction of Transmission Corridor with Fine-Scale
Power Line
Reprinted from: *Remote Sens.* **2021**, *13*, 4097, doi:10.3390/rs13204097 . . . . **75**

Liyang Zhou, Zhuang Zhang, Hanqing Jiang, Han Sun, Hujun Bao and Guofeng Zhang
DP-MVS: Detail Preserving Multi-View Surface Reconstruction of Large-Scale Scenes
Reprinted from: *Remote Sens.* **2021**, *13*, 4569, doi:10.3390/rs13224569 . . . . **99**

Yunsheng Zhang, Chi Zhang, Siyang Chen and Xueye Chen
Automatic Reconstruction of Building Façade Model from Photogrammetric Mesh Model
Reprinted from: *Remote Sens.* **2021**, *13*, 3801, doi:10.3390/rs13193801 . . . . **119**

**Haiqing He, Jing Yu, Penggen Cheng, Yuqian Wang, Yufeng Zhu, Taiqing Lin and Guoqiang
Da**
Automatic, Multiview, Coplanar Extraction for CityGML Building Model Texture Mapping
Reprinted from: *Remote Sens.* **2022**, *14*, 50, doi:10.3390/rs14010050 . . . . **135**

**Jingyuan Wang, Xinli Hu, Qingyan Meng, Linlin Zhang, Chengyi Wang, Xiangchen Liu and
Maofan Zhao**
Developing a Method to Extract Building 3D Information from GF-7 Data
Reprinted from: *Remote Sens.* **2021**, *13*, 4532, doi:10.3390/rs13224532 . . . . **155**

Yuntao Wang, Lin Zhao, Liman Liu, Huaifei Hu and Wenbing Tao
URNet: A U-Shaped Residual Network for Lightweight Image Super-Resolution
Reprinted from: *Remote Sens.* **2021**, *13*, 3848, doi:10.3390/rs13193848 . . . . **175**

Liman Liu, Jinjin Yu, Longyu Tan, Wanjuan Su, Lin Zhao and Wenbing Tao
Semantic Segmentation of 3D Point Cloud Based on Spatial Eight-Quadrant Kernel Convolution
Reprinted from: *Remote Sens.* **2021**, *13*, 3140, doi:10.3390/rs13163140 . . . . **197**

Shuhao Ran, Xianjun Gao, Yuanwei Yang, Shaohua Li, Guangbin Zhang and Ping Wang
Building Multi-Feature Fusion Refined Network for Building Extraction from High-Resolution
Remote Sensing Images
Reprinted from: *Remote Sens.* **2021**, *13*, 2794, doi:10.3390/rs13142794 . . . . **213**

Pingbo Hu, Yiming Miao and Miaole Hou
Reconstruction of Complex Roof Semantic Structures from 3D Point Clouds Using Local
Convexity and Consistency
Reprinted from: *Remote Sens.* **2021**, *13*, 1946, doi:10.3390/rs13101946 . . . . **237**

Junliang Zheng, Wanqiang Yao, Xiaohu Lin, Bolin Ma and Lingxiao Bai
An Accurate Digital Subsidence Model for Deformation Detection of Coal Mining Areas Using
a UAV-Based LiDAR
Reprinted from: *Remote Sens.* **2022**, *14*, 421, doi:10.3390/rs14020421 . . . . **263**

About the Editors

Wanshou Jiang

Wanshou Jiang received his bachelor and master degrees in photogrammetry and remote sensing from Wuhan Technical University of Surveying and Mapping respectively in 1989 and 1996. In 2004, he received the PhD degree in photogrammetry and remote sensing from Wuhan University. He started his research career in 1989 as a software developer in analytical photogrammetry. In 2000, he joined the LIESMARS (the State Key Laboratory of Information Engineering in Surveying, Mapping and Remote Sensing) as an associate researcher and then he got the tenure position of researcher in 2005. His research interest includes image registering, image classification, change detection, 3D reconstruction, etc. He made a lot of contribution to the famous digital photogrammetric workstation VirtuoZo and designed a software platform, named OpenRS, for remote sensing image processing.

San Jiang

San Jiang received the B.S. degree in remote sensing science and technology from Wuhan University in 2010, and the M.Sc. and Ph.D. degrees in photogrammetry and remote sensing from Wuhan Univeristy in 2012 and 2018, respectively. From 2012 to 2014, he worked as an assistant engineer in Tianjin Institute of Surveying and Mapping. From 2014 to 2015, he joined the LIESMARS (State Key Laboratory of Information Engineering in Surveying, Mapping and Remote Sensing of Wuhan Univeristy) as a research assistant. Currently, he is an associate professor in the School of Computer Science at China University of GeoSciences (Wuhan). His research interests include image matching, SfM-based aerial triangulation, and 3D reconstruction.

Xiongwu Xiao

Xiongwu Xiao received the B.S. degree in geographic information system from Yangtze University in 2011, and the Ph.D. degree in photogrammetry and remote sensing from Wuhan Univeristy in 2018. From 2016 to 2017, he worked as a visiting scholar and joint PhD student in applied mathematics at the University of California at Irvine. In 2018, he joined the LIESMARS (State Key Laboratory of Information Engineering in Surveying, Mapping and Remote Sensing of Wuhan Univeristy) as an assistant research fellow, and he is an associate research professor at the LIESMARS currently. His research interests include SLAM and real-time photogrammetry, multi-source data fusion, 3D reconstruction, building extraction and intelligent 3D mapping. He designed a real-time mapping system, named DirectMap, for UAV image real-time processing and the real-time generation of DOM in April 2019.

Preface to "Techniques and Applications of UAV-Based Photogrammetric 3D Mapping"

In the last decade, unmanned aerial vehicles (UAVs) have become one of the most important remote sensing data sources for photogrammetric 3D mapping. This special issue focuses on the techniques for UAV-based 3D mapping and hopes to give a review of recent development, especially for trajectory planning for UAV data acquisition in complex environments, recent algorithms for feature matching of aerial-ground images, SfM and SLAM for efficient image orientation, as well as the usage of DL techniques in the 3D mapping pipeline. For the construction of this special issue, we really appreciate the authors who contribute their valuable work and the Editors for their Passionate assistance, which are the base for the successful organization of this special issue.

Wanshou Jiang, San Jiang, and Xiongwu Xiao
Editors

remote sensing

Editorial

Editorial on Special Issue "Techniques and Applications of UAV-Based Photogrammetric 3D Mapping"

Wanshou Jiang [1,*], San Jiang [2] and Xiongwu Xiao [1]

[1] State Key Laboratory of Information Engineering in Surveying, Mapping and Remote Sensing, Wuhan University, Wuhan 430072, China
[2] School of Computer Science, China University of Geosciences, Wuhan 430074, China
* Correspondence: jws@whu.edu.cn

Citation: Jiang, W.; Jiang, S.; Xiao, X. Editorial on Special Issue "Techniques and Applications of UAV-Based Photogrammetric 3D Mapping". *Remote Sens.* **2022**, *14*, 3804. https://doi.org/10.3390/rs14153804

Received: 5 August 2022
Accepted: 5 August 2022
Published: 7 August 2022

Publisher's Note: MDPI stays neutral with regard to jurisdictional claims in published maps and institutional affiliations.

1. Introduction

Recently, 3D mapping has begun to play an increasingly important role in photogrammetric applications. In the last decade, unmanned aerial vehicle (UAV) images have become one of the most critical remote sensing data sources because of the high flexibility of UAV platforms and the extensive usage of low-cost cameras. The techniques and applications of UAV-based photogrammetric 3D mapping are undergoing explosive development, which can be observed from the adopted cutting-edge techniques, including SfM (Structure from Motion) for offline image orientation, SLAM (Simultaneous Localization and Mapping) for online UAV navigation, and the deep learning (DL) embedded 3D reconstruction pipeline.

This Special Issue includes a collection of papers that mainly focus on the techniques and applications of UAV-based 3D mapping. There are a total of 13 papers published in this Special Issue, which range from review papers on recent techniques to research papers for feature detection and matching, false match removal, camera self-calibration, SfM-based image orientation, MVS-based (Multi-view Stereo) dense point cloud generation, building façade model reconstruction, and other related applications in varying fields. The details of each paper will be described in the following section.

2. Overview of Contributions

Yao et al. [1] gave a review of recently reported learning-based methods for wide-baseline image matching, which includes approaches involving feature detection, feature description, and end-to-end image matching. By using benchmark datasets, some typical methods have also been evaluated in this study. The paper reveals that no algorithm can adapt to all wide-baseline images and the generalization ability of learning-based methods should be improved by expanding training data or combining different model design strategies.

For robust and accurate image orientation, Huang et al. [2] proposed a camera self-calibration solution for long-corridor UAV images, such as transmission lines. The proposed solution combines two novel strategies for parameter initialization and high-precision GNSS fusion, in which the former is implemented by an iterative camera parameter optimization algorithm, and the latter is achieved by inequality constrained bundle adjustment. The validation of the proposed solution was verified by using four UAV images that are recorded from transmission corridors. The experimental results demonstrate that the proposed solution can alleviate the "bowl effect" for weakly structured long-corridor UAV images and achieve high precision in absolute orientation when compared with other methods.

In SfM-based image orientation, match pair selection is a key step, which can improve the efficiency of feature matching and decrease the involvement of false matches. In the work of Xiao et al. [3], a progressive structure-from-motion technique was designed to cope with false match pairs retained from repetitive patterns and short baseline images, which iteratively selects initial matches by extracting minimum spanning trees and cycle consistency inference. They verified the validation of the proposed algorithm by using UAV images.

After SfM-based image orientation, multi-view stereo is used to resume dense point clouds. Considering the 3D reconstruction of fine-scale power lines, Huang et al. [4] designed an efficient PatchMatch-based dense matching algorithm, which improves the steps of random red–black checkerboard propagation, matching cost computation, and depth map fusion. When compared with the traditional PatchMatch algorithm, speedup ratios ranging from 4 to 7 were achieved in the tests for transmission corridor UAV images. In addition, the proposed algorithm can improve the completeness of reconstructed power towers and lines.

In contrast to dense matching of normal objects, Zhou et al. [5] proposed a dense matching algorithm, termed DP-MVS, for detail-preserving 3D reconstruction. DP-MVS is achieved by using detail-preserving PatchMatch for the depth estimation of individual images and detail-aware surface meshing to reconstruct final models. The proposed algorithm can cope with the 3D modeling of thin objects, such as communication towers and transmission corridors, and it is 4 times faster than other methods in the dense matching of benchmark datasets.

Zhang et al. [6] presented a newly developed method for automatically generating 3D regular building façade models from the photogrammetric mesh model using the contour as the main cue. The contours tracked on the mesh are grouped into trees and segmented into groups to represent a topological relationship of building components. Then, each component of the mesh is iteratively abstracted into cuboids and the parameters of each cuboid are adjusted to be close to the original mesh model.

Wang et al. [7] proposed a U-Shaped Residual Network for Lightweight Image Super-Resolution (URNet), which applies to low-computing-power or portable devices. Firstly, a more effective feature distillation pyramid residual group (FDPRG) is proposed to extract features from low-resolution images. Then, a step-by-step fusion strategy is utilized to fuse the features of different blocks and further refine the learned features. To capture the global context information, a lightweight asymmetric non-local residual block is introduced. In addition, to alleviate the problem of smoothing image details caused by pixel-wise loss, a simple but effective high-frequency loss function is designed to help optimize the model.

In their study, Wang et al. [8] developed a workflow to extract building 3D information from GF-7 multi-view images. The workflow consists of four main steps, namely building footprint extraction from multi-spectral images, point cloud generation from the stereo image pair with SGM matching, normalized digital surface model (nDSM) generated from the point cloud, and building height calculation. Among the four steps, the main contribution is the multi-stage attention U-Net (MSAU-Net) designed for building footprint extraction. The experiments based on a study area in Beijing show the RMSE between the estimated building height and the reference building height is 5.42 m, and the MAE is 3.39 m.

The study by He et al. [9] proposed a novel approach to achieve CityGML building model texture mapping by multi-view coplanar extraction from UAV or terrestrial images. They first utilized a deep convolutional neural network to filter out object occlusion (e.g., pedestrians, vehicles, and trees) and obtain building-texture distribution. Then, point-line-based features are extracted to characterize multi-view coplanar textures in a 2D space under the constraint of a homography matrix, and geometric topology is subsequently conducted to optimize the boundary of textures by combining Hough-transform and iterative least-squares methods. This approach can map the texture of 2D terrestrial images to building façades without the requirement of exterior orientation information.

To deal with the problem that some existing semantic segmentation networks for 3D point clouds generally have poor performance on small objects, Liu et al. [10] presented a Spatial Eight-Quadrant Kernel Convolution (SEQKC) algorithm to enhance the ability of the network for extracting fine-grained features from 3D point clouds. Based on the

SEQKC, they designed a downsampling module for point clouds, and embed it into classical semantic segmentation networks (PointNet++, PointSIFT, and PointConv) for semantic segmentation. As a result, the semantic segmentation accuracy of small objects in indoor scenes can be improved.

Ran et al. [11] presented a building multi-feature fusion refined network (BMFR-Net) to extract buildings accurately and completely. BMFR-Net was based on an encoding and decoding structure, mainly consisting of two parts: the continuous atrous convolution pyramid (CACP) module and the multiscale output fusion constraint (MOFC) structure. The CACP module was positioned at the end of the contracting path and the MOFC structure performed predictive output at each stage of the expanding path and integrated the results into the network.

Hu et al. [12] presented an automated modeling approach that could semantically decompose and reconstruct the complex building light detection and ranging (LiDAR) point clouds into simple parametric structures, and each generated structure was an un-ambiguous roof semantic unit without overlapping planar primitive. The method begins by extracting roof planes using a multi-label energy minimization solution, followed by constructing a roof connection graph associated with proximity, similarity, and consis-tency attributes. Then, a progressive decomposition and reconstruction algorithm was introduced to generate explicit semantic subparts and hierarchical representation of an isolated building.

Zheng et al. [13] made a digital subsidence model (DSuM) for deformation detection in coal mining areas based on airborne light detection and ranging (LiDAR). Noise points were removed by multi-scale morphological filtering, and the progressive triangulation filtering classification (PTFC) algorithm was used to obtain the ground point cloud. The DEM was generated from the clean ground point cloud, and an accurate DSuM was obtained through multiple periods of DEM difference calculations. Then, data mining was conducted based on the DSuM to obtain parameters such as the maximum surface subsidence value, a subsidence contour map, the subsidence area, and the subsidence boundary angle.

3. Conclusions

This Special Issue aims to attract a collection of papers that focus on the recent tech-niques for UAV-based 3D mapping, especially for trajectory planning for data acquisition in complex environments, recent algorithms for feature matching, SfM and SLAM for efficient image orientation, the usage of DL techniques in 3D mapping, and the applications of UAV-based 3D mapping. Furthermore, this Special Issue hopes to promote and inspire further research in the field of UAV-based photogrammetric 3D mapping.

Funding: This research received no external funding.

Acknowledgments: The authors would like to thank the authors who have contributed their work to this Special Issue and reviewers who have paid their valuable time and effort to ensure the quality of all accepted papers. Meanwhile, our heartfelt thanks goes to the Editors of this Special Issue for careful and patient help in its organization.

Conflicts of Interest: The author declares no conflict of interest.

References

1. Yao, G.; Yilmaz, A.; Meng, F.; Zhang, L. Review of Wide-Baseline Stereo Image Matching Based on Deep Learning. *Remote Sens.* **2021**, *13*, 3247. [CrossRef]
2. Huang, W.; Jiang, S.; Jiang, W. Camera Self-Calibration with GNSS Constrained Bundle Adjustment for Weakly Structured Long Corridor UAV Images. *Remote Sens.* **2021**, *13*, 4222. [CrossRef]
3. Xiao, T.; Yan, Q.; Ma, W.; Deng, F. Progressive Structure from Motion by Iteratively Prioritizing and Refining Match Pairs. *Remote Sens.* **2021**, *13*, 2340. [CrossRef]
4. Huang, W.; Jiang, S.; He, S.; Jiang, W. Accelerated Multi-View Stereo for 3D Reconstruction of Transmission Corridor with Fine-Scale Power Line. *Remote Sens.* **2021**, *13*, 4097. [CrossRef]
5. Zhou, L.; Zhang, Z.; Jiang, H.; Sun, H.; Bao, H.; Zhang, G. DP-MVS: Detail Preserving Multi-View Surface Reconstruction of Large-Scale Scenes. *Remote Sens.* **2021**, *13*, 4569. [CrossRef]

6. Zhang, Y.; Zhang, C.; Chen, S.; Chen, X. Automatic Reconstruction of Building Façade Model from Photogrammetric Mesh Model. *Remote Sens.* **2021**, *13*, 3801. [CrossRef]
7. He, H.; Yu, J.; Cheng, P.; Wang, Y.; Zhu, Y.; Lin, T.; Dai, G. Automatic, Multiview, Coplanar Extraction for CityGML Building Model Texture Mapping. *Remote Sens.* **2022**, *14*, 50. [CrossRef]
8. Wang, J.; Hu, X.; Meng, Q.; Zhang, L.; Wang, C.; Liu, X.; Zhao, M. Developing a Method to Extract Building 3D Information from GF-7 Data. *Remote Sens.* **2021**, *13*, 4532. [CrossRef]
9. Wang, Y.; Zhao, L.; Liu, L.; Hu, H.; Tao, W. URNet: A U-Shaped Residual Network for Lightweight Image Super-Resolution. *Remote Sens.* **2021**, *13*, 3848. [CrossRef]
10. Liu, L.; Yu, J.; Tan, L.; Su, W.; Zhao, L.; Tao, W. Semantic Segmentation of 3D Point Cloud Based on Spatial Eight-Quadrant Kernel Convolution. *Remote Sens.* **2021**, *13*, 3140. [CrossRef]
11. Ran, S.; Gao, X.; Yang, Y.; Li, S.; Zhang, G.; Wang, P. Building Multi-Feature Fusion Refined Network for Building Extraction from High-Resolution Remote Sensing Images. *Remote Sens.* **2021**, *13*, 2794. [CrossRef]
12. Hu, P.; Miao, Y.; Hou, M. Reconstruction of Complex Roof Semantic Structures from 3D Point Clouds Using Local Convexity and Consistency. *Remote Sens.* **2021**, *13*, 1946. [CrossRef]
13. Zheng, J.; Yao, W.; Lin, X.; Ma, B.; Bai, L. An Accurate Digital Subsidence Model for Deformation Detection of Coal Mining Areas Using a UAV-Based LiDAR. *Remote Sens.* **2022**, *14*, 421. [CrossRef]

 *remote sensing*

Review

Review of Wide-Baseline Stereo Image Matching Based on Deep Learning

Guobiao Yao [1,2,*], Alper Yilmaz [2], Fei Meng [1] and Li Zhang [3]

1 School of Surveying and Geo-Informatics, Shandong Jianzhu University, No. 1000 Fengming Road, Jinan 250101, China; lzhmf@sdjzu.edu.cn
2 Photogrammetric Computer Vision Lab, The Ohio State University, Columbus, OH 43210, USA; yilmaz.15@osu.edu
3 Chinese Academy of Surveying & Mapping, No. 28 Lianhuachi West Road, Beijing 100830, China; zhangl@casm.ac.cn
* Correspondence: yao7837005@sdjzu.edu.cn; Tel.: +86-531-8636-1159

Abstract: Strong geometric and radiometric distortions often exist in optical wide-baseline stereo images, and some local regions can include surface discontinuities and occlusions. Digital photogrammetry and computer vision researchers have focused on automatic matching for such images. Deep convolutional neural networks, which can express high-level features and their correlation, have received increasing attention for the task of wide-baseline image matching, and learning-based methods have the potential to surpass methods based on handcrafted features. Therefore, we focus on the dynamic study of wide-baseline image matching and review the main approaches of learning-based feature detection, description, and end-to-end image matching. Moreover, we summarize the current representative research using stepwise inspection and dissection. We present the results of comprehensive experiments on actual wide-baseline stereo images, which we use to contrast and discuss the advantages and disadvantages of several state-of-the-art deep-learning algorithms. Finally, we conclude with a description of the state-of-the-art methods and forecast developing trends with unresolved challenges, providing a guide for future work.

Keywords: wide-baseline stereo image; deep learning; convolutional neural network; affine invariant feature; image matching

Citation: Yao, G.; Yilmaz, A.; Meng, F.; Zhang, L. Review of Wide-Baseline Stereo Image Matching Based on Deep Learning. *Remote Sens.* **2021**, *13*, 3247. https://doi.org/10.3390/rs13163247

Academic Editor: Riccardo Roncella

Received: 7 July 2021
Accepted: 11 August 2021
Published: 17 August 2021

Publisher's Note: MDPI stays neutral with regard to jurisdictional claims in published maps and institutional affiliations.

1. Introduction

Wide-baseline image matching is the process of automatically extracting corresponding features from stereo images with substantial changes in viewpoint. It is the key technology for reconstructing realistic three-dimensional (3D) models [1–3] based on two-dimensional (2D) images [4–6]. Wide-baseline stereo images provide rich spectral, real texture, shape, and context information for detailed 3D reconstruction. Moreover, they have advantages with respect to spatial geometric configuration and 3D reconstruction accuracy [7]. However, because of the significant change in image viewpoint, there are complex distortions and missing content between corresponding objects in regard to scale, azimuth, surface brightness, and neighborhood information, which make image matching very challenging [8]. Hence, many scholars in the fields of digital photogrammetry and computer vision have intensely explored the deep-rooted perception mechanism [9] for wide-baseline images, and have successively proposed many classic image-matching algorithms [10].

Based on the recognition mechanism, existing wide-baseline image-matching methods can be divided into two categories [11–13]: Handcrafted matching and deep-learning matching. Inspired by professional knowledge and intuitive experience, several researchers have proposed handcrafted matching methods that can be implemented by intuitive computational models and their empirical parameters according to the image-matching task [14–18].This category of methods is also referred to as traditional matching, the classical representative of which is the scale invariant feature transform (SIFT) algorithm [14].Traditional matching has many problems [15–18] such as repetition in wide-baseline image feature extraction or the reliability of the feature descriptors and matching measures. Using multi-level convolutional neural network (CNN) architecture, learning-based methods perform iterative optimization by back-propagation and model parameter learning from a large amount of annotated matching data to develop the trained image-matching CNN model [19]. A representative deep-learning model under this category can be chosen, such as MatchNet [20]. Methods under this category offer a different approach to solving the problem of wide-baseline image matching, but they are currently limited by the number and scope of training samples, and it is difficult to learn the optimal model parameters that are suitable for practical applications [21–25]. Learning-based image matching is essentially a method that is driven by prior knowledge. In contrast to the traditional handcrafted methods, it can avoid the need for many manual interventions with respect to feature detection [26], feature description [27], model design [28], and network parameter assignment [29]. Moreover, it can adaptively learn the deep representation and correlation of the topographic features directly from large-scale sample data. According to the scheme used for model training, the wide-baseline matching methods can be further divided into two types [30]: Multi-stage training with (1) step-by-step [31] and (2) end-to-end training [32]. The former focuses on the concrete issues of each stage, such as feature detection, neighborhood direction estimation, and descriptor construction, and it can be freely integrated with handcrafted methods [33]; whereas the latter considers the multiple stages of feature extraction, description, and matching as a whole and achieves the global optimum by jointly training with various matching stages [34]. In recent years, with the growth of training datasets and the introduction of transfer learning [35], deep-learning-based image matching has been able to perform most wide-baseline image-matching tasks [36], and its performance can, in some cases, surpass that of traditional handcrafted algorithms. However, the existing methods still need to be further studied in terms of network structure [37], loss function [38], matching metric [39], and generalization ability [40], especially for typical image-matching problems such as large viewpoint changes [41], surface discontinuities [42], terrain occlusion [43], shadows [44], and repetitive patterns [45–47].

On the basis of a review of the image-matching process, we incrementally organize, analyze, and summarize the characteristics of proposed methods in the existing research, including the essence of the methods as well as their advantages and disadvantages. Then, the classical deep-learning models are trained and tested on numerous public datasets and wide-baseline stereo images. Furthermore, we compare and evaluate the state-of-the-art methods and determine their unsolved challenges. Finally, possible future trends in the key techniques are discussed. We hope that research into wide-baseline image matching will be stimulated by the review work of this article.

The main contributions of this article are summarized as follows. First, we conduct a complete review for the learning-based matching methods, from the feature detection to end-to-end matching, which involves the essences, merits, and defects of each method for wide-baseline images. Second, we construct various combined methods to evaluate the representative modules fairly and uniformly by using numerous qualitative and quantitative tests. Third, we reveal the root cause for struggling to produce high-quality matches across wide-baseline stereo images and present some feasible solutions for the future work.

In Section 2, this article reviews the most popular learning-based matching methods, including the feature detection, feature description, and end-to-end strategies. The results and discussion are presented in Section 3. The following summary and outlook are given in Section 4. Finally, Section 5 draws the conclusions of this article.

2. Deep-Learning Image-Matching Methodologies

At present, the research on deep-learning methods for wide-baseline image matching mainly focuses on three topics: Feature detection, feature description, and end-to-end matching (see Figure 1). Therefore, this section provides a review and summary of the related work in these research topics below.

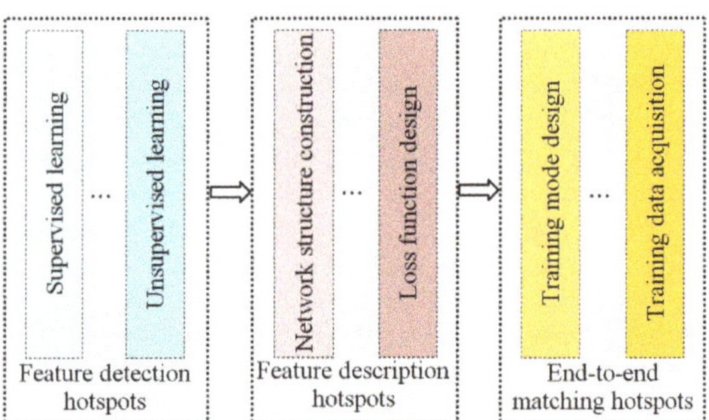

Figure 1. Focus of this review: Topics of deep-learning methods for wide-baseline stereo image matching.

2.1. Deep-Learning-Based Feature Detection

Figure 2 summarizes the progress in deep-learning feature-detection methods. Based on the implemented learning mode, the mainstream deep-learning feature-detection algorithms can be divided into two types: Supervised learning [48] and unsupervised learning [49]. Supervised learning feature detection takes the feature points extracted by traditional methods as "anchor points", and then trains a regression neural network to predict the location of more feature points; whereas the unsupervised learning strategy uses a neural network directly to train the candidate points and their response-values, and then takes the candidate points at the top or bottom of the ranking as the final feature points.

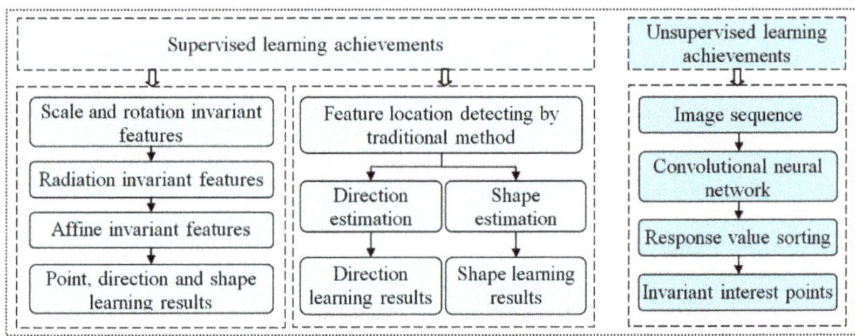

Figure 2. Development of feature detection with deep learning.

The basis of wide-baseline image matching is the extraction of local invariant features, which are local features that remains table between the stereo images under geometric or radiometric distortions, such as viewpoint change or illumination variation. In recent years, researchers have focused on exploring feature detection schemes for deep learning with enhancing network [50]. Using the supervised learning strategy as an example, Lenc et al. first proposed a local invariant feature loss function $L_{cov}(x)$ [51].

$$L_{cov}(x) = \min\|g\phi(x) - \phi(gx)q\|_F^2 \tag{1}$$

where $\|\cdot\|_F^2$ is F-norm, x is the image block to be processed, g is the random geometric transformation, gx is the random transformation result of x, $\phi(\cdot)$ is the transformation matrix output by the neural network, and q is the complementary residual transformation of g. On this basis, this algorithm employs the Siamese neural network DetNet to learn the invariant feature geometric transformations. Moreover, it uses image control points as anchor points and treats potential feature points as certain transformation forms of these anchor points. In the training phase, the images with anchor points are input to the regression neural network, and the optimal transformation is learned iteratively. Then, the weights of the regression neural network are adjusted according to the loss function and finally interpolated to obtain more feature positions, directions, and shapes. This method created a precedent for deep-learning invariant feature detection, and the detected features are equipped with good scale and rotation invariance.

Zhang et al. [52] used the illumination-invariant feature TILDE [30] of deep learning as an anchor point, which solved the problem of image matching under strong illumination changes; on this basis, Doiphode et al. [53] used a triple network [54] and introduced an affine invariant constraint to learn stable and reliable affine invariant features. The above methods give the target features a certain geometric and radiation invariance, but the geometric relationship between the image blocks must be roughly known before training the model; this invisibly increases the workload of the training dataset production.

Yi et al. [55] further studied the Edge Foci (EF) [56] and SIFT [14] features to detect the location of key points and learned the neighborhood direction of features based on a CNN; Mishkin et al. [57] used a multi-scale Hessian to detect the initial feature points and estimate the affine invariant region based on the triplet network AffNet. This method combines traditional feature extraction algorithms with deep-learning invariant features, which substantially improves the efficiency and reliability of feature detection.

In addition to the above-mentioned features for supervised learning, Savinov et al. [58] also proposed a classic feature-learning strategy with unsupervised idea. This method transforms the learning problem of feature detection into a learning problem of response-value sorting of image interest points. The response function of the image point is denoted by $H(p|w)$, where p represents the image point, and H and w represent the CNN to be trained and the weight vector of the network, respectively. The image point response-value sorting model is then expressed as follows:

$$\begin{cases} H(p_d^i|w) > H(p_d^j|w) \ \& \ H(p_{t(d)}^i|w) > H(p_{t(d)}^j|w) \\ \qquad\qquad\qquad\text{or} \\ H(p_d^i|w) < H(p_d^j|w) \ \& \ H(p_{t(d)}^i|w) < H(p_{t(d)}^j|w) \end{cases} \tag{2}$$

where d represents one scene target in the image and p is located on d; i and j are the indexes of p, and $i \neq j$; $p_{t(d)}^i$ and $p_{t(d)}^j$ are generated respectively by transformation t of p_d^i and p_d^j. Therefore, all points p on target d are sorted according to the response-value function and Equation (2), and the image points with the response-values in the top or bottom ranks are retained as feature points. The key purpose of this method is to learn the invariant response function of the image point using the neural network. The feature points maintain good invariance to the perspective transformation of the images; additional experiments in Reference [58] demonstrate that the proposed method may outperform the Difference of Gaussian (DoG) strategy [14] regarding feature repeatability for view-change images. However, the existing methods still have many shortcomings with respect to feature point detection repeatability and stability for wide-baseline images with large view changes.

As mentioned above, the most learning-based methods for feature detection are categorized as supervised learning achievements. Such mainstream methods can handily surpass the unsupervised strategies in invariant feature learning because the supervised methods may directly and separately produce the geometric covariant frames for wide-baseline images, while the unsupervised methods need to simultaneously cope with the locations of interest points and their invariance during learning process.

2.2. Deep-Learning Feature Description

Deep-learning feature description [59] has been widely applied in professional tasks [60] such as image retrieval, 3D reconstruction, face recognition, interest point detection, and target positioning and tracking. Specific research on this topic mainly focuses on network structure construction and loss function design, as shown in Figure 3. Among them, the network structure of deep learning directly determines the discrimination and reliability of the feature descriptors, while the loss function affects the training performance of the model by controlling the iterative update frequency of the model parameters and optimizing the quantity and quality of the sample input.

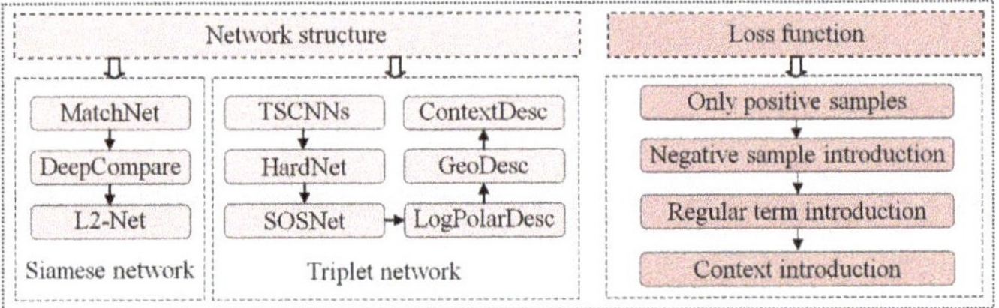

Figure 3. Development of deep-learning feature description.

The key to high-quality feature description is to consider both similarity and discrimination. "Similarity" refers to the ability of corresponding feature descriptors to maintain good invariance to signal noise, geometric distortion, and radiation distortion, thereby retaining a high degree of similarity. In contrast, "discrimination" refers to the idea that there should be a large difference between any non-matching feature descriptors. To generate high-quality descriptors, the learning-based method departs from the paradigm of traditional algorithms and builds Siamese network or triplet network, which emulates the cognitive structure of human visual nerves. The Siamese network, also known as the dual-channel network, is a coupled architecture based on a binary branch network, whereas the triple network has one more branch than the Siamese network, and thus it can be adapted to a scenario in which three samples are input simultaneously.

Figure 4 shows the evolution of several typical feature-description networks. Among them, a representative approach is MatchNet [20], which uses the original Siamese network and is composed of two main parts: A feature coding network and a similarity measurement network. The two branches of the feature network maintain dynamic weight sharing and extract the feature patches from stereo images through a convolution layer [58], a maximum pooling layer [61], and other layers. Furthermore, it calculates the similarity between image blocks though a series connecting to the top fully connected network [62], and then determines the matching blocks based on the similarity score. Subsequently, Zagoruyko et al. [63] further explored the role of the central-surround two-stream network (CSTSNet) [64] and the spatial pyramid pooling net (SPPNet) [65] in the feature description. CSTSNet combines a low-resolution surround stream with a high-resolution center stream, which not only use the multi-resolution information of the image, but also emphasize the information of the center pixels, thus substantially improving the matching performance. In contrast, SPPNet inherits the good characteristics of the Siamese network, then it enhances the adaption to image block data of different sizes by introducing a spatial pyramid pooling layer. To apply SPPNetto the description of features in satellite images, Fan et al. [66] designed a dual-channel description network based on a spatial-scale convolutional layer to improve the accuracy of satellite image matching.

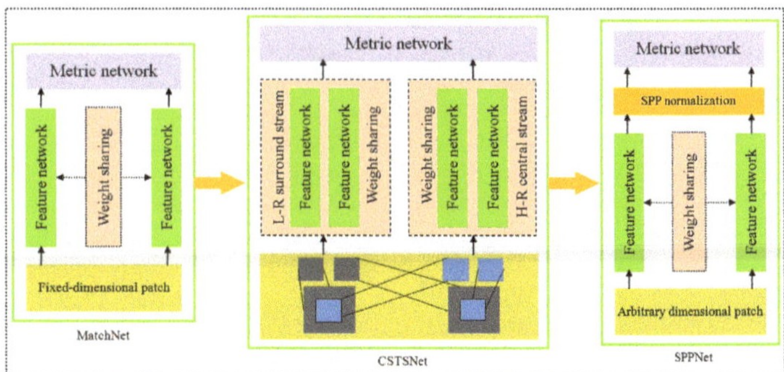

Figure 4. Evolution of representativefeature description networks.

These descriptor measurement networks belong to the fully connected category of networks, which consume a large amount of computing resources during training and testing, and hence have low matching efficiency. To address this, Tian et al. proposed a feature description model called L2-Net [67] with a full convolutional network representation. This method inherits the idea of SIFT descriptors, namely, it adjusts the dimension of network output to 128 and uses the L2 norm measure of Euclidean distance instead of a metric network to evaluate the similarity of the feature descriptors. The basic structure of the L2-Net network is shown in Figure 5. This network consists of seven convolutional layers and a local response normalization layer (LRN). In the figure, the term "3 × 3 Conv" in the convolutional layer refers to convolution, batch normalization, and linear activation operations in the series, and "8 × 8 Conv" represents the convolution and batch normalization processing operations. Moreover, "32" represents a 32-dimensional convolution with a step size of 1 and "64/2" refers to a 64-dimensional convolution operation with a step size of 2. The final output layer LRN is used to generate unit descriptor vectors while accelerating network convergence and enhancing model generalization.

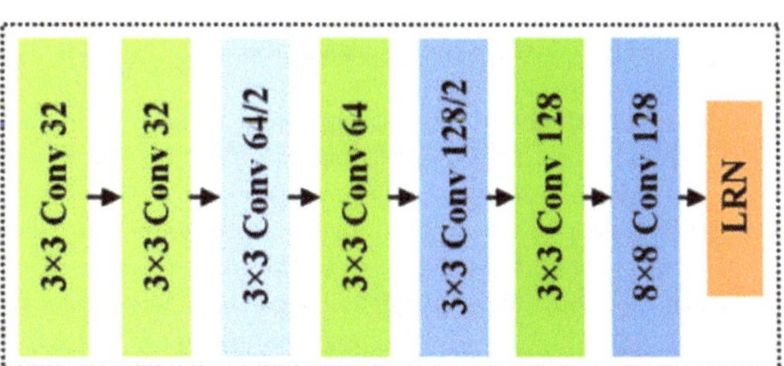

Figure 5. Basic architecture of L2-Net.

The results on the open-source dataset Brown [68], Oxford [10], and HPatches [69] training and testing datasets show that L2-Net has good generalization ability, and its performance is better than the existing traditional descriptors. Moreover, L2-Net performs well with respect to image feature classification as well as wide-baseline stereo image feature description and matching, and thus many researchers regard it as a classic feature description network and have extended it with improvements in network structure. Balntas et al. [34] found that one disadvantage of L2-Net is that it ignores the contribution of negative samples to the loss function value. Hence, they proposed the triplets and shallow CNN (TSCNN). This method simplifies the L2-Net network layer and the number of channels, then incorporates negative samples into the network training, and hence that the modified model can reduce the distance between matching feature descriptors while increasing the distance between non-matching feature descriptors. However, the negative samples are input into TSCNNs using random sampling strategy, and as a result, most negative samples do not sufficiently contribute to the model training, which limits the improvements in descriptor discrimination. In view of this, HardNet [70] incorporates the most difficult negative sample, namely the nearest non-matching descriptor, into the training of the model, which substantially enhances the training efficiency and matching performance. The triplet margin loss (TML) function used by this model is as follows:

$$L = \frac{1}{m} \sum_{i=1}^{m} \max(0.1 + \mathrm{d}(\boldsymbol{a}_i, \boldsymbol{p}_i) - \min(\mathrm{d}(\boldsymbol{a}_i, \boldsymbol{n}_{j_{\min}}), \mathrm{d}(\boldsymbol{n}_{k_{\min}}, \boldsymbol{p}_i))) \tag{3}$$

where m is the batch size, $\mathrm{d}()$ is the Euclidean distance between two descriptors, $\boldsymbol{a}_i$ and $\boldsymbol{p}_i$ are an arbitrary pair of matching descriptors, and $\boldsymbol{n}_{j_{\min}}$ and $\boldsymbol{n}_{k_{\min}}$ represent the closest non-matching descriptors to $\boldsymbol{a}_i$ and $\boldsymbol{p}_i$, respectively.

On the basis of the L2-Net network structure, the HardNet descriptor model employs the nearest neighbor negative sample sampling strategy and the TML loss function, which is another important advance in the descriptor network model. Inspired by HardNet, some notable deep-learning models for feature description have been further explored. For example, LogPolarDesc [71] uses a polar transform network to extract corresponding image blocks with higher similarity to improve the quality and efficiency of model training; SOSNet [72] introduces the second-order similarity regularization into the loss function to prevent over-fitting of the model and substantially improve the utilization of the descriptors. To generate a descriptor with both global and local geometric invariance, some researchers have proposed making full use of the geometry or the visual context information of an image. The representative approach GeoDesc [73] employs cosine similarity to measure the matching degree of descriptors. It also sets self-adaptive distance thresholds to handle different training image blocks and then introduces a geometric loss function to enhance the geometric invariance of the descriptor, which is expressed by the following equation:

$$E_{\text{geometric}} = \sum_i \max(0, \beta - s_{i,i}), \beta = \begin{cases} 0.7 & s_{\text{patch}} \geq 0.5 \\ 0.5 & 0.2 \leq s_{\text{patch}} < 0.5 \\ 0.2 & \text{otherwise} \end{cases} \tag{4}$$

where β represents the adaptive threshold; $s_{i,i}$ represents the cosine similarity between corresponding features descriptors; and s_{patch} represents the similarity of the correspondingimage blocks. On this basis, ContextDesc [74] integrates geometry and visual context perception into the process of network model construction, thus improving the utilization of image geometry and visual context information. Finally, many data tests show that the ContextDesc adapts well to the geometric and radiation distortions of different scenes.

In short, feature description plays a vital role in image matching, as the high-quality descriptor can absorb the local and global information from the feature neighborhoods, which may provide adequate knowledge for recognizing the unique feature from extensive false candidates. Based on the aforementioned, the triple networks can perform better than the Siamese or sole model, because multi-branch networks can be efficient in learning the uniqueness of features and make full use of context information.

2.3. Deep-Learning End-to-End Matching

The end-to-end matching strategy integrates three different stages of image feature extraction, description, and matching into one system for training, which is beneficial for learning the globally optimal model parameters, and adaptively improves the performance of each stage [75]. Figure 6 summarizes the development of end-to-end deep-learning matching. Most end-to-end methods focus on the design of training modes and the automatic acquisition of training data [76]. The design of training modes is intended to obtain high-quality image features and descriptors in a more concise and efficient way; the aim of automatic acquisition of data is to achieve fully automatic training by means of a classical feature detection algorithm and spatial multi-scale sampling strategy.

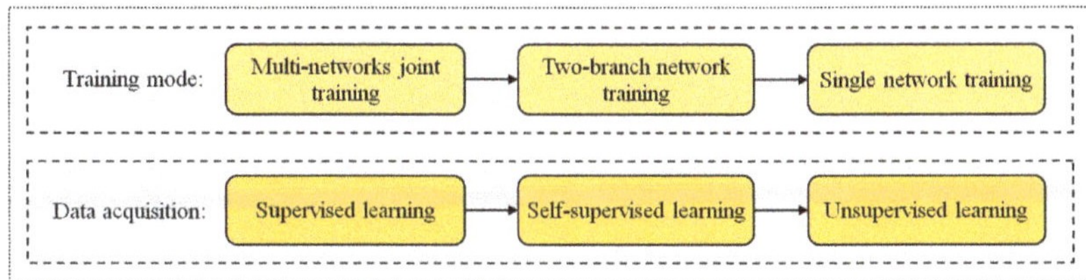

Figure 6. Development of end-to-end matching with learning-based methods.

Yi et al. proposed the learned invariant feature transform (LIFT) network structure [77]. This network first integrates feature detection, direction estimation, and feature description into one pipeline based on the Transformer (ST) [78] and softargmax algorithm [79]. The end-to-end training is carried out by back propagation. The complete training and testing process of this method is shown in Figure 7.

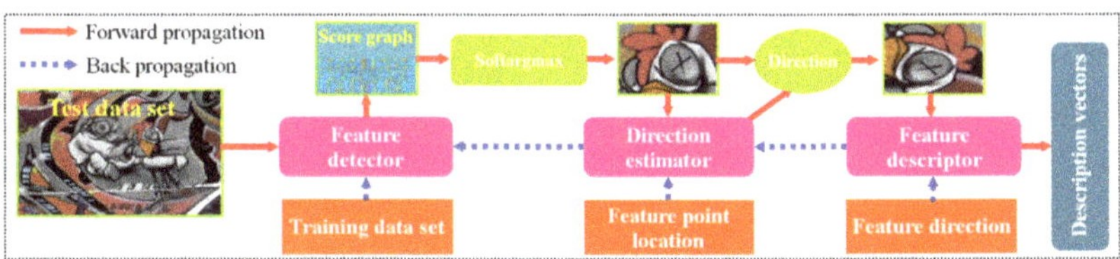

Figure 7. Training and testing of the LIFT pipeline.

The back propagation-based training process of LIFT can be briefly described as follows. First, the feature location and principal direction can be extracted using the structure from motion (SFM) algorithm [80], and then the feature descriptor is trained. Second, guided by the feature descriptor, the direction estimator is trained based on the feature location and its neighborhood ST. Finally, the feature descriptor and direction estimator are united to train the feature detector based on the training dataset. After the LIFT has been trained, the corresponding test process proceeds as follows. First, the feature score map of a multi-scale image is obtained based on the feature detector. Second, scale-space non-maximum suppression is performed using the softargmax function and then the scale invariant feature region is extracted. Finally, the feature region is further normalized and then the description vectors are extracted by the feature descriptor.

Although LIFT belongs to the category of end-to-end network models, a back propagation-based multi-stage training mode is adopted in the network training, which reduces the training efficiency and practicality of the model; additionally, LIFT employs an SFM strategy and random spatial transformation to provide matching image blocks for training, which limits the discrimination of descriptors. In view of this, DeTone et al. [81] proposed a self-supervised network model called MagicPoint instead of SFM to label training data. They then use the SuperPoint model to learn feature points and extract their descriptors for end-to-end training.

SuperPoint realizes the joint training of feature detection and description through the encoding structure [82] and decoding structure [83]. The encoding structure is used for image feature extraction, whereas the decoding structure can not only output the position of the feature point, but also output the descriptor vector. Similarly, Revaud et al. [84] proposed the Siamese decoding structure R2D2, which focuses more on the repetitive and discriminative expression of training features than SuperPoint.

The learning-based method of MagicPoint can replace the handcrafted labeling of feature points, but a small amount of handcrafted labeling data is still required when obtaining the pre-trained model. Ono et al. [85] proposed LF-Net, which is an end-to-end model that uses unsupervised training. This method directly uses the stereo images obtained by a metric camera, an image depth map, the camera position, and orientation data, and other prior information to complete the end-to-end model training, which greatly reduces the need for manual intervention and promotes the automated process of deep-learning matching. In addition, Dusmanu et al. proposed a combination of feature detection and descriptor extraction that can make more effective use of high-level semantic information. They then proposed a simplified end-to-end model D2Net [86]. The difference between this model and the traditional model is depicted in Figure 8. Figure 8a shows the traditional "detect-then-describe" model, that is, SuperPoint [81], which is a representative model of this type, and Figure 8b shows the D2Net "describe-and-detect" model. In contrast to a Siamese or multi-branch network structure [87], D2Net adopts a single-branch network architecture, and the feature location and descriptor information of the image are stored in high-dimensional feature channels, which is thus more conducive to obtaining stable and efficient matches. However, D2Net must extract dense descriptors in the process of using high-level semantic information, which reduces the accuracy and efficiency of feature detection.

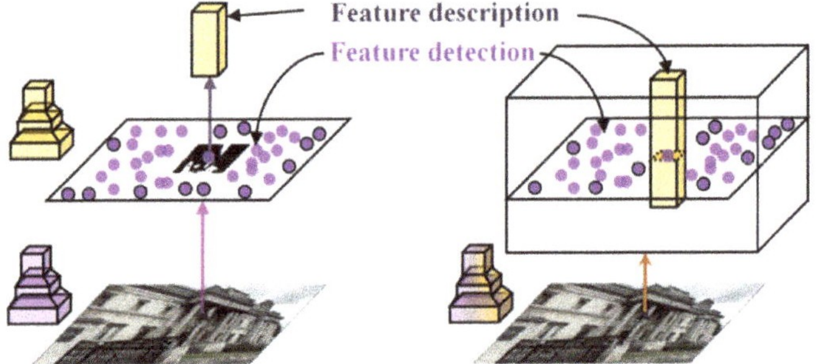

(a) Traditional model: detection, then description (b) D2Net: description and detection

Figure 8. Difference between D2Net and the traditional model.

All in all, the end-to-end strategy is prone to train the optimal parameters for image matching. Multi-networks with complex architecture need to input more training samples than a single network. Considering the available scale of training data [76], the self-supervised learning mode is the best choice for current practical applications.

3. Results and Discussion

3.1. Representative Algorithms and Experimental Data

To evaluate the performance, advantages, and disadvantages of deep-learning stereo matching algorithms, we selected a total five categories of 10 well-performed algorithms for the experiments, including deep-learning end-to-end matching, deep-learning feature detection and description, deep-learning feature detection and handcrafted feature description, handcrafted feature detection and deep-learning feature description, and handcrafted image matching, as shown in Table 1. In addition, the key source code of each algorithm can be obtained from the corresponding link in this table. The above methods were selected due to the following reasons. First, as the representatives of deep-learning end-to-end matching, SuperPoint [81] and D2Net [86] were published recently, and they have been widely applied [88–90] in the fields of photogrammetry and computer vision. Second, the deep-learning feature detectors AffNet [57] andDetNet [51], deep-learning feature descriptors HardNet [70], SOSNet [72], and ContextDesc [74], were all proposed for wide-baseline image matching, and were often used as benchmarks [12]. Third, the classical handcrafted methods are used here to verify the strength of deep-learning methods. Finally, all selected methods are effective and well-performed in previous reports, and the source codes are open to public.

Table 1. Representative algorithms and their references.

Categories	Algorithms	Code links
Deep learning end-to-end matching	①SuperPoint [81] ②D2Net [86]	https://github.com/rpautrat/SuperPoint https://github.com/mihaidusmanu/d2-net
Deep learning feature detection and description	③AffNet [57] + HardNet [70] ④AffNet [57] + SOSNet [72] ⑤DetNet [51] + Contexdesc [74] ⑥DetNet [51] + HardNet [70]	https://github.com/DagnyT/hardnet https://github.com/scape-research/SOSNet https://github.com/lzx551402/contextdesc https://github.com/lenck/ddet
Deep learning feature detection and handcrafted feature description	⑦AffNet [57] + SIFT [14]	https://github.com/ducha-aiki/affnet
Handcrafted feature detection and deep learning feature description	⑧Hessian [16] + HardNet [70]	https://github.com/doomie/HessianFree
Handcrafted matching	⑨MSER [17] + SIFT [14] ⑩ASIFT [18]	https://github.com/idiap/mser https://github.com/search?q=ASIFT

The datasets used to train each deep-learning algorithm are as follows: SuperPoint using MSCOCO [91]; D2Net using MegaDepth [92]; AffNet using UBC Phototour [68]; both HardNet and SOSNet using HPatches [69]; ContextDesc using GL3D [93]; DetNet using DTU-Robots [94]. According to their corresponding literatures [68,69,91–94], the characters of each dataset would be discussed and summarized as follows. MSCOCO was proposed with the goal of advancing the state-of-the-art in scene understanding and object detection. In contrast to the popular datasets, MSCOCO involves fewer common object categories but more instances per category, which would be useful for learning complex scenes. MegaDepth was created to exploit multi-view internet images and produce a large amount of training data by the SFM method. It performs well for challenging environments such as offices and close-ups, but MegaDepth is biased towards outdoor scenes. UBC Phototour initially proposed patch verification as an evaluation protocol. There is large number of patches available in this dataset, which is particularity suited for deep-learning-based detectors. The images in this dataset have notable variations in illumination and view changes, but most of these images only focus on three scenes: Liberty, Notre-Dame, and Yosemite. HPatches presented a new large-scale dataset special for training local descriptors, aiming to eliminate the ambiguities and inconsistencies in scene understanding. It has the superiorities of diverse scenes and notable viewpoint changes. GL3D designed a large-scale database for 3D surface reconstruction and geometry-related learning issues. This dataset covered many different scenes, including rural area, urban, and scenic spots taken from multiple scales and viewpoints. The DTU-Robots dataset involves real images of 3D scenes, shot using a robotic arm in rigorous laboratory conditions, which is suitable for certain application but of limited size and variety in the data.

The representative wide-baseline test data are presented in Figure 9, and the corresponding data descriptions are listed in Table 2. Algorithms ①, ②, and ⑩ can directly output the corresponding features, and for the descriptors output by algorithms ③–⑨, we adopt the nearest-neighbor and second nearest-neighbor distance ratio to obtain the matches. Finally, each algorithm employs the random sample consensus (RANSAC) strategy to eliminate outliers. The performance of the algorithms is objectively evaluated according to the number of matching points, matching accuracy, and matching spatial distribution indexes.

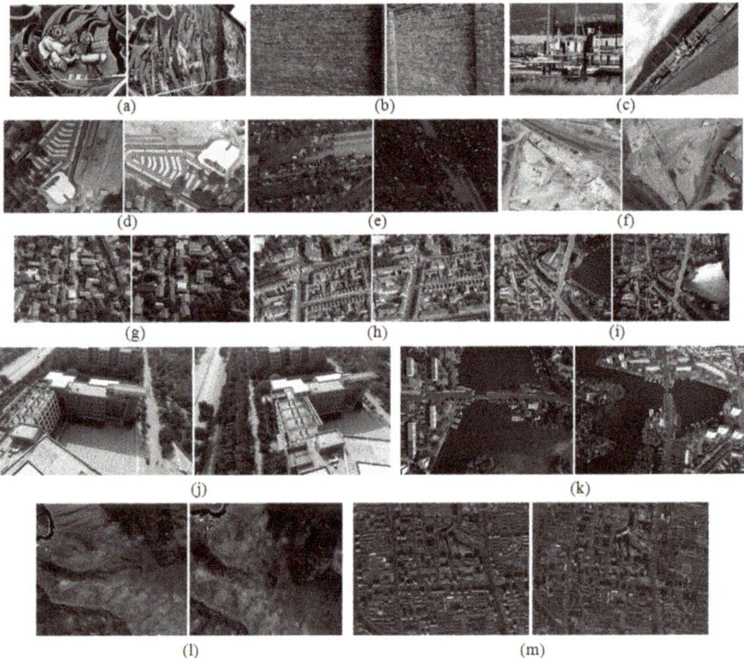

Figure 9. (a–m) Wide-baseline test data. These data are carefully selected from different platforms of ground close-range, UAV, and satellite, respectively. They cover various terrains and have significant viewpoint changes.

3.2. Experimental Results

For the 13 sets of wide-baseline stereo images and 10 representative algorithms, the results are as follows: Table 3 presents the number of image matches obtained by each algorithm, and the bold number is the maximum number of matches in each group of test data; Figure 10 shows the matching errors of each algorithm, where the matching error ε is estimated by the following equations [95]:

$$\left.\begin{array}{l} \varepsilon_H = \sqrt{\frac{1}{N}\sum_{j=1}^{N}\left\|x'_j - Hx_j\right\|^2} \\ \varepsilon_F = \sqrt{\frac{1}{N}\sum_{j=1}^{N}\left(x'^{\mathrm{T}}_j Fx_j\right)^2 / (Fx_j)_1^2 + (Fx_j)_2^2} \end{array}\right\} \tag{5}$$

where N is the number of matches, x_j and x'_j are an arbitrary pair of matching point coordinates, and H and F are the known true perspective transformation matrix and true fundamental matrix, respectively. The matching errors of test data (a)–(f),which consist of planar or approximately planar scenes, are evaluated by ε_H (pixel), and the matching errors of test data (g)–(m), which consist of non-planar scenes, are evaluated by ε_F (pixel). Figure 11 shows the image-matching results of each algorithm. Because of the limited space, this figure only exhibits the matching results of algorithms ①, ③, ④, ⑤, and ⑩ based on test data (a), (f), (g), (i), (j), and (m), where the matching points are indicated by red dots and joined by yellow lines, and the most matches in each row of the figure are marked by a green frame. For each algorithm, Figure 12 shows the matching distribution quality *Dis*, which is estimated by the following equation [96]:

$$Dis = \sqrt{\sum_{i=1}^{n}\left[(A_i/\overline{A}) - 1\right]/(n-1)} \times \sqrt{\sum_{i=1}^{n}(S_i - 1)/(n-1)}, \quad \overline{A} = \frac{1}{n}\sum_{i=1}^{n}A_i, \quad S_i = 3\max(J_i)/\pi \tag{6}$$

where n represents the total number of Delaunay triangles generated by the matching points, A_i denotes the area of the i-th triangle, $\max(J_i)$ represents the radian value of the maximum internal angle, and $\overline{A}$ represents the average area of the triangle. The value of Dis can reveal the consistency and uniformity of the spatial distribution of the triangle network, and a smaller Dis value indicates that the matches have a higher spatial distribution quality.

Table 2. Description of the wide-baseline test data.

	Testdata	Left Image (Pixels)	Right Image (Pixels)	Description for Image Pair	True Perspective Transform Matrix H or True Fundamental Matrix F
Ground close-ranges data	a	800 × 640	800 × 640	Close-range stereo images with 60 deg viewpoint change	H is provided by Reference [10]
	b	1000 × 700	880 × 680	Close-range stereo images with repetitive patterns and 60 deg viewpoint change	H is provided by Reference [10]
	c	850 × 680	850 × 680	Close-range stereo images with about 45 deg rotation and 2.5 times scale transform	H is provided by Reference [10]
Low attitude data	d	900 × 700	900 × 700	UAV stereo images with 90 deg rotation and significant oblique viewpoint change	H is estimated by manual work
	e	800 × 600	800 × 600	UAV stereo images with 90 deg rotation, large oblique view change, and radiometric distortion	H is estimated by manual work
	f	900 × 700	900 × 700	UAV stereo images with rare texture, large view change, and radiometric distortion	H is estimated by manual work
	g	800 × 600	800 × 600	UAV stereo images with significant scale deformation, oblique view change, radiometric distortion, and numerous 3D scenes	F is estimated by manual work
	h	800 × 600	800 × 600	UAV stereo images with large oblique view change, and numerous 3D scenes	F is estimated by manual work
	i	1084 × 814	1084 × 814	UAV stereo images with significant oblique view change, radiometric distortion, and complex 3D scenes	F is estimated by manual work
	j	5472 × 3468	5472 × 3468	UAV stereo images with significant view change, surface discontinuity, object occlusion, and rare texture	F is estimated by manual work
	k	4200 × 3154	4200 × 3154	UAV stereo images with about 90 deg rotation, significant oblique view change, single texture, and large area of water	F is estimated by manual work

Table 2. *Cont.*

Testdata		Left Image (Pixels)	Right Image (Pixels)	Description for Image Pair	True Perspective Transform Matrix H or True Fundamental Matrix F
Sallite data	l	2316 × 2043	2316 × 2043	Satellite optical stereo image with notable rotation, significant topography variation, and rare texture	F is estimated by manual work
	m	2872 × 2180	2872 × 2180	Satellite optical stereo images with significant surface discontinuity, radiometric distortion, dense 3D buildings, and single texture	F is estimated by manual work

Table 3. The contrast of ten algorithms in the aspect of number of matches. Bold font denotes the best results.

Algorithms	a	b	c	d	e	f	g	h	i	j	k	l	m
①SuperPoint	65	277	0	0	0	5	**337**	**498**	**523**	**618**	0	419	**3856**
②D2Net	7	118	0	0	0	0	36	38	32	23	0	33	114
③AffNet + HardNet	239	414	54	617	229	200	147	152	178	147	141	198	62
④AffNet + SOSNet	263	421	39	690	233	**208**	152	151	178	120	134	237	58
⑤DetNet + Contexdesc	201	540	**119**	939	**607**	152	102	207	339	22	18	79	489
⑥DetNet + HardNet	7	0	29	15	48	7	45	67	90	7	144	38	21
⑦AffNet + SIFT	33	59	6	131	36	7	7	16	24	16	0	49	7
⑧Hessian + HardNet	180	313	64	620	191	176	124	131	154	144	142	188	53
⑨MSER + SIFT	37	185	6	54	29	6	17	31	50	16	8	224	0
⑩ASIFT	**855**	**2339**	22	**1287**	223	175	153	188	348	528	**275**	**1304**	2580

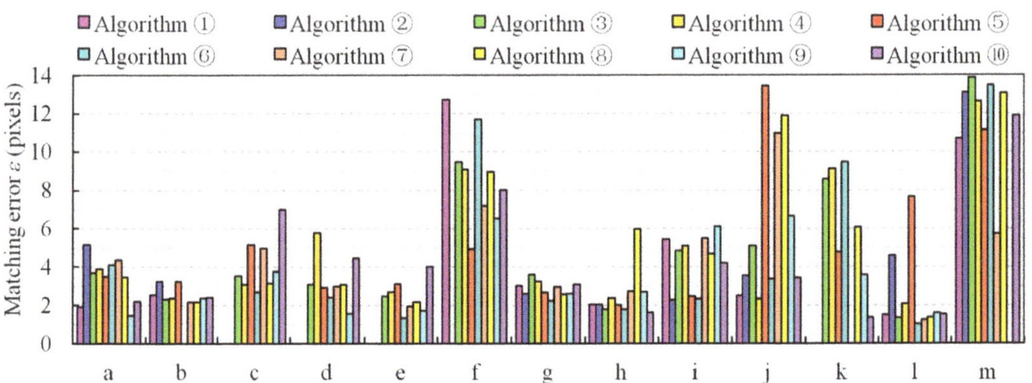

Figure 10. Comparison of the matching error results of the ten algorithms.

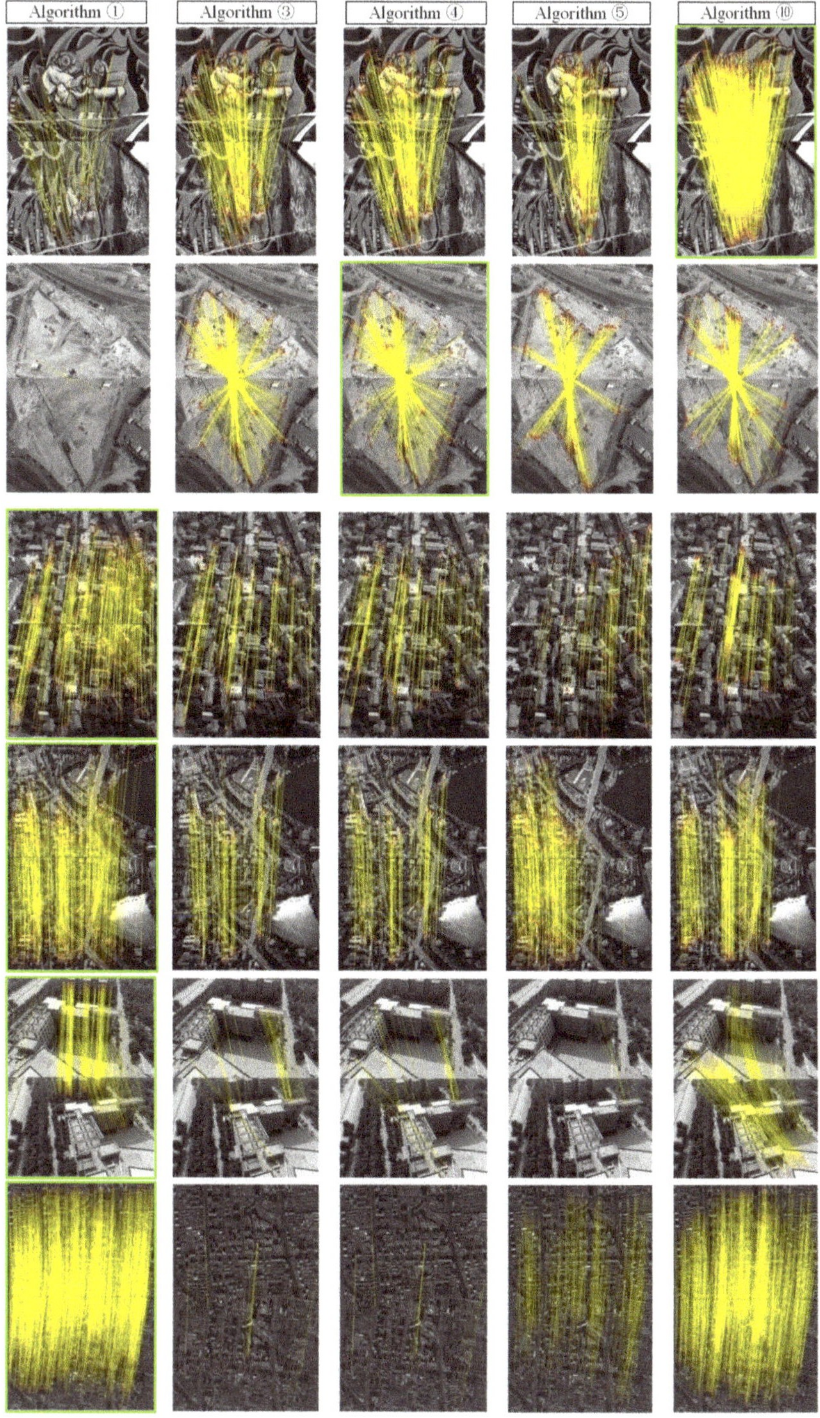

Figure 11. Matching results of algorithms ①, ③, ④, ⑤, and ⑩ on test data (a), (f), (g), (i), (j), and (m).

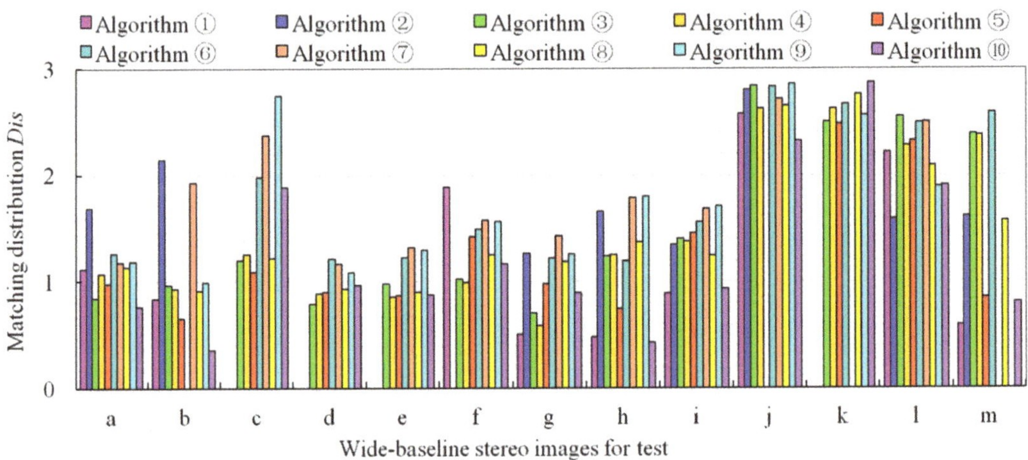

Figure 12. Comparison of the matching distribution quality of the ten algorithms.

3.3. Analysis and Discussion

First, we discuss the test results of compared methods as a whole. The results in Table 3 and Figures 11 and 12 show that no single algorithm always obtains the best performance on stereo images with different platform types, different viewpoint changes, and various texture structures. As a typical representative of handcrafted algorithms, ASIFT can achieve affine invariant stereo matching through a multi-level sampling strategy in 3D space; however, compared with the deep-learning algorithms, the test results of ASIFT show that its advantageis in the number of matches in close-range images with planar scene or satellite images. In contrast, the deep-learning algorithms DetNet + Contexdesc, AffNet + SOSNet, and SuperPointcan perform better on close-range stereo images with large rotations and scale changes, low-altitude stereo images with approximately planar scenes, and high-altitude stereo images with complex 3D scenes. This is because handcrafted algorithms tend to adopt the global spatial geometry rectification or a single segmentation model, which is more suitable for simple stereo images with planar scenes; whereas deep-learning algorithms build deep convolutional layers or fully connected neural network models from the perspective of emulating human visual cognition, and they iteratively learn the optimal network model parameters based on a large number of training samples, which can theoretically approximate any complex geometric or radiometric transform model, and hence this type of algorithm is more suitable for matching wide-baseline images with complex scenes. For test data (a), (b), (h), and (j), ASIFT yields better matching distribution quality; the algorithms DetNet + Contexdesc and AffNet + HardNet respectively perform well on data (c) and (d) with respect to matching distribution, whereas SuperPoint performs well on data (g) and (m) with respect to matching distribution. All compared algorithms consistently achieve poor matching distribution quality for data (c), (j), (k), and (l). This is mainly because the traditional problems of digital photogrammetry, such as large-scale deformation of images, lack of texture, terrain occlusion, and surface discontinuity, are still difficult for the available algorithms to handle. On this topic, we suggest that handcrafted algorithms may expand the search range of geometric transform parameters to enhance adaptability to large-scale deformation data, whereas deep-learning algorithms may also improve the matching compatibility of complex terrain by increasing the number of samples in such areas.

Second, we discuss the CNN architectures combining the used training datasets. Deep-learning wide-baseline image matching is mainly limited by the structure of the neural network model and the size of the training dataset. Table 3 and Figure 11 show that the SuperPoint algorithm can obtain the most matches from the complex 3D scene (data (g)–(j) and (m)) for UAV oblique stereo images (data (d)–(i), (j), and (k)) or satellite wide-baseline images (data (l) and (m)), but it almost fails on simple ground scenes (data (d)–(f)). Although the MSCOCO training dataset used by SuperPoint contains large-scale independent structural objects, it lacks ground scene annotation instances with a single texture, and hence this training dataset limits the matching performance of SuperPoint on the ground scenes. The AffNet+SOSNet algorithm can obtain a sufficient number of matches from wide-baseline images (data (d)–(f)) with ground scenes and poor texture, where the spatial distribution of the matches is relatively uniform, as presented in Figures 11 and 12. The reason is that the UBC Phototour and HPatches datasets cover a large number of homogeneous structures such as ground, wall, and sculpture structures, which enables the algorithm to enhance its perception of some scenes with a single texture. A comparison of the matching results of algorithms 3 and 4 shows that, even with the same training dataset, the feature description performance of SOSNet is substantially better than that of HardNet. Reviewing the structures of the two networks shows that on the basis of the HardNet, SOSNet embeds a second-order similarity regularization term in the loss function to avoid over-fitting problems in the model training and further improve the similarity and discrimination of the descriptors. The ContextDesc algorithm integrates visual and geometric context encoding structures into the network model to improve the use of image context information. The test results show that it is particularly suitable for image matching in scenes with cluttered background (data (c)) or large radiometric distortion (data (e)).

Third, we further discuss the strengths and weaknesses of integrating methods for the difficult test data. Although algorithms ③, ④, and ⑦ all adopt AffNet to extract affine invariant features, the test results of algorithms ③ and ④ are substantially better. We speculate the reason is that the deep-learning descriptors of algorithms ③ and ④ perform better than the handcrafted descriptor SIFT of algorithm 7. Figure 10 shows that the matching of both the deep learning and handcrafted algorithms is not able to achieve sub-pixel accuracy. The main reason is that the two stages of feature detection and feature matching are relatively independent, which makes it difficult for the corresponding points to be accurately aligned. The complete UAV dataset, which is larger in size and resolution (data (j) and (k)) was also used for testing. It should theoretically be beneficial for each algorithm to obtain more matches; however, Table 3 shows that the number of matches did not increase substantially as a result. We believe that a high resolution will exacerbate the lack of local texture in the image, and larger images tend to introduce more occluded regions. Specifically, data (j) contain more occluded scenes and homogenous textures, whereas data (k) involve a large area of water and scenes with viewpoint changes. Additionally, the ratio of the corresponding regions in the larger images is lower. Thus, it would become more difficult to obtain the corresponding features in the absence of prior knowledge or initial matches. For satellite wide-baseline images with various mountainous and urban areas, both the deep-learning approach SuperPoint and the handcrafted ASIFT method can obtain a significant number of matches.

4. Summary and Outlook

For wide-baseline image-matching problems, this paper systematically organized, analyzed, and summarized the existing deep-learning image invariant feature detection, description, and end-to-end matching models. In addition, the matching performances of the representative algorithms were evaluated and compared through comprehensive experiments on wide-baseline images. According to the above test results and discussion, future research and challenges can be summarized as follows.

(1) The current deep-learning invariant feature detection approach continues to reveal its potential, and the research on invariant features and their applications has increasingly developed, from the scale invariant feature learning of Reference [52] to the affine invariant feature training of Reference [53]. Experiments have shown that learning-based methods have better potential than handcrafted detection algorithms such as DoG [14] and pixel watershed [17]. In addition, the strategy of combining handcrafted methods with learning-

based methods [55] to extract invariant features has become a good option, but this type of method obviously depends on the accurate extraction of the image features by the handcrafted algorithms. In short, although the feature detection methods based on deep learning tend to show abilities beyond the traditional methods, this approach is not yet fully mature, especially for the matching problem of wide-baseline images with complex scenes and various textures, and it still faces great challenges. Therefore, extracting invariant features with high repeatability and stability needs further study.

(2) Deep-learning feature description is essentially metric learning; this kind of method is mainly focused on network model construction and loss function design. From the MatchNet Siamese network [20] to the SOSNet triplet structure [72], the model parameters are gradually simplified, and the performance is correspondingly improved. However, most network backbones still inherit the structure of the classic L2-Net [67]. Especially for the affine invariant feature description network structure, we suggest introducing a viewpoint transform module, which could enhance the transparency, perception, and generalization capabilities of existing models for wide-baseline images. Moreover, the loss function design is mainly used to select reasonable training samples. Although the existing functions focus on traditional problems such as the selection of positive and negative samples, they do not consider the inherent characteristics of wide-baseline images. Therefore, to improve the performance of the descriptors, future work could involve the construction of novel wide-baseline network structures or the design of universal loss functions.

(3) Recently, end-to-end learning-based methods such as back propagation-trained LIFT [77] and feature description-and-detection D2Net [86] have received increasing attention. This type of method has led to numerous innovations in terms of training mode and the automatic acquisition of training data. The research shows that the end-to-end method has a faster computation speed than other learning-based methods and can meet the performance requirements for simultaneous localization and mapping (SLAM) [68], SFM [80], and other real-time vision tasks. However, for wide-baseline image-matching tasks, it is difficult for this type of method to extract sufficient feature points. Therefore, in the field of wide-baseline image matching, we should further explore the end-to-end learning of unconventional and complex distortions as well as the image features of various textures and structures.

(4) Image matching based on deep learning is a data-driven image-matching method that must automatically learn the deep expression mechanism of ground surface features from a large amount of image data. Therefore, the key for deep-learning image matching is to build a diverse and massive training dataset. At present, the main training datasets for deep-learning wide-baseline image matching are UBC Phototour [68] and HPatches [69]. The UBC Phototour dataset contains a large number of artificial statues, whereas the HPatches dataset mainly consists of simple facade configurations. These available training datasets are very different from the data captured by aerial photography or satellite remote sensing, which stop the affine invariant network models, such as AffNet [57], HardNet [70], and SOSNet [72], from achieving the optimal matches in wide-baseline remote sensing images. Therefore, it is an urgent task to establish a large wide-baseline dataset of multi-source, multi-platform, and multi-spectrum data through crowd sourcing or sharing mechanisms.

(5) The existing studies have shown that the comprehensive performance of a model can be substantially improved through transfer learning, which has been widely applied in the fields of target recognition, image classification, and change detection. However, there are few published reports on transfer learning in the field of deep-learning wide-baseline image matching, specifically for feature detection, feature description, and end-to-end methods. Therefore, on the basis of establishing a wide-baseline image dataset, further work should focus on training a network for wide-baseline matching using a transfer learning strategy to achieve high-quality matching for wide-baseline images. In addition, for the original observation of matching points, the positioning errors must be fully considered

in the field of digital photogrammetry. However, the corresponding points across wide-baseline images cannot be registered precisely by learning-based methods. Hence, the matching accuracy could be improved by some optimization strategies, such as least-squares image matching or the Newton iteration method, which remains as future work.

5. Conclusions

In this paper, based on a review of the image-matching stages, we organized and summarized the development status and trends of existing learning-based methods. Moreover, the matching performance, advantages, and disadvantages of typical algorithms were evaluated through comprehensive experiments on the representative wide-baseline images. The results reveal that there is no algorithm that can adapt to all types of wide-baseline images with various viewpoint changes and texture structures. Therefore, the currently urgent task is to enhance the generalization ability of the models by combining a mixed model with more extensive training datasets. Moreover, it was suggested that a critical task is to construct deep network models with elastic receptive field and self-adaptive loss functions based on wide-baseline imaging properties and typical problems in image matching. It is our hope that the review work of this paper will act as a reference for future research.

Author Contributions: Conceptualization, A.Y. and G.Y.; data curation, L.Z. and G.Y.; validation, G.Y. and F.M.; writing—original draft preparation, G.Y.; writing—review and editing, A.Y.; supervision, A.Y. All authors have read and agreed to the published version of the manuscript.

Funding: This work is supported by the National Natural Science Foundation of China with Project No. 41601489, the Shandong Provincial Natural Science Foundation with Project No. ZR2015DQ007, the Postgraduate Education and Teaching Reform Foundation of Shandong Province with Project No. SDYJG19115, and the Open Fund of Key Laboratory of Geographic Information Science (Ministry of Education, East China Normal University) with Project No. KLGIS2021A02.

Acknowledgments: The authors would like to thank Karel Lenc, Dmytro Mishkin, and Daniel DeTonefor providing their key algorithms. We are very grateful also for the valuable comments and contributions of anonymous reviewers and members of the editorial team.

Conflicts of Interest: The authors declare no conflict of interest.

References

1. Cao, M.; Gao, H.; Jia, W. Stable image matching for 3D reconstruction in outdoor. *Int. J. Circuit Theory Appl.* **2021**, *49*, 2274–2289. [CrossRef]
2. Yao, J.; Qi, D.; Yao, Y.; Cao, F.; He, Y.; Ding, P.; Jin, C.; Jia, T.; Liang, J.; Deng, L.; et al. Total variation and block-matching 3D filtering-based image reconstruction for single-shot compressed ultrafast photography. *Opt. Lasers Eng.* **2020**, *139*, 106475. [CrossRef]
3. Park, S.-W.; Yoon, R.; Lee, H.; Lee, H.-J.; Choi, Y.-D.; Lee, D.-H. Impacts of Thresholds of Gray Value for Cone-Beam Computed Tomography 3D Reconstruction on the Accuracy of Image Matching with Optical Scan. *Int. J. Environ. Res. Public Health* **2020**, *17*, 6375. [CrossRef] [PubMed]
4. Zhang, Y.; Zhang, Z.; Gong, J. Generalized photogrammetry of spaceborne, airborne and terrestrial multi-source remote sensing datasets. *Acta Geod. Cartogr. Sin.* **2021**, *50*, 1–11. [CrossRef]
5. Chen, M.; Zhu, Q.; He, H.; Yan, S.; Zhao, Y. Structure adaptive feature point matching for urban area wide-baseline images with viewpoint variation. *Acta Geod. Cartogr. Sin.* **2019**, *48*, 1129–1140. [CrossRef]
6. Zhang, L.; Ai, H.; Xu, B.; Sun, Y.; Dong, Y. Automatic tie-point extraction based on multiple-image matching and bundle adjustment of large block of oblique aerial images. *Acta Geod. Cartogr. Sin.* **2017**, *46*, 554–564. [CrossRef]
7. Yao, G.; Deng, K.; Zhang, L.; Ai, H.; Du, Q. An algorithm of automatic quasi-dense matching and three-dimensional reconstruction for oblique stereo images. *Geomat. Informat. Sci. Wuhan Univ.* **2014**, *39*, 843–849.
8. Jin, Y.; Mishkin, D.; Mishchuk, A.; Matas, J.; Fua, P.; Yi, K.M.; Trulls, E. Image Matching across Wide Baselines: From Paper to Practice. *Int. J. Comput. Vis.* **2020**, *129*, 517–547. [CrossRef]
9. Sarlin, P.-E.; DeTone, D.; Malisiewicz, T.; Rabinovich, A. SuperGlue: Learning Feature Matching with Graph Neural Networks. In Proceedings of the IEEE/CVF Conference on Computer Vision and Pattern Recognition (CVPR), Seattle, WA, USA, 14–19 June 2020. [CrossRef]
10. Mikolajczyk, K.; Tuytelaars, T.; Schmid, C.; Zisserman, A.; Matas, J.; Schaffalitzky, F.; Kadir, T.; Van Gool, L. A Comparison of Affine Region Detectors. *Int. J. Comput. Vis.* **2005**, *65*, 43–72. [CrossRef]

11. Kasongo, S.M.; Sun, Y. A deep learning method with wrapper based feature extraction for wireless intrusion detection system. *Comput. Secur.* **2020**, *92*, 101752. [CrossRef]
12. Ma, J.; Jiang, X.; Fan, A.; Jiang, J.; Yan, J. Image Matching from Handcrafted to Deep Features: A Survey. *Int. J. Comput. Vis.* **2020**, *129*, 23–79. [CrossRef]
13. Chen, L.; Rottensteiner, F.; Heipke, C. Feature detection and description for image matching: From hand-crafted design to deep learning. *Geo-Spat. Inf. Sci.* **2020**, *24*, 58–74. [CrossRef]
14. Lowe, D.G. Distinctive Image Features from Scale-Invariant Keypoints. *Int. J. Comput. Vis.* **2004**, *60*, 91–110. [CrossRef]
15. Yao, G.; Cui, J.; Deng, K.; Zhang, L. Robust Harris Corner Matching Based on the Quasi-Homography Transform and Self-Adaptive Window for Wide-Baseline Stereo Images. *IEEE Trans. Geosci. Remote Sens.* **2017**, *56*, 559–574. [CrossRef]
16. Mikolajczyk, K. Scale & Affine Invariant Interest Point Detectors. *Int. J. Comput. Vis.* **2004**, *60*, 63–86. [CrossRef]
17. Matas, J.; Chum, O.; Urban, M.; Pajdla, T. Robust wide-baseline stereo from maximally stable extremal regions. *Image Vis. Comput.* **2004**, *22*, 761–767. [CrossRef]
18. Morel, J.-M.; Yu, G. ASIFT: A New Framework for Fully Affine Invariant Image Comparison. *SIAM J. Imaging Sci.* **2009**, *2*, 438–469. [CrossRef]
19. Zhang, Y.; Xia, G.; Wang, J.; Lha, D. A Multiple Feature Fully Convolutional Network for Road Extraction from High-Resolution Remote Sensing Image Over Mountainous Areas. *IEEE Geosci. Remote Sens. Lett.* **2019**, *16*, 1600–1604. [CrossRef]
20. Han, X.; Leung, T.; Jia, Y.; Sukthankar, R.; Berg, A.C. MatchNet: Unifying feature and metric learning for patch-based matching. In Proceedings of the IEEE Conference on Computer Vision and Pattern Recognition (CVPR), Boston, MA, USA, 7–12 June 2015; pp. 3279–3286. [CrossRef]
21. Sangwan, D.; Biswas, R.; Ghattamaraju, N. An effective analysis of deep learning based approaches for audio based feature extraction and its visualization. *Multimedia Tools Appl.* **2018**, *78*, 23949–23972. [CrossRef]
22. Yu, Y.; Li, X.; Liu, F. Attention GANs: Unsupervised Deep Feature Learning for Aerial Scene Classification. *IEEE Trans. Geosci. Remote Sens.* **2019**, *58*, 519–531. [CrossRef]
23. Alshaikhli, T.; Liu, W.; Maruyama, Y. Automated Method of Road Extraction from Aerial Images Using a Deep Convolutional Neural Network. *Appl. Sci.* **2019**, *9*, 4825. [CrossRef]
24. Saeedimoghaddam, M.; Stepinski, T.F. Automatic extraction of road intersection points from USGS historical map series using deep convolutional neural networks. *Int. J. Geogr. Inf. Sci.* **2019**, *34*, 947–968. [CrossRef]
25. Jiang, X.; Ma, J.; Xiao, G.; Shao, Z.; Guo, X. A review of multimodal image matching: Methods and applications. *Inf. Fusion* **2021**, *73*, 22–71. [CrossRef]
26. Cosgriff, C.V.; Celi, L.A. Deep learning for risk assessment: All about automatic feature extraction. *Br. J. Anaesth.* **2020**, *124*, 131–133. [CrossRef] [PubMed]
27. Maggipinto, M.; Beghi, A.; McLoone, S.; Susto, G.A. DeepVM: A Deep Learning-based approach with automatic feature extraction for 2D input data Virtual Metrology. *J. Process. Control* **2019**, *84*, 24–34. [CrossRef]
28. Sun, Y.; Yen, G.G.; Yi, Z. Evolving Unsupervised Deep Neural Networks for Learning Meaningful Representations. *IEEE Trans. Evol. Comput.* **2018**, *23*, 89–103. [CrossRef]
29. Lee, K.; Lim, J.; Ahn, S.; Kim, J. Feature extraction using a deep learning algorithm for uncertainty quantification of channelized reservoirs. *J. Pet. Sci. Eng.* **2018**, *171*, 1007–1022. [CrossRef]
30. Verdie, Y.; Yi, K.M.; Fua, P.; Lepetit, V. TILDE: A Temporally Invariant Learned DEtector. In Proceedings of the IEEE Conference on Computer Vision and Pattern Recognition (CVPR), Boston, MA, USA, 7–12 June 2015; pp. 5279–5288. [CrossRef]
31. Shukla, S.; Arac, A. A Step-by-Step Implementation of DeepBehavior, Deep Learning Toolbox for Automated Behavior Analysis. *J. Vis. Exp.* **2020**, e60763. [CrossRef]
32. Yan, M.; Li, Z.; Yu, X.; Jin, C. An End-to-End Deep Learning Network for 3D Object Detection From RGB-D Data Based on Hough Voting. *IEEE Access* **2020**, *8*, 138810–138822. [CrossRef]
33. Laguna, A.B.; Riba, E.; Ponsa, D.; Mikolajczyk, K. Key.Net: Keypoint Detection by Handcrafted and Learned CNN Filters. In Proceedings of the IEEECVF International Conference on Computer Vision (ICCV), Seoul, Korea, 27–29 October 2019. [CrossRef]
34. Balntas, V.; Riba, E.; Ponsa, D.; Mikolajczyk, K. Learning local feature descriptors with triplets and shallow convolutional neural networks. In Proceedings of the British Machine Vision Conference, York, UK, 19–22 September 2016. [CrossRef]
35. Zheng, X.; Pan, B.; Zhang, J. Power tower detection in remote sensing imagery based on deformable network and transfer learning. *Acta Geod. Cartogr. Sin.* **2020**, *49*, 1042–1050. [CrossRef]
36. Yao, Y.; Park, H.S. Multiview co-segmentation for wide baseline images using cross-view supervision. In Proceedings of the IEEE/CVF Winter Conference on Applications of Computer Vision (WACV), Snowmass, CL, USA, 1–5 March 2020; pp. 1942–1951.
37. Liu, J.; Wang, S.; Hou, X.; Song, W. A deep residual learning serial segmentation network for extracting buildings from remote sensing imagery. *Int. J. Remote Sens.* **2020**, *41*, 5573–5587. [CrossRef]
38. Zhu, Y.; Zhou, Z.; Liao, G.; Yuan, K. New loss functions for medical image registration based on VoxelMorph. In *Image Processing of Medical Imaging, Proceedings of the SPIE Medical Imaging, Houston, TX, USA, 15–20 February 2020*; p. 11313. [CrossRef]
39. Cao, Y.; Wang, Y.; Peng, J.; Zhang, L.; Xu, L.; Yan, K.; Li, L. DML-GANR: Deep Metric Learning with Generative Adversarial Network Regularization for High Spatial Resolution Remote Sensing Image Retrieval. *IEEE Trans. Geosci. Remote Sens.* **2020**, *58*, 8888–8904. [CrossRef]

40. Yang, Y.; Li, C. Quantitative analysis of the generalization ability of deep feedforward neural networks. *J. Intell. Fuzzy Syst.* **2021**, *40*, 4867–4876. [CrossRef]

41. Wang, L.; Qian, Y.; Kong, X. Line and point matching based on the maximum number of consecutive matching edge segment pairs for large viewpoint changing images. *Signal Image Video Process.* **2021**, 1–8. [CrossRef]

42. Zheng, B.; Qi, S.; Luo, G.; Liu, F.; Huang, X.; Guo, S. Characterization of discontinuity surface morphology based on 3D fractal dimension by integrating laser scanning with ArcGIS. *Bull. Int. Assoc. Eng. Geol.* **2021**, *80*, 2261–2281. [CrossRef]

43. Ma, Y.; Peng, S.; Jia, Y.; Liu, S. Prediction of terrain occlusion in Change-4 mission. *Measures* **2020**, *152*. [CrossRef]

44. Zhang, X.; Zhu, X. Efficient and de-shadowing approach for multiple vehicle tracking in aerial video via image segmentation and local region matching. *J. Appl. Remote Sens.* **2020**, *14*, 014503. [CrossRef]

45. Yuan, X.; Yuan, W.; Xu, S.; Ji, Y. Research developments and prospects on dense image matching in photogrammetry. *Acta Geod. Cartogr. Sin.* **2019**, *48*, 1542–1550.

46. Liu, J.; Ji, S. Deep learning based dense matching for aerial remote sensing images. *Acta Geod. Cartogr. Sin.* **2019**, *48*, 1141–1150. [CrossRef]

47. Chen, X.; He, H.; Zhou, J.; An, P.; Chen, T. Progress and future of image matching in low-altitude photogrammetry. *Acta Geod. Cartogr. Sin.* **2019**, *48*, 1595–1603. [CrossRef]

48. Li, Y.; Huang, X.; Liu, H. Unsupervised Deep Feature Learning for Urban Village Detection from High-Resolution Remote Sensing Images. *Photogramm. Eng. Remote Sens.* **2017**, *83*, 567–579. [CrossRef]

49. Chen, Q.; Liu, T.; Shang, Y.; Shao, Z.; Ding, H. Salient Object Detection: Integrate Salient Features in the Deep Learning Framework. *IEEE Access* **2019**, *7*, 152483–152492. [CrossRef]

50. Xu, D.; Wu, Y. FE-YOLO: A Feature Enhancement Network for Remote Sensing Target Detection. *Remote Sens.* **2021**, *13*, 1311. [CrossRef]

51. Lenc, K.; Vedaldi, A. Learning Covariant Feature Detectors. In Proceedings of the ECCV Workshop on Geometry Meets Deep Learning, Amsterdam, The Netherlands, 31 August–1 September 2016; pp. 100–117. [CrossRef]

52. Zhang, X.; Yu, F.X.; Karaman, S.; Chang, S.-F. Learning Discriminative and Transformation Covariant Local Feature Detectors. In Proceedings of the IEEE Conference on Computer Vision and Pattern Recognition (CVPR), Honolulu, HI, USA, 21–26 July 2017; pp. 4923–4931. [CrossRef]

53. Doiphode, N.; Mitra, R.; Ahmed, S.; Jain, A. An Improved Learning Framework for Covariant Local Feature Detection. In Proceedings of the Asian Conference on Computer Vision (ACCV), Perth, Australia, 2–6 December 2019; pp. 262–276. [CrossRef]

54. Hoffer, E.; Ailon, N. Deep Metric Learning Using Triplet Network. In Proceedings of the International Conference on Learning Representations, San Diego, CA, USA, 7–9 May 2015; pp. 84–92. [CrossRef]

55. Yi, K.M.; Verdie, Y.; Fua, P.; Lepetit, V. Learning to Assign Orientations to Feature Points. In Proceedings of the IEEE Conference on Computer Vision and Pattern Recognition (CVPR), Las Vegas, NV, USA, 26 June–1 July 2016; pp. 107–116. [CrossRef]

56. Zitnick, C.L.; Ramnath, K. Edge foci interest points. In Proceedings of the IEEE International Conference on Computer Vision (ICCV), Barcelona, Spain, 6–13 November 2011; pp. 359–366. [CrossRef]

57. Mishkin, D.; Radenović, F.; Matas, J. Repeatability Is Not Enough: Learning Affine Regions via Discriminability. In *European Conference on Computer Vision (ECCV)*; Springer: Cham, Switzerland, 2018; pp. 287–304. [CrossRef]

58. Savinov, N.; Seki, A.; Ladicky, L.; Sattler, T.; Plooeleys, M. Quad-networks: Unsupervised learning to rank for interest point detection. In Proceedings of the IEEE Conference on Computer Vision and Pattern Recognition, Honolulu, HI, USA, 21–26 July 2017; pp. 1822–1830.

59. De Vos, B.D.; Berendsen, F.F.; Viergever, M.A.; Sokooti, H.; Staring, M.; Išgum, I. A deep learning framework for unsupervised affine and deformable image registration. *Med. Image Anal.* **2019**, *52*, 128–143. [CrossRef] [PubMed]

60. Abdullah, T.; Bazi, Y.; Al Rahhal, M.M.; Mekhalfi, M.L.; Rangarajan, L.; Zuair, M. TextRS: Deep Bidirectional Triplet Network for Matching Text to Remote Sensing Images. *Remote Sens.* **2020**, *12*, 405. [CrossRef]

61. Wei, X.; Zhang, Y.; Gong, Y.; Zheng, N. Kernelized subspace pooling for deep local descriptors. In Proceedings of the IEEE/CVF Conference on Computer Vision and Pattern Recognition (CVPR), Salt Lake City, UT, USA, 18–23 June 2018.

62. Simo-Serra, E.; Trulls, E.; Ferraz, L.; Kokkinos, I.; Fua, P.; Moreno-Noguer, F. Discriminative Learning of Deep Convolutional Feature Point Descriptors. In Proceedings of the IEEE International Conference on Computer Vision (ICCV), Santiago, Chile, 7–13 December 2015; pp. 118–126. [CrossRef]

63. Zagoruyko, S.; Komodakis, N. Learning to compare image patches via convolutional neural networks. In Proceedings of the IEEE Conference on Computer Vision and Pattern Recognition (CVPR), Boston, MA, USA, 7–12 June 2015; pp. 4353–4361.

64. Simonyan, K.; Zisserman, A. Very deep convolutional networks for large-scale image recognition. In Proceedings of the IEEE Conference on Computer Vision and Pattern Recognition (CVPR), Columbus, OH, USA, 23–28 June 2014. *IEEE Trans. Pattern Anal. Mach. Intell.* **2014**, 346–361. [CrossRef]

65. He, K.; Zhang, X.; Ren, S.; Sun, J. Spatial pyramid pooling in deep convolutional networks for visual recognition. *IEEE T. Pattern. Anal.* **2014**, *37*, 1904–1916. [CrossRef] [PubMed]

66. Fan, D.; Dong, Y.; Zhang, Y. Satellite image matching method based on deep convolution neural network. *Acta Geod. Cartogr. Sin.* **2018**, *47*, 844–853. [CrossRef]

67. Tian, Y.; Fan, B.; Wu, F. L2-Net: Deep Learning of Discriminative Patch Descriptor in Euclidean Space. In Proceedings of the IEEE Conference on Computer Vision and Pattern Recognition (CVPR), Honolulu, HI, USA, 21–26 July 2017; pp. 6128–6136. [CrossRef]

68. A Brown, M.; Hua, G.; Winder, S. Discriminative Learning of Local Image Descriptors. *IEEE Trans. Pattern Anal. Mach. Intell.* **2010**, *33*, 43–57. [CrossRef] [PubMed]
69. Balntas, V.; Lenc, K.; Vedaldi, A.; Mikolajczyk, K. HPatches: A Benchmark and Evaluation of Handcrafted and Learned Local Descriptors. In Proceedings of the IEEE Conference on Computer Vision and Pattern Recognition (CVPR), Honolulu, HI, USA, 21–26 July 2017; pp. 3852–3861. [CrossRef]
70. Mishchuk, A.; Mishkin, D.; Radenovic, F. Working hard to know your neighbor's margins: Local descriptor learning loss. In Proceedings of the Neural Information Processing Systems, Long Beach, CA, USA, 4–9 December 2017; pp. 4826–4837.
71. Ebel, P.; Mishchuk, A.; Yi, K.M.; Fua, P.; Trulls, E. Beyond cartesian representations for local descriptors. In Proceedings of the IEEE International Conference on Computer Vision, Seoul, Korea, 27–28 October 2019; pp. 253–262.
72. Tian, Y.; Yu, X.; Fan, B.; Wu, F.; Heijnen, H.; Balntas, V. SOSNet: Second Order Similarity Regularization for Local Descriptor Learning. In Proceedings of the Computer Vision and Pattern Recognition, Long Beach, CA, USA, 16–20 June 2019; pp. 11008–11017. [CrossRef]
73. Luo, Z.; Shen, T.; Zhou, L.; Zhu, S.; Zhang, R.; Yao, Y.; Fang, T.; Quan, L. GeoDesc: Learning Local Descriptors by Integrating Geometry Constraints. In Proceedings of the European Conference on Computer Vision (ECCV), Munich, Germany, 8–14 September 2018; pp. 170–185. [CrossRef]
74. Luo, Z.; Shen, T.; Zhou, L.; Zhang, J.; Yao, Y.; Li, S.; Fang, T.; Quan, L. ContextDesc: Local Descriptor Augmentation with Cross-Modality Context. In Proceedings of the IEEE Conference on Computer Vision and Pattern Recognition (CVPR), Long Beach, CA, USA, 16–20 June 2019; pp. 2522–2531. [CrossRef]
75. Yao, G.; Yilmaz, A.; Zhang, L.; Meng, F.; Ai, H.; Jin, F. Matching Large Baseline Oblique Stereo Images Using an End-To-End Convolutional Neural Network. *Remote Sens.* **2021**, *13*, 274. [CrossRef]
76. Mahapatra, D.; Ge, Z. Training data independent image registration using generative adversarial networks and domain adaptation. *Pattern Recognit.* **2019**, *100*, 107109. [CrossRef]
77. Yi, K.M.; Trulls, E.; Lepetit, V.; Fua, P. LIFT: Learned Invariant Feature Transform. In Proceedings of the European Conference on Computer Vision (ECCV), Amsterdam, The Netherlands, 11–14 October 2016; pp. 467–483. [CrossRef]
78. Jaderberg, M.; Simonyan, K.; Zisserman, A. Spatial transformer networks. *Adv. Neural Inf. Process. Syst.* **2015**, *28*, 2017–2025.
79. Chapelle, O.; Wu, M. Gradient descent optimization of smoothed information retrieval metrics. *Inf. Retr.* **2009**, *13*, 216–235. [CrossRef]
80. Zhu, S.; Zhang, R.; Zhou, L.; Shen, T.; Fang, T.; Tan, P.; Quan, L. Very Large-Scale Global SfM by Distributed Motion Averaging. In Proceedings of the IEEE/CVF Conference on Computer Vision and Pattern Recognition, Salt Lake City, UT, USA, 18–23 June 2018; pp. 4568–4577. [CrossRef]
81. DeTone, D.; Malisiewicz, T.; Rabinovich, A. SuperPoint: Self-Supervised Interest Point Detection and Description. In Proceedings of the IEEE Conference on Computer Vision and Pattern Recognition Workshops, Salt Lake City, UT, USA, 18–22 June 2018; pp. 337–33712. [CrossRef]
82. Li, H.; Li, F. Image Encode Method Based on IFS with Probabilities Applying in Image Retrieval. In Proceedings of the Fourth Global Congress on Intelligent Systems (GCIS), Hong Kong, China, 2–3 December 2013; pp. 291–295. [CrossRef]
83. Lie, W.-N.; Gao, Z.-W. Video Error Concealment by Integrating Greedy Suboptimization and Kalman Filtering Techniques. *IEEE Trans. Circuits Syst. Video Technol.* **2006**, *16*, 982–992. [CrossRef]
84. Revaud, J.; Weinzaepfel, P.; De, S. R2D2: Repeatable and reliable detector and descriptor. *arXiv* **2019**, arXiv:1906.06195.
85. Ono, Y.; Trulls, E.; Fua, P.; Mooyi, K. LF-Net: Learning local features from images. In Proceedings of the Neural Information Processing Systems, Montreal, QC, Canada, 3–8 December 2018; pp. 6234–6244.
86. Dusmanu, M.; Rocco, I.; Pajdla, T.; Pollefeys, M.; Sivic, J.; Torii, A.; Sattler, T. D2-Net: A Trainable CNN for Joint Description and Detection of Local Features. In Proceedings of the IEEE Conference on Computer Vision and Pattern Recognition (CVPR), Long Beach, CA, USA, 16–20 June 2019; pp. 8084–8093. [CrossRef]
87. Xu, X.-F.; Zhang, L.; Duan, C.-D.; Lu, Y. Research on Inception Module Incorporated Siamese Convolutional Neural Networks to Realize Face Recognition. *IEEE Access* **2019**, *8*, 12168–12178. [CrossRef]
88. Li, J.; Xie, Y.; Li, C.; Dai, Y.; Ma, J.; Dong, Z.; Yang, T. UAV-Assisted Wide Area Multi-Camera Space Alignment Based on Spatiotemporal Feature Map. *Remote Sens.* **2021**, *13*, 1117. [CrossRef]
89. Hasheminasab, S.M.; Zhou, T.; Habib, A. GNSS/INS-Assisted Structure from Motion Strategies for UAV-Based Imagery over Mechanized Agricultural Fields. *Remote Sens.* **2020**, *12*, 351. [CrossRef]
90. Lee, S.-H.; Yoo, J.; Park, M.; Kim, J.; Kwon, S. Robust Extrinsic Calibration of Multiple RGB-D Cameras with Body Tracking and Feature Matching. *Sensors* **2021**, *21*, 1013. [CrossRef] [PubMed]
91. Lin, T.-Y.; Maire, M.; Belongie, S.; Hays, J.; Perona, P.; Ramanan, D.; Dollár, P.; Zitnick, C.L. Microsoft COCO: Common Objects in Context. In Proceedings of the European Conference on Computer Vision, Zurich, Switzerland, 6–12 September 2014; Springer International Publishing: Cham, Switzerland; pp. 740–755. [CrossRef]
92. Li, Z.; Snavely, N. MegaDepth: Learning Single-View Depth Prediction from Internet Photos. In Proceedings of the 2018 IEEE/CVF Conference on Computer Vision and Pattern Recognition (CVPR), Salt Lake City, UT, USA, 18–23 June 2018; pp. 2041–2050. [CrossRef]

93. Shen, T.; Luo, Z.; Zhou, L.; Zhang, R.; Zhu, S.; Fang, T.; Quan, L. Matchable Image Retrieval by Learning from Surface Reconstruction. In *Computer Vision–ACCV, Proceedings of the 14th Asian Conference on Computer Vision, Perth, Australia, 2–6 December 2019*; Springer: Berlin/Heidelberg, Germany, 2019; pp. 415–431. [CrossRef]
94. Aanæs, H.; Jensen, R.R.; Vogiatzis, G.; Tola, E.; Dahl, A.B. Large-Scale Data for Multiple-View Stereopsis. *Int. J. Comput. Vis.* **2016**, *120*, 153–168. [CrossRef]
95. Yao, G.; Deng, K.; Zhang, L.; Yang, H.; Ai, H. An automated registration method with high accuracy for oblique stereo images based on complementary affine invariant features. *Acta Geod. Cartogr. Sin.* **2013**, *42*, 869–876. [CrossRef]
96. Zhu, Q.; Wu, B.; Xu, Z.-X.; Qing, Z. Seed Point Selection Method for Triangle Constrained Image Matching Propagation. *IEEE Geosci. Remote Sens. Lett.* **2006**, *3*, 207–211. [CrossRef]

remote sensing

MDPI

Article

Camera Self-Calibration with GNSS Constrained Bundle Adjustment for Weakly Structured Long Corridor UAV Images

Wei Huang [1], San Jiang [2,*] and Wanshou Jiang [1,3]

[1] State Key Laboratory of Information Engineering in Surveying, Mapping and Remote Sensing, Wuhan University, 129 Luoyu Road, Wuhan 430079, China; hw1006@whu.edu.cn (W.H.); jws@whu.edu.cn (W.J.)
[2] School of Computer Science, China University of Geosciences, Wuhan 430074, China
[3] Collaborative Innovation Center of Geospatial Technology, Wuhan University, 129 Luoyu Road, Wuhan 430079, China
*   Correspondence: jiangsan@cug.edu.cn

Abstract: Camera self-calibration determines the precision and robustness of AT (aerial triangulation) for UAV (unmanned aerial vehicle) images. The UAV images collected from long transmission line corridors are critical configurations, which may lead to the "bowl effect" with camera self-calibration. To solve such problems, traditional methods rely on more than three GCPs (ground control points), while this study designs a new self-calibration method with only one GCP. First, existing camera distortion models are grouped into two categories, i.e., physical and mathematical models, and their mathematical formulas are exploited in detail. Second, within an incremental SfM (Structure from Motion) framework, a camera self-calibration method is designed, which combines the strategies for initializing camera distortion parameters and fusing high-precision GNSS (Global Navigation Satellite System) observations. The former is achieved by using an iterative optimization algorithm that progressively optimizes camera parameters; the latter is implemented through inequality constrained BA (bundle adjustment). Finally, by using four UAV datasets collected from two sites with two data acquisition modes, the proposed algorithm is comprehensively analyzed and verified, and the experimental results demonstrate that the proposed method can dramatically alleviate the "bowl effect" of self-calibration for weakly structured long corridor UAV images, and the horizontal and vertical accuracy can reach 0.04 m and 0.05 m, respectively, when using one GCP. In addition, compared with open-source and commercial software, the proposed method achieves competitive or better performance.

Keywords: digital photogrammetry; camera self-calibration; Brown model; polynomial model; aerial triangulation

Citation: Huang, W.; Jiang, S.; Jiang, W. Camera Self-Calibration with GNSS Constrained Bundle Adjustment for Weakly Structured Long Corridor UAV Images. *Remote Sens.* **2021**, *13*, 4222. https://doi.org/10.3390/rs13214222

Academic Editors: Francesco Nex and Marco Piras

Received: 14 September 2021
Accepted: 18 October 2021
Published: 21 October 2021

1. Introduction

With the advantages of flexible data acquisition and ease of use, UAV has become one of the most important remote sensing platforms for the photogrammetry and remote sensing community [1]. The UAVs have the characteristics of small size, autonomous vertical take-off and landing with low site requirements, high flight safety performance, and the flexibility to adjust the direction of flight, making them widely used for the regular inspection of high-voltage transmission line corridors [2–6]. However, the UAV platforms are often equipped with consumer-grade, non-metric digital cameras, mainly due to the limitations of the platform's load capacity. These cameras have non-ignored lens distortions when compared with metric sensors, which influences the robustness and precision of AT. The long corridor structure of UAV images is a critical configuration, and the reconstructed model would be bending with the inaccurately estimated distortion parameters. For example, Figure 1 illustrates the "bowl effect" of self-calibration with commercial software and open-source software for long corridor UAV images because of the failure in camera distortion parameter estimation. Thus, it is very critical to accurately estimate the distortion parameters of cameras for high-precision AT.

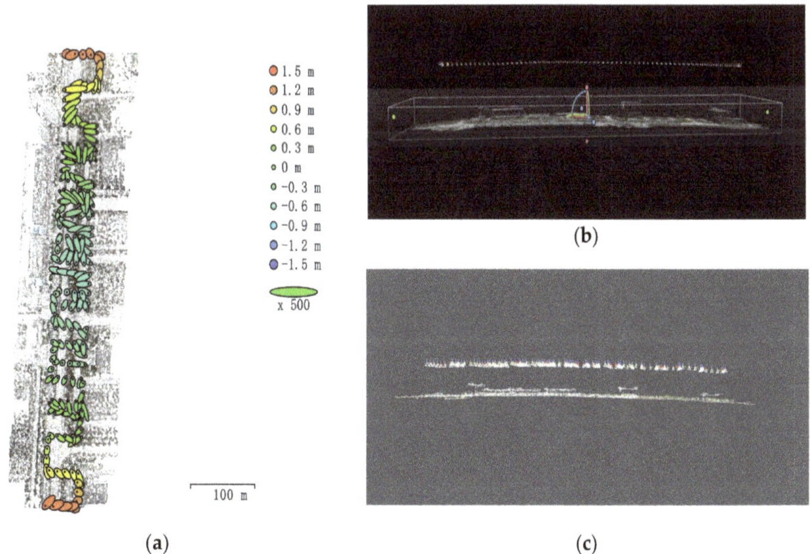

Figure 1. The "bowl effect" of self-calibration BA for long corridor UAV images. (**a**) The result of commercial software AgiSoft Photoscan V1.4.1, Z error is represented by the ellipse color, and X, Y errors are represented by the ellipse shape. (**b**) The result of commercial software Reality Capture V1.2, and (**c**) the result of open-source software AliceVision.

In photogrammetry and computer vision fields, the purpose of camera self-calibration is to solve the internal orientation parameters and camera lens distortion parameters, which affect the precision of 3D reconstruction. At present, the existing camera calibration methods can be categorized into two groups: pre-calibration and self-calibration. The pre-calibration method usually depends on either the indoor calibration board with fixed patterns [7,8] or the outer door large-scale 3D calibration test field [9]. However, this kind of method has many disadvantages applied in UAV image calibration. On the one hand, the indoor calibration errors of the calibration board with special patterns would be expanded with the increase of UAV flying height, and the focal length of the camera is usually set to be fixed depending on the flying height of UAVs, which is not suitable for indoor camera pre-calibration; on the other hand, although the outer door 3D calibration test field could improve the precision of camera calibration, it requires a lot of manpower and consumes time. Compared with the drawbacks of pre-calibration, the self-calibration process is simpler and more convenient without any known calibration targets.

With the above considerations, the relevant scholars have performed in-depth research on the camera self-calibration methods with different camera distortion models, including the physical model and mathematical model. Among the camera distortion models, the Brown model [10] and its improved model [11] are the most classical physical models. However, it is a challenge for camera self-calibration to occur in the critical configuration, due to the high correlation between the distortion parameters of the Brown model [12]. In the field of computer vision, the division model is another kind of physical model commonly used [13], which can fit simple camera distortion. Recently, many researchers have combined the division model with the fundamental matrix or the essential matrix to solve the camera distortion parameters by establishing polynomial equations [14–17]. However, it cannot fit complex camera distortion and is not suitable for UAV camera self-calibration. The physical camera model cannot describe the complex distortion precisely and may not work when the pattern of camera distortion is not apparent and the precise knowledge on distortion is unavailable. Based on this consideration, the mathematical model tries to use function approximation theory to accurately fit complex camera distortion, such as the quadratic orthogonal polynomial [18] and quartic orthogonal polynomial model [19]. Tang et al. [20,21] proposed the orthogonal polynomial models based on Legendre and Fourier polynomials and applied the models in an aerial camera for self-calibration. Subsequently, Babapour et al. [22] presented the Chebyshev–Fourier and Jacobi–Fourier camera models, which significantly improved the horizontal and elevation accuracy of aerial images. However, few studies have applied the Legendre- and Fourier-based distortion models to camera self-calibration for long corridor UAV images. This is the first key research content of this paper.

For the camera self-calibration with a long corridor structure, the related research can be divided into three categories: the research on the theoretic analysis [23,24], the research on the strategies of self-calibration [25,26], and the accuracy verification with such structures [27–32]. Wu et al. [23] analyzed the motion field of images with radial distortion and proved the ambiguous reconstruction with the "bowl effect" of camera self-calibration under weak structures and configuration through mathematical theory. Zhou et al. [24] discussed the impact of the focal length parameter estimation of camera self-calibration with a flat, corridor configuration. The research on the theoretic analysis of camera self-calibration with a long corridor structure has focused on investigating the causes and influencing factors, but has not presented any solutions to solve the problems. For the long corridor structure, Tournadre et al. [25] presented a 7th-order polynomial combined with radial camera distortion model (F15P7) and verified the accuracy of the orientations with a weak configuration using ground control points (GCP). Although this method can alleviate the "bowl effect", it relied on more than three GCPs for absolute image orientation. Polic et al. [26] proposed an uncertainty-based camera model selection method to reduce the "bowl effect", but this method did not consider the newest mathematical-based distortion models and high-precision GNSS observations. Griffiths et al. [27] analyzed the accuracy of 3D reconstruction from long corridor structure UAV images in detail, and experiments show that the more complex distortion model can improve the accuracy of camera self-calibration. The related works of [28–32] are mainly focused on accessing the accuracy of the DSM (Digital Surface Model), DTM (Digital Terrain Model), the influence of the distribution of GCPs, and on giving suggestions for data collection without improvement of the strategies for camera self-calibration. Compared to the existing literature about self-calibration with a long corridor structure, the proposed paper extends the scope of research on camera distortion models and investigates the accuracy of the recently proposed orthogonal polynomial model with the strategies for camera parameter initialization and high-precision GNSS fusion in a long corridor structure.

For the traditional aerial photogrammetry in surveying and mapping, the UAV platform generally collects image data with the regular region and often has multiple parallel and overlapping stable structures. However, in the application of UAV inspection for power lines, only rectangle or S-shaped strip flight trajectory is adopted to collect image data, due to cost considerations. Since the constraints between the long corridor structure are reduced to the minimum, the correlation of camera intrinsic parameters and external parameters cannot be restricted with the stability structure of images, which leads to the "bowl effect" phenomenon and affects the relative and absolute accuracy of 3D reconstruction. At present, most UAVs are equipped with centimeter-level high-precision differential GNSS, which can provide better initial position parameters to constrain the camera projection centers [33,34]. Traditional technology for fusing high-precision GNSS locations and oriented images of SfM is to minimize a weighted sum of image and GNSS errors. However, when the structure of images is degenerated and unstable, the oriented images bend after camera self-calibration. In this situation, the traditional technology of fusing GNSS locations and SfM would not align the projection centers of the image to the GNSS locations, which cannot eliminate the "bowl effect". How to use the high-precision

GNSS information to significantly alleviate the "bowl effect" is the second key research content of this paper.

To overcome the problem, this paper firstly investigates the classical physical model and orthogonal polynomial model for camera self-calibration. Then, a new strategy combined with the parameter initialization of UAV images and high-precision GNSS observation fusion is proposed for camera self-calibration with the physical model and orthogonal polynomial model. Finally, four UAV image datasets are used in the experiments for camera self-calibration, which illustrates the feasibility of this strategy. Compared with the related works of long corridor camera self-calibration, the proposed method has the following contributions: (1) the accuracy of the newest mathematical camera distortion models is investigated and verified, which can achieve better accuracy in the vertical direction; (2) a new strategy of camera self-calibration for long corridor UAV images is proposed, which can alleviate the "bowl effect" with the high-precision GNSS locations for direct georeferencing; (3) compared with the traditional method, which needs more than three GCPs to solve the problem of "bowl effect", the proposed method achieves competitive accuracy with only one GCP constraint, which is meaningful for the UAV photogrammetric community.

This paper is organized as follows. Section 2 presents the camera distortion model and the proposed camera self-calibration method with the camera parameters' initialization and high-precision GNSS fusion in detail. In Section 3, UAV datasets and experimental results are presented and discussed. Section 4 concludes the results of this study and presents future work.

2. Methodology

The proposed method mainly studies camera self-calibration in the incremental SfM framework, which is used to accurately estimate the image orientation and camera intrinsic parameters. Firstly, the most commonly used camera distortion models are analyzed in Section 2.1, including physical models and mathematical models. Secondly, the BA with inequality constraint is investigated in detail in Section 2.2. Finally, the camera self-calibration strategy is introduced for the long corridor structure of UAV images in Section 2.3. Figure 2 describes our main research contents. The camera distortion model introduced in Section 2.1 is applied in the proposed camera self-calibration method, and the bundle adjustment with inequality constraint described in Section 2.2 is applied in the absolute orientation of the proposed camera self-calibration method.

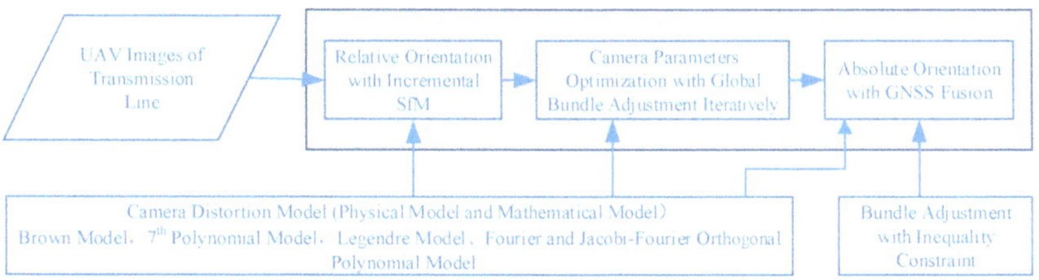

Figure 2. The camera self-calibration workflow for long corridor UAV images.

2.1. Camera Distortion Model

The classical Brown camera model in the field of photogrammetry and computer vision, the 7th polynomial model, and the Legendre, Fourier, and Jacobi–Fourier orthogonal polynomial models are compared and analyzed. The mathematical form of each camera distortion model is described below.

2.1.1. Brown Model

The Brown distortion model [10] has been widely used in the photogrammetry and computer vision field. The parameters of the Brown model mainly include symmetrical radial distortion, tangential distortion, and in-plane distortion [35,36]. The mathematical equations are as shown below.

$$\Delta x = \overline{x}(k_1 r^2 + k_2 r^4 + k_3 r^6) + p_1(r^2 + 2\overline{x}^2) + 2p_2\overline{xy} + b_1\overline{x} + b_2\overline{y}$$
$$\Delta y = \overline{y}(k_1 r^2 + k_2 r^4 + k_3 r^6) + p_1(r^2 + 2\overline{y}^2) + 2p_2\overline{xy} \tag{1}$$
$$r = \sqrt{(x - x_0)^2 + (y - y_0)^2} = \sqrt{\overline{x}^2 + \overline{y}^2}$$

where x_0, y_0 is the principal point of the image, k_1, k_2, k_3 are the radial distortion coefficients, p_1, p_2 are the tangential distortion coefficients, and b_1, b_2 are the in-plane distortion coefficients, which are named as the affinity and the shear terms. The in-plane coefficients represent different scaling factors and non-orthogonal pixels in the image along the x- and y-axis.

2.1.2. 7th Polynomial Model

The 7th polynomial camera distortion model is provided by the open-source software MicMac [25], and it contains a total of 66 parameters. The distortion functions are as follows:

$$
\begin{aligned}
\Delta x = {} & a_0\overline{x} + a_1\overline{y} - 2a_2\overline{x}^2 + a_3\overline{xy} + a_4\overline{y}^2 + \\
& a_6\overline{x}^3 + a_7\overline{x}^2\overline{y} + a_8\overline{xy}^2 + a_9\overline{y}^3 + \\
& a_{14}\overline{x}^4 + a_{15}\overline{x}^3\overline{y} + a_{16}\overline{x}^2\overline{y}^2 + a_{17}\overline{xy}^3 + a_{18}\overline{y}^4 + \\
& a_{24}\overline{x}^5 + a_{25}\overline{x}^4\overline{y} + a_{26}\overline{x}^3\overline{y}^2 + a_{27}\overline{x}^2\overline{y}^3 + a_{28}\overline{xy}^4 + a_{29}\overline{y}^5 + \\
& a_{36}\overline{x}^6 + a_{37}\overline{x}^5\overline{y} + a_{38}\overline{x}^4\overline{y}^2 + a_{39}\overline{x}^3\overline{y}^3 + a_{40}\overline{x}^2\overline{y}^4 + a_{41}\overline{xy}^5 + a_{42}\overline{y}^6 + \\
& a_{50}\overline{x}^7 + a_{51}\overline{x}^6\overline{y} + a_{52}\overline{x}^5\overline{y}^2 + a_{53}\overline{x}^4\overline{y}^3 + a_{54}\overline{x}^3\overline{y}^4 + a_{55}\overline{x}^2\overline{y}^5 + a_{56}\overline{xy}^6 + a_{57}\overline{y}^7 \\
\Delta y = {} & -a_0\overline{y} + a_1\overline{x} + a_2\overline{xy} - 2a_3\overline{y}^2 + a_5\overline{x}^2 + \\
& a_{10}\overline{x}^3 + a_{11}\overline{x}^2\overline{y} + a_{12}\overline{xy}^2 + a_{13}\overline{y}^3 + \\
& a_{19}\overline{x}^4 + a_{20}\overline{x}^3\overline{y} + a_{21}\overline{x}^2\overline{y}^2 + a_{22}\overline{xy}^3 + a_{23}\overline{y}^4 + \\
& a_{30}\overline{x}^5 + a_{31}\overline{x}^4\overline{y} + a_{32}\overline{x}^3\overline{y}^2 + a_{33}\overline{x}^2\overline{y}^3 + a_{34}\overline{xy}^4 + a_{35}\overline{y}^5 + \\
& a_{43}\overline{x}^6 + a_{44}\overline{x}^5\overline{y} + a_{45}\overline{x}^4\overline{y}^2 + a_{46}\overline{x}^3\overline{y}^3 + a_{47}\overline{x}^2\overline{y}^4 + a_{48}\overline{xy}^5 + a_{49}\overline{y}^6 + \\
& a_{58}\overline{x}^7 + a_{59}\overline{x}^6\overline{y} + a_{60}\overline{x}^5\overline{y}^2 + a_{61}\overline{x}^4\overline{y}^3 + a_{62}\overline{x}^3\overline{y}^4 + a_{63}\overline{x}^2\overline{y}^5 + a_{64}\overline{xy}^6 + a_{65}\overline{y}^7
\end{aligned} \tag{2}
$$

where $\overline{x}, \overline{y}$ are the same as the definition in the Brown model, and $a_0, a_1, \ldots, a_{65}$ are the coefficients of the 7th polynomial. Six coefficients are eliminated to reduce the correlation between those coefficients.

2.1.3. Legendre Model

The Legendre model is composed of a series of orthogonal polynomials, which greatly reduce the correlation between distortion coefficients. This model has been used in the camera self-calibration for professional digital mapping cameras. However, the feasibility of a consumer-grade camera applicated in camera self-calibration needs to be verified. Therefore, the Legendre orthogonal polynomial model with 66 parameters is introduced in this paper. The mathematical expression is presented in Formula (3), where $P_{m,n} = 0^{-6} l_m(\overline{x}) l_n(\overline{y})$, $m, n \in [0, 5]$; $\overline{x} = \frac{x - w/2}{w}$, $\overline{y} = \frac{y - h/2}{h}$; x, y are the pixel coordinates in the image; and w and h represent the width and height of the image, respectively. $l_m(\overline{x})$ and $l_n(\overline{y})$ mean the Legendre polynomials and $a_0, a_1, \ldots, a_{65}$ are the coefficients. Similar to the 7th polynomial distortion model, six coefficients are eliminated in the Legendre model.

$$\begin{aligned}
\Delta x = {} & a_0 p_{1,0} + a_1 p_{0,1} + a_2 p_{2,0} + a_3 p_{1,1} + a_4 p_{0,2} + a_5 p_{3,0} + a_6 p_{2,1} + \\
& a_7 p_{1,2} + a_8 p_{0,3} + a_9 p_{4,0} + a_{10} p_{3,1} + a_{11} p_{2,2} + a_{12} p_{1,3} + a_{13} p_{0,4} + \\
& a_{14} p_{5,0} + a_{15} p_{4,1} + a_{16} p_{3,2} + a_{17} p_{2,3} + a_{18} p_{1,4} + a_{19} p_{0,5} + a_{20} p_{5,1} + \\
& a_{21} p_{4,2} + a_{22} p_{3,3} + a_{23} p_{2,4} + a_{24} p_{1,5} + a_{25} p_{5,2} + a_{26} p_{5,3} + a_{27} p_{3,4} + \\
& a_{28} p_{2,5} + a_{29} p_{5,3} + a_{30} p_{4,4} + a_{31} p_{3,5} + a_{32} p_{5,4} + a_{33} p_{4,5} + a_{34} p_{5,5} \\
\Delta y = {} & a_1 p_{1,0} - a_0 p_{0,1} + a_{35} p_{2,0} - a_2 p_{1,1} - a_3 p_{0,2} + a_{36} p_{3,0} + a_{37} p_{2,1} + \\
& a_{38} p_{1,2} + a_{39} p_{0,3} + a_{40} p_{4,0} + a_{41} p_{3,1} + a_{42} p_{2,2} + a_{43} p_{1,3} + a_{44} p_{0,4} + \\
& a_{45} p_{5,0} + a_{46} p_{4,1} + a_{47} p_{3,2} + a_{48} p_{2,3} + a_{49} p_{1,4} + a_{50} p_{0,5} + a_{51} p_{5,1} + \\
& a_{52} p_{4,2} + a_{53} p_{3,3} + a_{54} p_{2,4} + a_{55} p_{1,5} + a_{56} p_{5,2} + a_{57} p_{5,3} + a_{58} p_{3,4} + \\
& a_{59} p_{2,5} + a_{60} p_{5,3} + a_{61} p_{4,4} + a_{62} p_{3,5} + a_{63} p_{5,4} + a_{64} p_{4,5} + a_{65} p_{5,5}
\end{aligned} \tag{3}$$

2.1.4. Fourier Model

The mathematical formula of 16 parameters in the first-order orthogonal polynomial distortion model based on the two-dimensional Fourier series [21] is as follows:

$$\begin{aligned}
\Delta x_f &= a_0 c_{1,0} + a_1 c_{0,1} + a_2 c_{1,-1} + a_3 c_{1,1} + a_4 s_{1,0} + a_5 s_{0,1} + a_6 s_{1,-1} + a_7 s_{1,1} \\
\Delta y_f &= a_8 c_{1,0} + a_9 c_{0,1} + a_{10} c_{1,-1} + a_{11} c_{1,1} + a_{12} s_{1,0} + a_{13} s_{0,1} + a_{14} s_{1,-1} + a_{15} s_{1,1}
\end{aligned} \tag{4}$$

where $c_{m,n} = 10^{-6} \cos(m\overline{x}_f + n\overline{y}_f)$, $s_{m,n} = 10^{-6} \sin(m\overline{x}_f + n\overline{y}_f)$, $\overline{x}_f = \frac{x - w/2}{w}\pi$, and $\overline{y}_f = \frac{y - h/2}{h}\pi$; x, y are the pixel coordinates in the image; w and h represent the width and height of the image, respectively; and $a_0, a_1, \ldots, a_{15}$ are the coefficients. When there is significant radial distortion in the image, it needs to be employed together with the radial distortion model. Therefore, the radial distortion and quadratic polynomial are applied.

$$\begin{aligned}
\Delta x_{rg} &= \overline{x}_r (k_1 r^2 + k_2 r^4 + k_3 r^6) + \\
& b_0 \overline{x}_g + b_1 \overline{y}_g - 2b_2 \overline{x}^2{}_g + b_3 \overline{x}_g \overline{y}_g + b_4 \overline{y}^2{}_g \\
\Delta y_{rg} &= \overline{y}_r (k_1 r^2 + k_2 r^4 + k_3 r^6) - \\
& b_0 \overline{y}_g + b_1 \overline{x}_g + b_2 \overline{x}_g \overline{y}_g - 2b_3 \overline{y}^2{}_{gg} + b_5 \overline{x}_g{}^2
\end{aligned} \tag{5}$$

where $\overline{x}_r, \overline{y}_r$ are consistent with $\overline{x}, \overline{y}$ in the Brown model; $\overline{x}_g = \overline{x}_f / \pi$, $\overline{y}_g = \overline{y}_f / \pi$; k_1, k_2, k_3 are the radial coefficients; and $b_0, b_1, \ldots, b_5$ are the coefficients of a quadratic polynomial. The final hybrid distortion model is as follows:

$$\begin{aligned}
\Delta x &= \Delta x_f + \Delta x_{rg} \\
\Delta y &= \Delta y_f + \Delta y_{rg}
\end{aligned} \tag{6}$$

2.1.5. Jacobi–Fourier Model

Compared with the Fourier model, the Jacobi–Fourier model has a higher horizontal and vertical accuracy [22]. In this paper, the Jacobi–Fourier model is adopted, and the function is presented in Formula (7), where $J_n(\alpha, \beta, r)$ is the Jacobi polynomial and the mathematical expression is as shown in Formula (8); $\overline{x}, \overline{y} \in [0, 1]$, represented as normalized image coordinates; r is the distance from the normalized pixel coordinate to the origin, $r^2 = \overline{x}^2 + \overline{y}^2$; N_J, M_F, N_F are the variable parameters of Jacobi and Fourier, respectively; and $a_{i,m,n}, b_{i,m,n}, a'_{i,m,n}, b'_{i,m,n}$ are the coefficients of the polynomial.

$$\begin{aligned}
\Delta x_{jf} ={} & \sum_{i=0}^{N_J} \sum_{m=0}^{M_F} \sum_{n=1}^{N_F} a_{i,m,n} J_i(\alpha, \beta, r) \sin(m\pi\overline{x} + n\pi\overline{y}) \\
& + \sum_{i=0}^{N_J} \sum_{m=0}^{M_F} \sum_{n=1}^{N_F} b_{i,m,n} J_i(\alpha, \beta, r) \cos(m\pi\overline{x} + n\pi\overline{y}) \\
\Delta y_{jf} ={} & \sum_{i=0}^{N_J} \sum_{m=0}^{M_F} \sum_{n=1}^{N_F} a'_{i,m,n} J_i(\alpha, \beta, r) \sin(m\pi\overline{x} + n\pi\overline{y}) \\
& + \sum_{i=0}^{N_J} \sum_{m=0}^{M_F} \sum_{n=1}^{N_F} b'_{i,m,n} J_i(\alpha, \beta, r) \cos(m\pi\overline{x} + n\pi\overline{y})
\end{aligned} \tag{7}$$

$$J_n(\alpha, \beta, \tau) = \sqrt{\frac{\omega(\alpha,\beta,\tau)}{b_n(\alpha,\beta)\cdot\tau}}\, G_n(\alpha, \beta, \tau)$$

$$G_n(\alpha, \beta, \tau) = \frac{n!(\beta-1)!}{(\alpha+n-1)!} \sum_{s=0}^{n} (-1)^s \frac{(\alpha+n+s-1)!}{(n-s)!s!(\beta+s-1)!} \tau^s \tag{8}$$

$$b_n(\alpha, \beta) = \frac{n![(\beta-1)!]^2(\alpha-\beta+n)!}{(\beta+n-1)!(\alpha+n-1)!(\alpha+2n)}$$

$$\omega(\alpha, \beta, \tau) = (1-\tau)^{\alpha-\beta}\tau^{\beta-1}$$

In Formula (8), α, β are set to 7 and 3, respectively, according to the suggestion of [22]; $\tau \in [0,1]$, and G_n, b_n, ω are the polynomial, normalizing constant, and weighting function, respectively. Similar to the Fourier model, the Jacobi–Fourier model is mixed with radial distortion and quadratic polynomial as given in the following.

$$\Delta x = \Delta x_{jf} + \Delta x_{rg}$$
$$\Delta y = \Delta y_{jf} + \Delta y_{rg} \tag{9}$$

2.2. GNSS-Aided Bundle Adjustment with Inequality Constraint

GNSS-aided BA is the commonly used method in the photogrammetry and computer vision fields. By minimizing the reprojection error, the traditional BA can optimize the internal and external parameters of the camera together with the 3D coordinates of the tie points. The error equation of GNSS-aided BA considers the deviation between the image projection center X_c and GNSS phase center X_{gps}. The jointly optimized error function is as described in Formula (10), where w is the weight of GNSS.

$$e_u = \sum_j \rho_j\left(\left\|\pi(P_c, X_k) - x_j\right\|_2^2\right) + \sum_n \rho_n(\|w(X_c - X_{gps}\|_2^2) \tag{10}$$

Different from the traditional weighted GNSS-aided BA, Maxime et al. [33] proposed an inequality constrained bundle adjustment (IBA) method with GNSS fusion to reduce the deviation error accumulation between the image projection center X_c and the GNSS location X_{gps} for long image sequences. The basic idea of IBA is to improve the absolute accuracy with GNSS-aided BA on the premise of appropriately increasing the reprojection error. Let $X^* = (X_c^T, X_a^T, X_k^T)$ be the optimal solution of standard BA without consideration of GNSS information, where X_c, X_a, X_k represent the projection center of image, rotation angle, and 3D coordinates of tie points, respectively. Further, let $e(X^*)$ be the minimum sum of square reprojection errors with the optimal solution, i.e., $\forall X, e(X^*) \leq e(X)$. Suppose e_t be a threshold that is slightly larger than the minimum reprojection error $e(X^*)$, i.e., $e(X^*) < e_t$. IBA assumes that the GNSS error is bounded and the reprojection error $e(X)$ of BA with GNSS constraint should be less than e_t, that is, $e(X) \leq e_t$. Under this condition, the optimized image projection center should be as close to the GNSS phase center as possible, i.e., $X_c \approx X_{gps}$.

Let $X_2 = (X_a, X_k)$, then the unknowns of BA can be expressed as $X = (X_c^T, X_2^T)$. Let $P = (I,0)$ be such that $X_c = PX$. IBA establishes the optimization equation by combining penalty function and inequality constraint, as shown in Formula (11).

$$e_I(X) = \frac{\gamma}{c_I(X)} + ||PX - X_{gps}||^2 \tag{11}$$

where γ is a user-defined weight and a positive number, $c_I(X) = e_t - e(X)$, and $c_I(X) > 0$. The objective function is iteratively minimized by this inequality and penalty function $\gamma/c_I(X)$ constraint. The penalty function value is close to positive infinity in the neighborhood of $c_I = 0$. The algorithm in C style is shown in Algorithm 1 and the parameter γ is set $\gamma = \frac{e_t - e(X^*)}{10}||PX^* - X_{gps}||^2$ according to [33]. More details should be referenced in [33].

Algorithm 1 The procedure of IBA

```
1: err = γ/(e_t − e(X)) + ||PX − X_gps||²
2: UpdateD = 1; λ = 0.001;
3: for (It = 0; It < It_max; It++) {
4: if (UpdateD) {
5:     UpdateD = 0;
6:     g = 2JᵀE(X); H = 2JᵀJ;
7:     g_I = γ/(e_t−e)² · g + 2Pᵀ(PX − X_gps); H = γ/(e_t−e)² · H + 2PᵀP;
8:     g̃ = √(2γ/(e_t−e)³ · g);
9: }
10: H̃ = H + λdiag(H + g̃g̃ᵀ)
11: solveH̃(a, b) = (−g, g̃);
12: Δ = a − (g̃ᵀa)/(1+g̃ᵀb) · b;
13: if (e(X + Δ) ≥ e_t) {
14:     λ = 10λ; continue;
15: }
16: err' = γ/(e_t−e(X+Δ)) + ||P(X + Δ) − X_gps||²;
17: if (err' < err) {
18:     X = X + Δ;
19:     if (0.9999err < err') break;
20:     err = err'; UpdateD = 1; λ = λ/10;
21: } else
22: λ = 10λ;
23:}
```

2.3. Camera Self-Calibration for Long Corridor UAV Images

The camera self-calibration for long corridor UAV images is realized under the framework of incremental SfM in ColMap. Firstly, the incremental SfM selects two seed images with enough matching feature points, which are of uniform distribution in the images and the intersection angle between the two-image pair should be large enough. Then the relative orientation and 3D coordinates of tie points in the initial image pair are calculated. Secondly, the next best images are selected, which are most fully connected with the existing reconstructed model. The image poses and 3D coordinates of tie points are recovered immediately. Finally, to eliminate accumulated error, the local and global BA optimization is carried out iteratively: (1) when the number of newly added images exceeds a given threshold, local BA optimization is performed for the local-oriented images; (2) when the percentage of registered images grows by a certain threshold, the reconstructed model is optimized by global BA [37]. The feature point extraction, exhaustive matching strategy, criteria for seed image selection, and the strategies of controlling local BA in ColMap are directly used in the proposed method without any changes.

For the long corridor UAV images, the existing incremental SfM framework has the following shortcomings:

1. From the perspective of camera self-calibration, the next best image selection does not consider whether the scene structure is degraded. If the structure of the seed image is poor and lacks height variation, the camera intrinsic parameters are unstable and may even lead to the failure of the final reconstruction.

2. At present, UAV images often record high-precision GNSS location information, which can alleviate the "bowl effect" of long corridor images. The existing incremental SfM framework of camera self-calibration does not take full advantage of GNSS information for absolute orientation.

3. The inaccurately estimated distortion parameters have an adverse impact on the 3D point clouds generated by dense matching technology. The power lines in UAV

images of high-voltage transmission are usually 1~2 pixels in width. When the distortion parameters are estimated inaccurately, the reconstructed point clouds of power lines are noisy and diverged around. Figure 3 shows the dense point clouds reconstructed by ColMap with inaccurate camera distortion parameters.

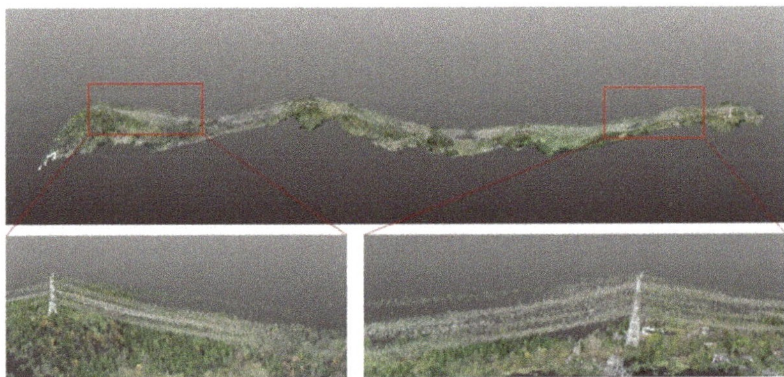

Figure 3. The influence of inaccurately estimated distortion parameters with dense matching for power lines.

Therefore, a camera self-calibration method is proposed, which combines the camera parameters initialization and high-precision differential GNSS position information fusion. The workflow is shown in Figure 4, and details of key steps are listed as follows:

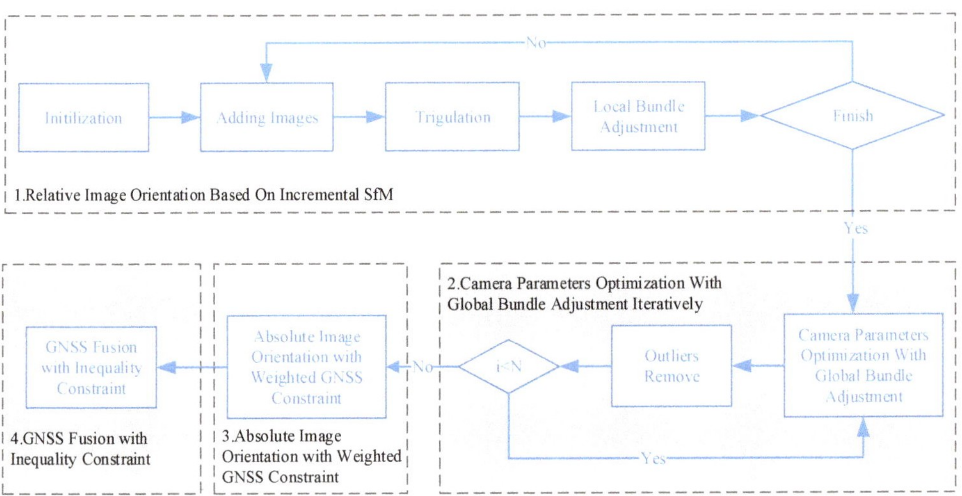

Figure 4. The camera self-calibration workflow for long corridor UAV images.

1. Image-relative orientation based on incremental SfM framework. Only the local BA is performed to reduce the error accumulation, and the focal length, principal point, and distortion parameters of the image are kept fixed to avoid the problem of the unstable and large variation of distortion parameters and focal length caused by the instability structure of image and scene degradation. The results of image-relative orientation are shown in Figure 5a.

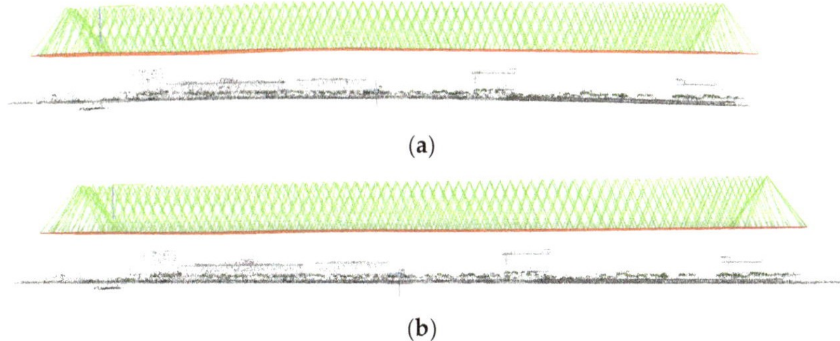

(a)

(b)

Figure 5. The reconstructed models with the proposed camera self-calibration strategy. (**a**) The result of image relative orientation, and (**b**) the result of image absolute orientation with the proposed strategy.

2. Global BA and iterative optimization of camera parameters. In the process of iterative global BA and gross error elimination, the optimization strategy of gradually freeing the intrinsic and distortion of camera parameters is adopted, that is, (a) distortion parameters, (b) focal length, and (c) principal point. This strategy can alleviate the correlation between focal length, principal point, and distortion parameters of the camera. According to the experiment, when the number of iterations is bigger than two, the camera intrinsic parameters become stable. In this paper, the global BA is iterated three times. In each iteration, the distortion, focal length, and principal point parameters of the camera are optimized using the strategy of gradually freeing parameters, to provide better initial values for GNSS-constrained BA.
3. Traditional weighted GNSS-constrained absolute orientation. At this time, the GNSS-constrained BA optimizes the focal length, principal point, and distortion parameters as unknowns together. Further, the error equation is shown in Formula (12).

$$e_u = \sum_j \rho_j \left(\left\| \pi(P_c, X_k, \theta) - x_j \right\|_2^2 \right) + \sum_n \rho_n (\|w(X_c - X_{gps}\|_2^2) \tag{12}$$

where $\theta = (f, c_x, c_y, Dist)$, including the focal length f, principal point c_x, c_y, and the distortion parameters $Dist$. The distortion parameters $Dist$ are determined by the selected camera distortion model introduced in Section 2.1. The weight of GNSS w is set to 10, keeping consistent with [33]. The cost function ρ is the Cauchy function with stronger noise resistance, as shown in Formula (13).

$$\rho(s) = \log(1 + s) \tag{13}$$

4. GNSS fusion based on IBA. This paper combines IBA to further fuse the GNSS. The main steps are as follows: (a) the camera focal length, principal point, and distortion parameters are used as unknowns to optimize together during camera self-calibration; (b) the initial input parameters $e(X^*)$ are the sum of the squares' reprojection error of weighted GNSS-constrained BA, and X^* are the minimum solvers; (c) all the image projection centers and the corresponding GNSS positions are used as constraints for global IBA to solve iteratively. The final reconstructed model with the proposed camera self-calibration strategy is shown in Figure 5b.

In summary, there are several differences between the proposed method with the incremental SfM in ColMap and the usage of IBA in [33]. In ColMap, the local BA is performed on the images that are connected with the most recently registered images, and the global BA is performed according to the growing percentage of the registered images. The camera intrinsic parameters are optimized during local BA and global BA. In the proposed method, only local BA is performed before all images are registered, and the global BA is literately performed after all images are registered. During the local BA, the camera intrinsic parameters are kept fixed. Further, during the global BA, the camera intrinsic parameters are gradually freed to get better initial values. Then the traditional weighted and inequality constrained BA with GNSS is performed for absolute orientation. In [33], the IBA is used in local BA to fuse the low-cost GNSS and image projection centers to refine the k-most recent images. The input-optimized initial parameters are the results of local BA and the 3D GNSS location of the corresponding most recent image. The camera intrinsic parameters are known and kept fixed. In the proposed method, the IBA is used in global BA to fuse high-precision GNSS locations and image projection centers. The camera intrinsic parameters are freed and optimized together with the GNSS constraint. The inputs of IBA in the proposed method are the optimized parameters of the weighted GNSS constraint BA and all the GNSS locations of registered images. The differences between the proposed method and the incremental SfM in ColMap and the IBA in [33] are listed in Tables 1 and 2, respectively.

Table 1. The differences between the incremental SfM in ColMap [38] and the proposed method.

Method	Incremental SfM in ColMap	The Proposed Method
Local BA	Camera intrinsic parameters are freed in local BA	Camera intrinsic parameters are fixed in local BA
Global BA	Global BA is performed after growing the registered images by a certain percentage. The camera intrinsic parameters are freed in global BA.	Global BA is performed after registering all images. Iteratively free distortion parameters, focal length, and principal point in global BA.

Table 2. The differences between the IBA used in [33] and the proposed method.

Method	IBA in [33]	The Proposed Method
Applied Stage	IBA is applied in Local BA	IBA is applied in Global BA
Initial Parameters	The input-optimized initial parameters are the results of local BA and the most recent GNSS location.	The input-optimized initial parameters are the results of weighted GNSS constraint BA and all the GNSS locations.
Camera Intrinsic Parameters	Keep fixed	Free

3. Results and Discussion

3.1. Test Sites and Datasets

Two datasets of long corridor transmission line UAV images were collected by DJI Phantom 4 RTK UAV, as shown in Figure 6. Figure 6a,b was collected using the rectangle closed-loop trajectory; Figure 6c,d was collected using the S-shaped strip trajectory. The rectangle and S-shaped trajectories were made by a third-party software developed based on DJI Mobile SDK, and during the flight, the standard control algorithm provided by DJI was applied to fly and take photos in autonomous mode along the trajectories. For the rectangle trajectories, the forward and side overlap ratio of images were set to 88% and 75%, respectively, and the flight speed was set to 4 m/s. For the S-shaped trajectories, the forward and side overlap ratio of images were set to 82% and 61%, respectively, and the flight speed was set to 6 m/s. The time interval for taking photos was 3 s for all flights. The camera focal length was kept fixed during image collection. The UAV flight height was set to 70 m, which is relative to the position from where the UAV takes off and the camera takes images vertically downward. The GSD (ground resolution distance) of images was 2.1 cm. The numbers of each image datasets were 140, 166, 165, and 132, respectively. To verify the absolute orientation accuracy of BA, the accurate ground coordinates in the two test sites were collected by Hi-Target iRTK2 GNSS receiver and FindCM CORS. The targets were marked with red paint manually on the road using the perpendicular lines with a width of about 10 cm, and the ground coordinates inside the right angle were measured. Finally, 15 targets were collected in test site 1 and 27 targets were collected in test site 2. The distribution of targets is listed in Figure 6e,f. The targets in test site 1 were labeled from A1 to A15, and the targets in test site 2 were labeled from B1 to B27. For the experiments of camera self-calibration without GCP constraint, all the targets were regarded as check points to evaluate the accuracy. For the experiments with one GCP constraint, the target of A14 in test site 1 and the target of B20 in test site 2 were regarded as control points and the rest of the targets were regarded as check points for accuracy evaluation. Both A14 and B20 were located in the middle of the long corridors.

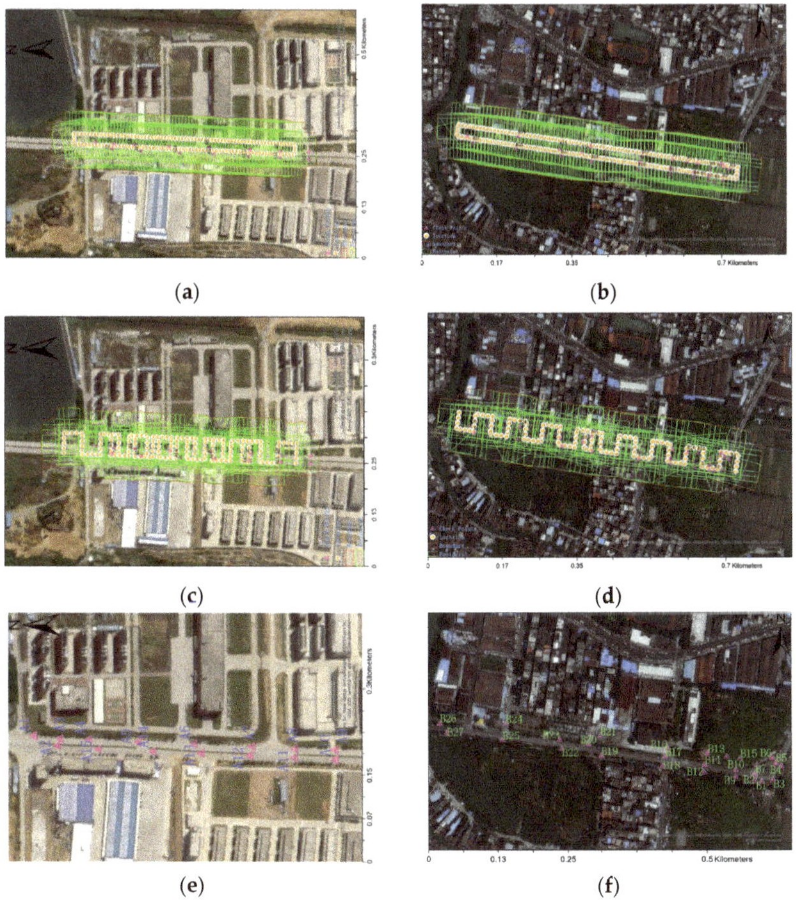

Figure 6. UAV images of the test sites. (**a**,**c**) The rectangle and S-shaped datasets of test site 1, respectively; (**b**,**d**) the rectangle and S-shaped datasets of test site 2, respectively. (**e**,**f**) the location distribution of GCPs in test site 1 and test site 2.

3.2. Analysis of the Influence of Camera Distortion Models

Firstly, the influence of different image acquisition modes and camera distortion models on the accuracy of bundle block adjustment was analyzed. For the hybrid Fourier and Jacobi–Fourier models, the radial and quadratic polynomial distortion parameters were first calculated with all images. Then, we kept these parameters fixed and calculated the parameters of Fourier and Jacobi–Fourier. For other distortion models, all the distortion parameters were calculated at one time. The mean, SD (standard deviation), and RMSE (root mean square error) were used to evaluate the accuracy of checkpoints.

The statistical results are listed in Table 3. Figures 7 and 8 show the residuals of check points for different distortion models. It can be seen that the accuracy of S-shaped image datasets was significantly better than that of rectangle image datasets (except for the vertical accuracy of the Brown model and hybrid Jacobi–Fourier model in test site 2). The main reason is that when the images are collected in the S-shaped method, the angle between the images is always changing, which can reduce the correlation between distortion and other parameters of the camera, and then the horizontal accuracy and vertical accuracy are improved.

Table 3. Statistical results of self-calibration for different camera distortion models.

Datasets		Camera Model	Mean(m)			SD(m)			RMSE(m)		
			X	Y	Z	X	Y	Z	X	Y	Z
Test Site 1	Rectangle	Brown	0.004	−0.001	0.174	0.011	0.036	0.037	0.012	0.036	0.178
		Poly7	0.008	0.018	2.456	0.018	0.050	0.032	0.020	0.053	2.456
		Legendre	−0.006	−0.002	0.829	0.026	0.082	0.032	0.027	0.082	0.830
		Fourier	0.001	0.000	0.755	0.013	0.054	0.032	0.013	0.054	0.755
		Jacobi–Four	−0.001	0.003	0.481	0.020	0.040	0.032	0.020	0.040	0.482
	S-Shaped	Brown	0.011	0.002	−0.348	0.014	0.018	0.022	0.018	0.018	0.349
		Poly7	0.004	0.001	0.084	0.012	0.028	0.025	0.012	0.028	0.088
		Legendre	0.003	−0.000	−0.061	0.012	0.030	0.026	0.012	0.030	0.066
		Fourier	0.010	0.007	0.054	0.013	0.013	0.018	0.016	0.015	0.057
		Jacobi–Four	0.019	0.007	0.031	0.020	0.014	0.022	0.028	0.015	0.038
Test Site 2	Rectangle	Brown	−0.018	0.028	0.753	0.030	0.025	0.036	0.035	0.037	0.754
		Poly7	0.001	0.014	0.240	0.022	0.032	0.028	0.022	0.035	0.242
		Legendre	−0.018	0.027	0.223	0.033	0.027	0.033	0.038	0.038	0.226
		Fourier	0.000	0.004	0.208	0.022	0.035	0.028	0.022	0.035	0.210
		Jacobi–Four	−0.011	0.024	0.104	0.023	0.025	0.029	0.026	0.035	0.108
	S-Shaped	Brown	0.016	−0.006	−1.259	0.012	0.012	0.041	0.020	0.013	1.260
		Poly7	0.004	0.003	0.058	0.016	0.012	0.018	0.017	0.012	0.061
		Legendre	0.003	0.004	0.125	0.017	0.012	0.018	0.017	0.013	0.127
		Fourier	0.005	0.005	0.098	0.014	0.012	0.015	0.013	0.013	0.099
		Jacobi–Four	0.022	0.003	−0.724	0.016	0.012	0.035	0.027	0.013	0.725

For further analysis, we can see the following: (1) For the rectangle dataset of test site 1, the horizontal and vertical accuracy with the Brown model was higher than other distortion models. However, for the other three datasets, the accuracy of the Brown model was the worst. (2) From the comparison of camera self-calibration between the Poly7 model and the Legendre model, it can be seen that the horizontal accuracy of Poly7 was better than the Legendre model for all the datasets. Further, the Poly7 model had better vertical accuracy in the S-shaped dataset of test site 2. However, the Legendre model had better vertical accuracy than the Poly7 model in the other three datasets. The main reason is that the orthogonality of the Legendre model can improve the accuracy of focal length for camera self-calibration, and then improve the vertical accuracy, but meanwhile, it loses the horizontal accuracy. (3) From the comparison of camera self-calibration between the hybrid Fourier model and hybrid Jacobi–Fourier model, it can be seen that the horizontal accuracy of the two models had little difference. However, the vertical accuracy of the hybrid Jacobi–Fourier model was better than the hybrid Fourier model in general (except for the S-shaped dataset in test site 2).

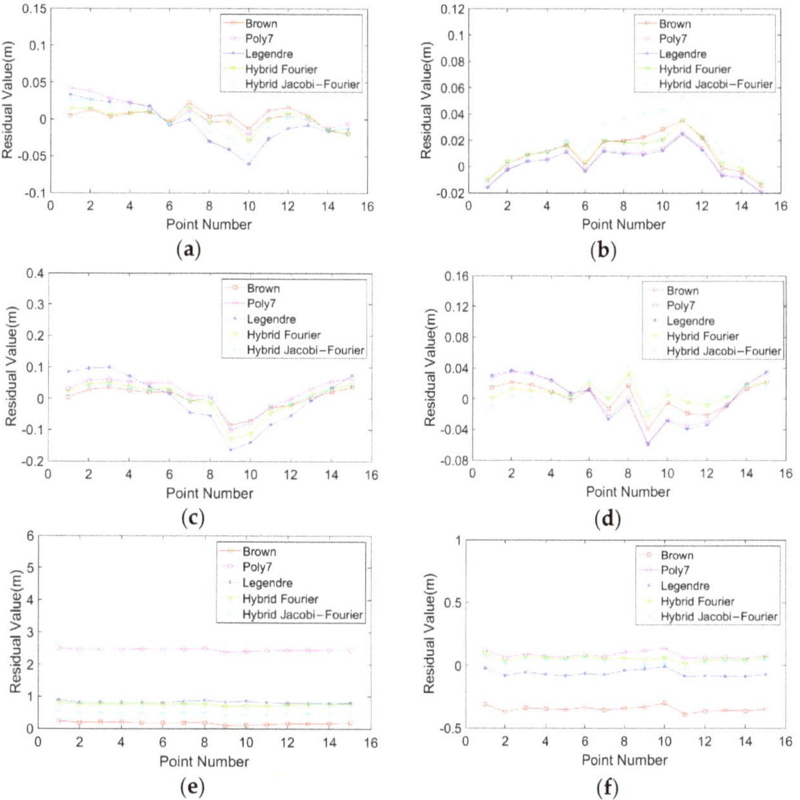

Figure 7. The residuals of check points after self-calibration for test site 1. (**a,c,e**) The residuals of check points of the rectangle dataset in the direction of X, Y, and Z, respectively; (**b,d,f**) the residuals of check points of the S-shaped dataset in the direction of X, Y, and Z, respectively.

In summary, the bundle block adjustment accuracy in the horizontal direction with the five different models can reach the centimeter level, while the vertical accuracy has great differences for the four datasets. No one distortion model can achieve the best accuracy among all four datasets. Overall, the horizontal and vertical accuracy of bundle block adjustment with mathematical distortion models was better than the physical model in the long corridor structure datasets, and the vertical accuracy of the hybrid Jacobi–Fourier model was generally better than the other three mathematical distortion models.

3.3. Analysis of the Performance of Proposed Self-Calibration

To verify the feasibility of the proposed strategy for camera self-calibration, this paper is compared with the scheme of ColMap [38]. Since ColMap does not implement the GNSS-constrained BA, the similarity transforms were applied for the projection center of images and GNSS after the final global BA in ColMap. Then, the weighted BA with GNSS as described in Formula (12) was conducted. Considering that ColMap does not provide any mathematical distortion models, the Brown model was adopted to make the experimental comparison.

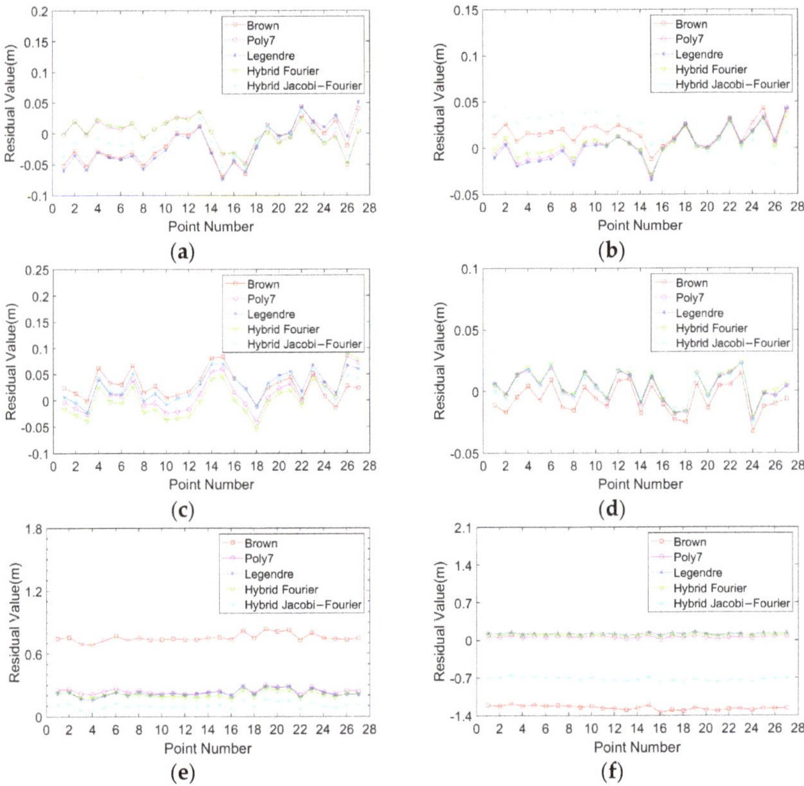

Figure 8. The residuals of check points after self-calibration for test site 2. (**a,c,e**) The residuals of check points of rectangle dataset in the direction of X, Y, and Z, respectively; (**b,d,f**) the residuals of check points of S-shaped dataset in the direction of X, Y, and Z, respectively.

The statistical results of check points after camera self-calibration are shown in Table 4. The results show that the horizontal accuracy with the proposed strategy is better than that of ColMap's in the four datasets of the two test sites. In the direction of elevation, the proposed strategy significantly improved the RMSE accuracy than ColMap (except for the rectangle dataset in test site 2). Due to the "bowl effect" in ColMap camera self-calibration, although its mean value was smaller than that of the proposed strategy in the S-shaped dataset of test site 1, it had a large standard deviation, which indicates that there is a large fluctuation in the vertical direction of the reconstructed model with ColMap, as shown in Figure 9b. In the rectangle dataset of test site 2, the elevation of ColMap had a higher accuracy in the mean and RMSE than that of the proposed strategy. The reason is that the initial image pair selected during the incremental SfM framework had a better structure, which led to a smaller variation range of focal length and a higher accuracy. However, it should be noted that the standard deviation of ColMap in this dataset was 0.055 larger than ours, which shows that the ColMap's fluctuation of elevation error is still large and there is a "bowl effect" in the reconstructed model with a certain bending phenomenon, as shown in Figure 9c. In conclusion, the proposed camera self-calibration strategy had advantages in horizontal accuracy and had better vertical accuracy than ColMap in three of the datasets.

Table 4. Statistical results of self-calibration for different calibration methods.

Datasets		Method	Mean(m)			SD(m)			RMSE(m)		
			X	Y	Z	X	Y	Z	X	Y	Z
Test Site 1	Rectangle	ColMap	0.006	−0.007	0.574	0.015	0.050	0.036	0.016	0.050	0.575
		Ours	0.004	−0.001	0.174	0.011	0.036	0.037	0.012	0.036	0.178
	S-Shaped	ColMap	0.107	0.103	0.055	0.040	0.320	0.404	0.114	0.337	0.407
		Ours	0.011	0.002	−0.348	0.014	0.018	0.022	0.018	0.018	0.349
Test Site 2	Rectangle	ColMap	−0.038	0.033	0.239	0.055	0.026	0.091	0.067	0.042	0.256
		Ours	−0.018	0.028	0.753	0.030	0.025	0.036	0.035	0.037	0.754
	S-Shaped	ColMap	0.012	−0.005	−1.380	0.018	0.011	0.019	0.021	0.012	1.381
		Ours	0.016	−0.006	−1.259	0.012	0.012	0.041	0.020	0.013	1.260

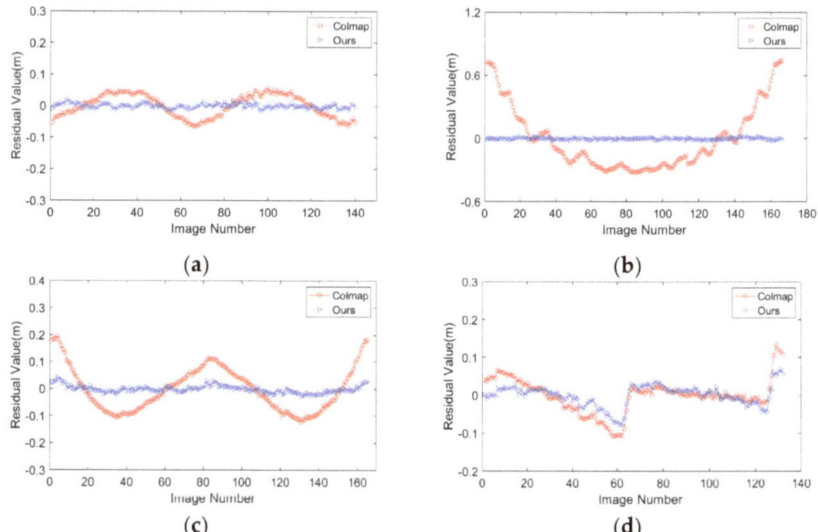

Figure 9. The coordinate offsets in the vertical direction between image projection center and differential GNSS location. (**a**,**b**) The offsets in the vertical direction with the rectangle dataset and S-shaped dataset in test site 1; (**c**,**d**) the offsets in the vertical direction with the rectangle dataset and S-shaped dataset in test site 2.

The DJI Phantom 4 RTK UAV records the differential GNSS location of the image with centimeter positioning accuracy. The relative accuracy between the projection center of the image after camera self-calibration with the proposed strategy and the corresponding GNSS location were analyzed. As the horizontal relative errors between the proposed method and ColMap had no obvious regularity and the vertical relative errors can indicate whether the "bowl effect" appears, the vertical offset distribution of the image projection centers after camera self-calibration and the corresponding GNSS locations in the four datasets are listed in Figure 9. It can be seen that, in the elevation direction, the reconstructed model of ColMap had obvious bending. It is a convex shape that is higher in the middle and lower on both sides in the rectangle dataset of test site 1. Further, it shows a concave shape that is lower in the middle and higher on both sides in both the S-shaped dataset of test site 1 and the rectangle dataset of test site 2. The offsets in the vertical direction of the reconstructed model with the proposed method are small, which significantly alleviates the "bowl effect". For the S-shaped dataset of test site 2, there are two broken jumps in the proposed method and ColMap. The main reason is that the large illumination leads to the increase of mismatch feature points, which affects the accuracy of bundle block adjustment. In summary, the vertical relative accuracy of the proposed method in the vertical direction is significantly improved compared with ColMap and the bending of the reconstructed model is reduced.

3.4. Comparison with State-Of-The-Art Software

The open-source software MicMac and the commercial software Pix4d Mapper were selected to compare and analyze with the proposed method. Based on the comparative experiments mentioned above with different distortion models, it was found that the overall performance of the hybrid Jacob–Fourier model was the best. Therefore, the hybrid Jacobi–Fourier distortion model was selected for further comparative analysis. The F15P7 distortion model was adopted in MicMac with the strategies proposed in [25]. The distortion model of Pix4d Mapper was unknown. In this section, self-calibration is conducted without and with GCPs.

3.4.1. Bundle Adjustment without GCP

For bundle adjustment without GCP, the experimental results are listed in Table 5. Figures 10 and 11 show the residual of check points with this software after camera self-calibration. From the analysis of horizontal accuracy, the following can be seen: (1) The proposed method had the smallest mean value in the two datasets of test site 1. The mean values of Pix4d in both datasets of test site 2 are the smallest. For the two datasets of test site 1, the standard deviation of MicMac in the Y direction is the largest, reaching 0.1 m, while Pix4d and the proposed method are both less than 0.07 m. Therefore, MicMac performs the worst overall. (2) The RMSE values in the X direction of Pix4d and the proposed method have little difference in the datasets except for the S-shaped dataset of test site 1. However, the RMSE values of Pix4d in the Y direction are 0.03 m, 0.015 m, and 0.010 m larger than the proposed method. Therefore, the horizontal accuracy of the proposed method is generally better than Pix4d.

Table 5. Statistical results of self-calibration for different software.

Datasets		Software	Mean(m)			SD(m)			RMSE(m)		
			X	Y	Z	X	Y	X	Y	Z	Z
Test site 1	Rectangle	MicMac	−0.007	−0.023	0.301	0.017	0.125	0.095	0.019	0.127	0.315
		Pix4d	−0.012	−0.024	−1.266	0.020	0.066	0.088	0.024	0.070	1.269
		Ours	−0.001	0.003	0.481	0.020	0.040	0.032	0.020	0.040	0.482
	S-shaped	MicMac	−0.029	−0.050	−0.075	0.013	0.155	0.157	0.031	0.163	0.174
		Pix4d	0.021	0.013	0.360	0.017	0.016	0.029	0.027	0.021	0.362
		Ours	0.019	0.007	0.031	0.020	0.014	0.022	0.028	0.015	0.038
Test site 2	Rectangle	MicMac	−0.016	0.047	0.716	0.037	0.043	0.056	0.041	0.064	0.718
		Pix4d	0.008	−0.013	−0.847	0.025	0.048	0.037	0.026	0.050	0.848
		Ours	−0.011	0.024	0.104	0.023	0.025	0.029	0.026	0.035	0.108
	S-shaped	MicMac	0.056	−0.023	−1.425	0.076	0.021	0.108	0.095	0.031	1.429
		Pix4d	0.011	0.005	−0.111	0.013	0.023	0.025	0.018	0.023	0.113
		Ours	0.022	0.003	−0.724	0.016	0.012	0.035	0.027	0.013	0.725

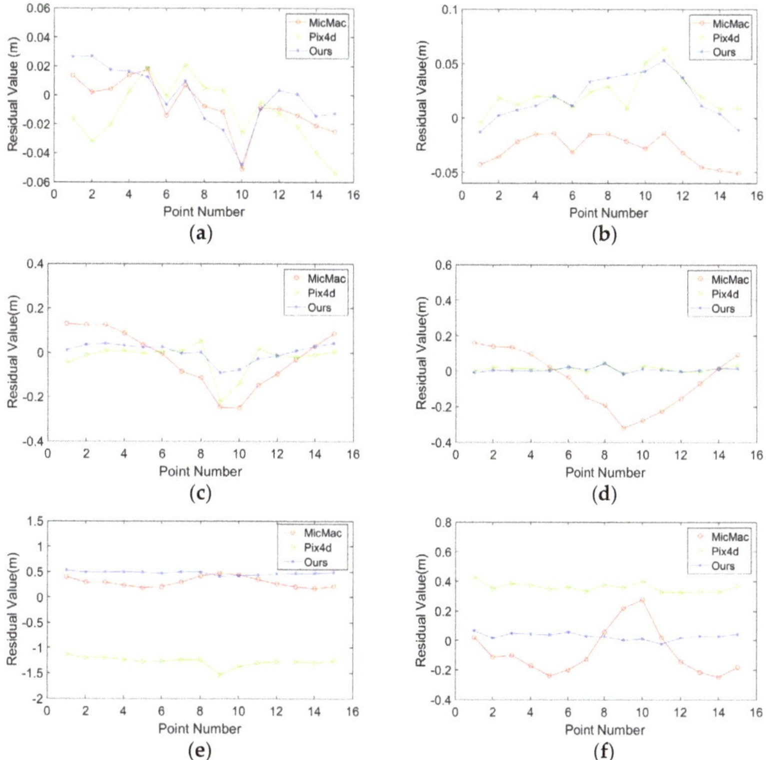

Figure 10. The residuals of check points after self-calibration with different software for test site 1. (**a**,**c**,**e**) The residuals of check points of rectangle dataset in the direction of X, Y, and Z, respectively; (**b**,**d**,**f**) the residuals of check points of S-shaped dataset in the direction of X, Y, and Z, respectively.

From the analysis of the vertical accuracy, the following can be seen: (1) The proposed method had the smallest standard deviation in the datasets except for the S-shaped dataset of test site 2. Pix4d had the smallest standard deviation in the S-shaped dataset of test site 2. MicMac had the largest standard deviation in the four datasets and the accuracy fluctuates greatly. (2) For test site 1, the RMSE values of MicMac were the smallest, but it had the largest standard deviation and the reconstructed model had obvious bending, as shown in Figure 12a,b. Pix4d had the smallest RMSE and standard deviation in the S-shaped dataset of test site 2. The possible reason is that the feature points matching of Pix4d is more robust with the large change of illumination. However, the RMSE values of Pix4d in the vertical direction were 0.787 m, 0.324 m, and 0.74 m larger than the proposed method, which indicates that the proposed has better accuracy in the vertical direction. To sum up, compared with MicMac and Pix4d, the proposed method still has certain advantages.

To evaluate the "bowl effect" with different software, the relative errors in the Z direction between the projection centers and the corresponding GNSS locations are shown in Figure 12. Compared with Pix4d and the proposed method, MicMac had the worst performance in the vertical relative accuracy between the projection centers and GNSS locations. The fluctuation range of vertical relative errors was between −0.2 m and 0.3 m with MicMac, which is much bigger than Pix4d and the proposed method. There was a "bowl effect" with MicMac in the four datasets except for the S-shaped dataset in test site 2, while the bending of the reconstructed models was significantly reduced with Pix4d and the proposed method. The vertical relative errors of the proposed method were close to Pix4d. The "bowl effect" was alleviated with the proposed method and Pix4d.

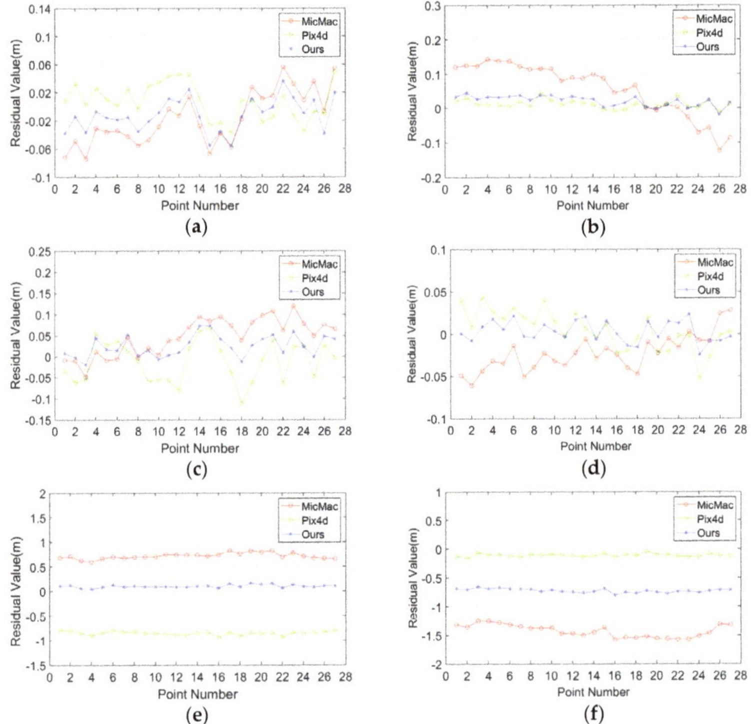

Figure 11. The residuals of check points after self-calibration with different software for test site 2. (**a**,**c**,**e**) The residuals of check points of rectangle dataset in the direction of X, Y, and Z, respectively; (**b**,**d**,**f**) the residuals of check points of S-shaped dataset in the direction of X, Y, and Z, respectively.

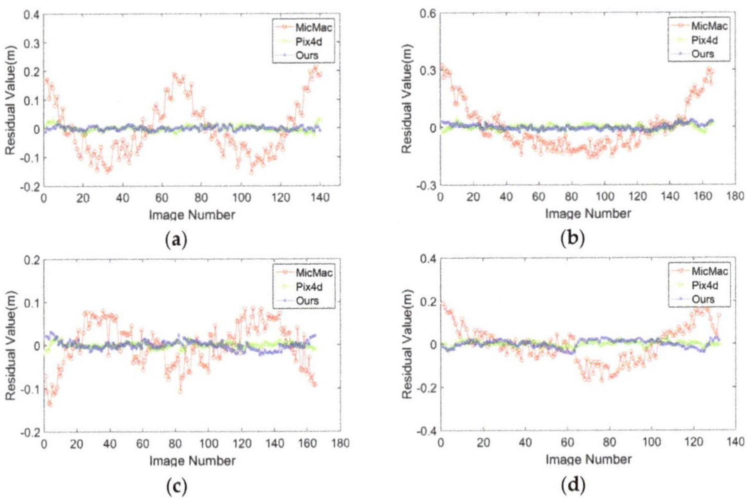

Figure 12. The coordinate offsets in the vertical direction between image projection center and differential GNSS location with different software. (**a**,**b**) The offsets in the vertical direction with the rectangle dataset and S-shaped dataset in test site 1; (**c**,**d**) the offsets in the vertical direction with the rectangle dataset and S-shaped dataset in test site 2.

3.4.2. Bundle Adjustment with GCP

For bundle adjustment with GCP, a single GCP was applied for the camera self-calibration experiment compared with MicMac and Pix4d Mapper. All the optimized parameters of Section 3.4.1 were used as the initial parameters for MicMac and the proposed method with one single GCP-constrained BA. The Brown and hybrid Jacobi–Fourier distortion models were selected with the proposed method. Table 6 lists the experimental results. Figures 13 and 14 show the residuals of check points after camera self-calibration with one single GCP. From the analysis of horizontal accuracy, the accuracy of the Brown and hybrid Jacobi–Fourier distortion models with the proposed strategy is comparable. The mean values in the horizontal direction of X and Y with MicMac were the largest in the datasets except for the S-shaped dataset of test site 1, and the horizontal RMSE values and standard deviation values were also the largest among the four datasets. In the two datasets of test site 1, the mean values and RMSE values of Pix4d in the X and Y horizontal directions were larger than the Brown and hybrid Jacobi–Fourier model. Further, in the two datasets of the test site 2, the mean values and RMSE values of Pix4d in the horizontal direction of X were smaller than Brown and hybrid Jacobi–Fourier model with the proposed strategy, but in the Y direction, they were larger than the two distortion models with the proposed strategy. In general, the horizontal accuracy of the proposed method is relatively close to Pix4d. The horizontal RMSE value of the two distortion models with the proposed method was better than 0.04 m, while MicMac was less than 0.5 m and Pix4d was less than 0.08 m.

Table 6. Statistical results of self-calibration for different software with one single ground control point constraint.

Datasets		Software	Mean(m)			SD(m)			RMSE(m)		
			X	Y	Z	X	Y	Z	X	Y	Z
Test site 1	Rectangle	Brown	−0.005	0.003	0.016	0.011	0.036	0.040	0.012	0.036	0.043
		Jacobi–Four	0.000	0.000	0.016	0.021	0.041	0.038	0.021	0.041	0.042
		MicMac	0.020	−0.081	0.169	0.019	0.200	0.119	0.027	0.216	0.207
		Pix4d	0.018	−0.019	0.018	0.020	0.073	0.083	0.027	0.076	0.085
	S-shaped	Brown	−0.011	−0.001	−0.014	0.014	0.018	0.022	0.018	0.018	0.026
		Jacobi–Four	−0.019	−0.004	−0.024	0.017	0.016	0.019	0.026	0.016	0.030
		MicMac	0.017	−0.029	0.164	0.017	0.044	0.142	0.024	0.053	0.217
		Pix4d	0.020	0.013	0.034	0.019	0.017	0.027	0.027	0.022	0.043
Test site 2	Rectangle	Brown	0.018	−0.027	0.005	0.030	0.026	0.029	0.035	0.038	0.029
		Jacobi–Four	0.012	−0.024	0.003	0.024	0.026	0.028	0.027	0.035	0.028
		MicMac	0.154	−0.071	0.016	0.214	0.067	0.105	0.264	0.098	0.106
		Pix4d	0.005	−0.010	−0.023	0.023	0.048	0.033	0.024	0.049	0.040
	S-shaped	Brown	−0.018	0.007	0.024	0.014	0.013	0.017	0.023	0.015	0.029
		Jacobi–Four	−0.023	−0.002	0.005	0.018	0.013	0.019	0.029	0.013	0.019
		MicMac	−0.283	0.057	0.022	0.381	0.053	0.190	0.475	0.078	0.191
		Pix4d	0.012	0.017	−0.017	0.013	0.021	0.024	0.017	0.027	0.030

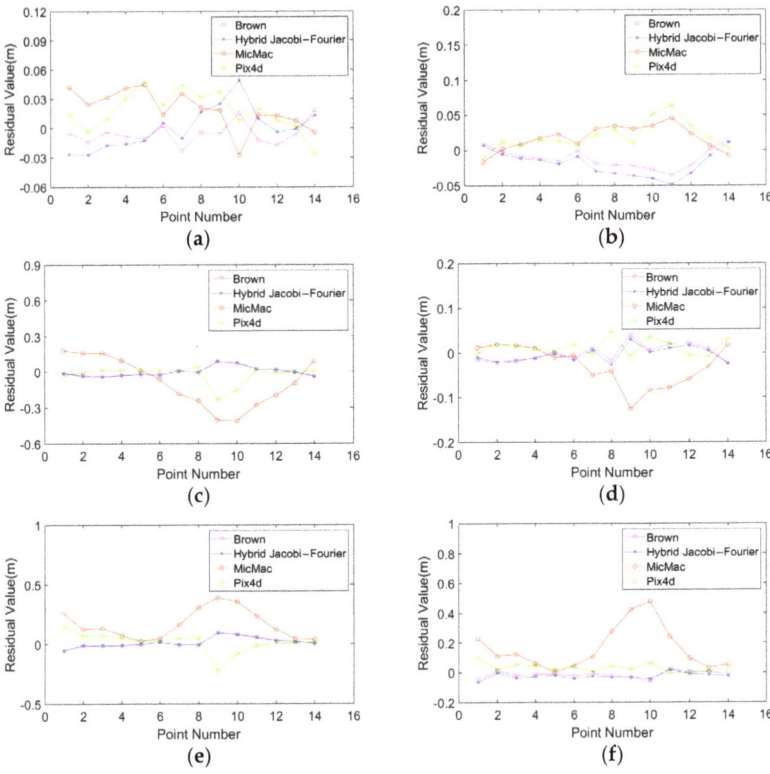

Figure 13. The residuals of check points after self-calibration with one single GCP for test site 1. (**a,c,e**) The residuals of check points of rectangle dataset in the direction of X, Y, and Z, respectively; (**b,d,f**) the residuals of check points of S-shaped dataset in the direction of X, Y, and Z, respectively.

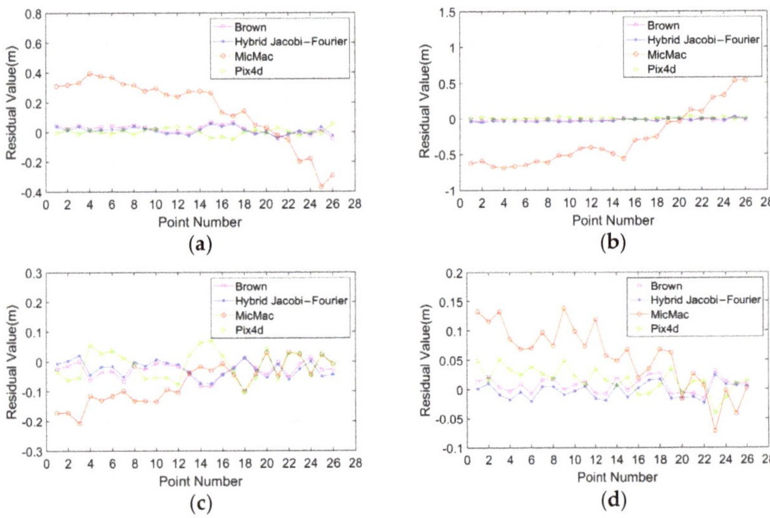

Figure 14. *Cont.*

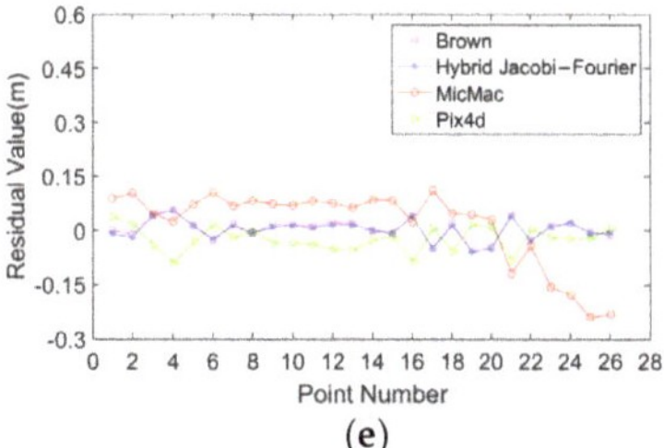 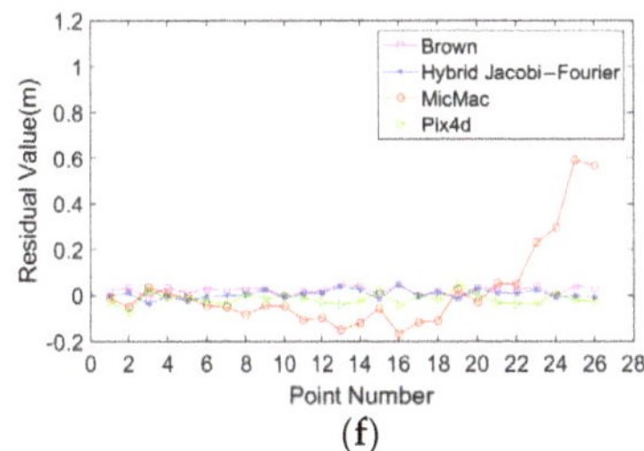

Figure 14. The residuals of check points after self-calibration with one single GCP for test site 2. (a,c,e) The residuals of check points of rectangle dataset in the direction of X, Y, and Z, respectively; (b,d,f) the residuals of check points of S-shaped dataset in the direction of X, Y, and Z, respectively.

From the analysis of elevation, the two distortion models with the proposed strategy achieved the best accuracy, while MicMac achieved the worse. The RMSE in the vertical direction of MicMac was greater than 0.1 m, while Pix4d was less than 0.1 m and the two distortion models with the proposed strategy were both less than 0.05 m. The vertical accuracy of the two distortion models was relatively close, and the difference between the first three datasets was only at the millimeter level, while the vertical RMSE of the hybrid Jacobi–Fourier model was 0.01 m better than the Brown in the S-shaped dataset of test site 2. Therefore, in the case of a single GCP constraint, the accuracy of the Brown model is comparable to the hybrid Jacobi–Fourier model, and the overall accuracy of camera self-calibration with the proposed strategy is better than that of MicMac and Pix4d. Additionally, compared with Table 5 in the vertical direction, MicMac had the smallest RMSE value in the rectangle dataset of test site 1 while Pix4d had the smallest RMSE value in the S-shaped dataset of test site 2. However, the vertical RMSE of Brown and hybrid Jacobi–Fourier models with the proposed strategy was better than MicMac and Pix4d after adding a single GCP. The reason is that the horizontal accuracy is ensured and the relative errors between image projection centers and GNSS locations are small with the experiments of direct georeferencing in Sections 3.3 and 3.4.1, which indicates that the proposed method can significantly reduce the bending of the reconstructed model and the image structures have litter distortion. The "bowl effect" is alleviated. The only problem is that the vertical accuracy is unstable because the focal length is highly correlated with other distortion parameters. With one GCP constraint, the correlation between focal length and other distortion parameters is reduced and the focal length can be accurately estimated. In this case, vertical accuracy can be ensured.

4. Conclusions

The UAV images collected on a linear axis and fixed height are critical configurations for camera self-calibration, which may lead to the "bowl effect". To solve such a tough problem, traditional methods rely on more than three GPCs, while the proposed method relies on only one GCP. The proposed new camera self-calibration method for long corridor UAV images in high transmission lines combines the initialization of the camera calibration parameters and the fusion of high-precision differential GNSS position information for long corridor UAV images in high transmission lines. Based on the comprehensive analysis of the physical and mathematical models of camera distortion, the new camera self-calibration method was designed, which takes full consideration of the initialization of the camera intrinsic parameters in long corridor UAV images and the fusion of differential GNSS with inequality constrained BA.

The UAV images of two test sites with two different acquisition modes were applied for camera self-calibration experiments. The experimental results show that the proposed camera self-calibration method can significantly alleviate the "bowl effect" for long corridor UAV images, reduce the bending of the reconstructed model, and improve the absolute accuracy. Compared to the accuracy using the physical distortion model without any GCPs, the mathematical distortion models achieve better horizontal and vertical accuracy in the weak structure datasets. Among them, the vertical accuracy of the hybrid Jacobi–Fourier distortion model is generally better than the other mathematical models. Furthermore, with only one single GCP constraint, the proposed method with Brown and hybrid Jacobi–Fourier distortion models achieved the best accuracy compared with open-source and commercial software. Compared with the open-source software MicMac, the RMSEs in the directions of X, Y, and Z improved the GSD value on average approximately 8.36, 4.02, and 7.07 times, respectively, in the four datasets with the Brown model, and improved the GSD value on average approximately 8.18, 4.05, and 7.12 times, respectively, with hybrid Jacobi–Fourier model. Compared with the commercial software Pix4d, the RMSEs in the directions of X, Y, and Z improved the GSD value on average approximately 0.08, 0.80, and 0.85 times, respectively in the four datasets with the Brown model. Although, in the direction of X, the RMSE lost 0.01 times the GSD value with the hybrid Jacobi–Fourier model, while the RMESs in the directions of Y and Z improved 0.82 and 0.94 times the GSD value on average, respectively. Considering that different distortion models perform differently with different scenes, this study focuses on how to select the appropriate distortion model according to the characteristics and uncertainty of the scene in future work.

Author Contributions: W.H. and W.J. conceived and designed the experiments; W.H. and S.J. performed the experiments; W.J. collected the data; S.J. and W.H. wrote the manuscript; W.J. revised the manuscript. All authors have read and agreed to the published version of the manuscript.

Funding: This research was funded by the National Natural Science Foundation of China (Grant No. 42001413), Nature Science Foundation of Hubei Province (Grant No. 2020CFB324), the Open Fund of State Laboratory of Information Engineering in Surveying, Mapping and Remote Sensing, Wuhan University (Grant No. 20E03).

Institutional Review Board Statement: Not applicable.

Informed Consent Statement: Not applicable.

Data Availability Statement: Data available on request due to restrictions, e.g., privacy or ethical. The data presented in this study are available on request from the corresponding author. The data are not publicly available due to [the policy to protect the coordinates of pylon].

Conflicts of Interest: The authors declare no conflict of interest.

References

1. Xiang, T.-Z.; Xia, G.-S.; Zhang, L. Mini-unmanned aerial vehicle-based remote sensing: Techniques, applications, and prospects. *IEEE Geosci. Remote Sens. Mag.* **2019**, *7*, 29–63. [CrossRef]
2. Jiang, S.; Jiang, W.; Huang, W.; Yang, L. UAV-Based Oblique Photogrammetry for Outdoor Data Acquisition and Offsite Visual Inspection of Transmission Line. *Remote Sens.* **2017**, *9*, 278. [CrossRef]
3. Jiang, S.; Jiang, W. UAV-based Oblique Photogrammetry for 3D Reconstruction of Transmission Line: Practices and Applications. In Proceedings of the International Archives of the Photogrammetry, Remote Sensing & Spatial Information Sciences, Enschede, The Netherlands, 10–14 June 2019; pp. 401–406.
4. Zhang, H.; Yang, W.; Yu, H.; Zhang, H.; Xia, G.-S. Detecting power lines in UAV images with convolutional features and structured constraints. *Remote Sens.* **2019**, *11*, 1342. [CrossRef]
5. Zhang, Y.; Yuan, X.; Li, W.; Chen, S. Automatic Power Line Inspection Using UAV Images. *Remote Sens.* **2017**, *9*, 824. [CrossRef]
6. Huang, W.; Jiang, S.; Jiang, W. A Model-Driven Method for Pylon Reconstruction from Oblique UAV Images. *Sensors* **2020**, *20*, 824. [CrossRef]
7. Triggs, B. Autocalibration from Planar Scenes. In Proceedings of the European Conference on Computer Vision, Berlin, Germany, 2–6 June 1998; pp. 89–105.
8. Zhang, Z. A Flexible New Technique for Camera Calibration. *IEEE Trans. Pattern Anal. Mach. Intell.* **2000**, *22*, 1330–1334. [CrossRef]
9. Oniga, E.; Pfeifer, N.; Loghin, A.-M. 3D Calibration Test-Field for Digital Cameras Mounted on Unmanned Aerial Systems (UAS). *Remote Sens.* **2018**, *10*, 2017. [CrossRef]
10. Duane, C.B. Close-Range Camera Calibration. *Photogramm. Eng.* **1971**, *37*, 855–866.
11. Fraser, C.S. Digital camera self-calibration. *ISPRS J. Photogramm. Remote Sens.* **1997**, *52*, 149–159. [CrossRef]
12. Luhmann, T.; Robson, S.; Kyle, S.; Harley, I.A. *Close Range Photogrammetry: Principles, Techniques and Applications*; Whittles Publishing: Dunbeath, Caithness, Scotland, 2006; Volume 3.

13. Fitzgibbon, A.W. Simultaneous Linear Estimation of Multiple View Geometry and Lens Distortion. In Proceedings of the IEEE Computer Society Conference on Computer Vision and Pattern Recognition, Kauai, HI, USA, 8–14 December 2001; p. I.
14. Kukelova, Z.; Pajdla, T. A Minimal Solution to the Autocalibration of Radial Distortion. In Proceedings of the IEEE Conference on Computer Vision and Pattern Recognition, Minneapolis, MN, USA, 17–22 June 2007; pp. 1–7.
15. Kukelova, Z.; Pajdla, T. A Minimal Solution to Radial Distortion Autocalibration. *IEEE Trans. Pattern Anal. Mach. Intell.* **2011**, *33*, 2410–2422. [CrossRef]
16. Jiang, F.; Kuang, Y.; Solem, J.E.; Åström, K. A Minimal Solution to Relative Pose with Unknown Focal Length and Radial Distortion. In Proceedings of the Asian Conference on Computer Vision, Singapore, 1–5 November 2014; pp. 443–456.
17. Kukelova, Z.; Heller, J.; Bujnak, M.; Fitzgibbon, A.; Pajdla, T. Efficient Solution to the Epipolar Geometry for Radially Distorted Cameras. In Proceedings of the 2015 IEEE International Conference on Computer Vision, Santiago, Chile, 11–18 December 2015; pp. 2309–2317.
18. Ebner, H. Self calibrating block adjustment. *Bildmess. Und Luftbildwessen* **1976**, *44*, 128–139.
19. Gruen, A. Accuracy, reliability and statistics in close-range photogrammetry. In Proceedings of the Inter-Congress Symposium of ISP Commission V, Stockholm, Sweden, 14–17 August 1978.
20. Tang, R.; Fritsch, D.; Cramer, M.; Schneider, W. A Flexible Mathematical Method for Camera Calibration in Digital Aerial Photogrammetry. *Photogramm. Eng. Remote Sens.* **2012**, *78*, 1069–1077. [CrossRef]
21. Tang, R.; Fritsch, D.; Cramer, M. New rigorous and flexible Fourier self-calibration models for airborne camera calibration. *ISPRS J. Photogramm. Remote Sens.* **2012**, *71*, 76–85. [CrossRef]
22. Babapour, H.; Mokhtarzade, M.; Valadan Zoej, M.J. Self-calibration of digital aerial camera using combined orthogonal models. *ISPRS J. Photogramm. Remote Sens.* **2016**, *117*, 29–39. [CrossRef]
23. Wu, C. Critical Configurations for Radial Distortion Self-Calibration. In Proceedings of the 2014 IEEE Conference on Computer Vision and Pattern Recognition, Columbus, OH, USA, 28 June 2014; pp. 25–32.
24. Zhou, Y.; Rupnik, E.; Meynard, C.; Thom, C.; Pierrot-Deseilligny, M. Simulation and Analysis of Photogrammetric UAV Image Blocks—Influence of Camera Calibration Error. *Remote Sens.* **2019**, *12*, 22. [CrossRef]
25. Tournadre, V.; Pierrot-Deseilligny, M.; Faure, P.H. UAV Linear Photogrammetry. In Proceedings of the International Archives of Photogrammetry, Remote Sensing and Spatial Information Sciences, La Grande Motte, France, 28 September–3 October 2015; p. 327.
26. Polic, M.; Steidl, S.; Albl, C.; Kukelova, Z.; Pajdla, T. Uncertainty based camera model selection. In Proceedings of the IEEE/CVF Conference on Computer Vision and Pattern Recognition, Seattle, WA, USA, 16–18 June 2020; pp. 5991–6000.
27. Griffiths, D.; Burningham, H. Comparison of pre- and self-calibrated camera calibration models for UAS-derived nadir imagery for a SfM application. *Prog. Phys. Geogr. Earth Environ.* **2018**, *43*, 215–235. [CrossRef]
28. Jaud, M.; Passot, S.; Le Bivic, R.; Delacourt, C.; Grandjean, P.; Le Dantec, N. Assessing the accuracy of high resolution digital surface models computed by PhotoScan® and MicMac® in sub-optimal survey conditions. *Remote Sens.* **2016**, *8*, 465. [CrossRef]
29. Salach, A.; Bakuła, K.; Pilarska, M.; Ostrowski, W.; Górski, K.; Kurczyński, Z. Accuracy assessment of point clouds from LiDAR and dense image matching acquired using the UAV platform for DTM creation. *ISPRS Int. J. Geo-Inf.* **2018**, *7*, 342. [CrossRef]
30. Jaud, M.; Passot, S.; Allemand, P.; Le Dantec, N.; Grandjean, P.; Delacourt, C. Suggestions to limit geometric distortions in the reconstruction of linear coastal landforms by SfM photogrammetry with PhotoScan® and MicMac® for UAV surveys with restricted GCPs pattern. *Drones* **2019**, *3*, 2. [CrossRef]
31. Nahon, A.; Molina, P.; Blázquez, M.; Simeon, J.; Capo, S.; Ferrero, C. Corridor mapping of sandy coastal foredunes with UAS photogrammetry and mobile laser scanning. *Remote Sens.* **2019**, *11*, 1352. [CrossRef]
32. Ferrer-González, E.; Agüera-Vega, F.; Carvajal-Ramírez, F.; Martínez-Carricondo, P. UAV Photogrammetry Accuracy Assessment for Corridor Mapping Based on the Number and Distribution of Ground Control Points. *Remote Sens.* **2020**, *12*, 2447. [CrossRef]
33. Maxime, L. Incremental Fusion of Structure-from-Motion and GPS Using Constrained Bundle Adjustments. *IEEE Trans. Pattern Anal. Mach. Intell.* **2012**, *34*, 2489–2495. [CrossRef]
34. Gopaul, N.S.; Wang, J.; Hu, B. Camera auto-calibration in GPS/INS/stereo camera integrated kinematic positioning and navigation system. *J. Glob. Position. Syst.* **2016**, *14*, 3. [CrossRef]
35. Snow, W.L.; Childers, B.A.; Shortis, M.R. The calibration of video cameras for quantitative measurements. In Proceedings of the 39th International Instrumentation Symposium, Albuquerque, NM, USA, 2–6 May 1993; pp. 103–130.
36. Wang, J.; Wang, W.; Ma, Z. Hybrid-model based Camera Distortion Iterative Calibration Method. *Bull. Surv. Mapp.* **2019**, *4*, 103–106.
37. Jiang, S.; Jiang, C.; Jiang, W. Efficient structure from motion for large-scale UAV images: A review and a comparison of SfM tools. *ISPRS J. Photogramm. Remote Sens.* **2020**, *167*, 230–251. [CrossRef]
38. Schonberger, J.L.; Frahm, J.M. Structure-from-Motion Revisited. In Proceedings of the IEEE Conference on Computer Vision & Pattern Recognition, Las Vegas, NV, USA, 27–30 June 2016; pp. 4104–4113.

remote sensing

Article

Progressive Structure from Motion by Iteratively Prioritizing and Refining Match Pairs

Teng Xiao, Qingsong Yan, Weile Ma and Fei Deng *

School of Geodesy and Geomatics, Wuhan University, Wuhan 430079, China; xiaoteng@whu.edu.cn (T.X.); yanqs_whu@whu.edu.cn (Q.Y.); maweile@whu.edu.cn (W.M.)
* Correspondence: fdeng@sgg.whu.edu.cn

Abstract: Structure from motion (SfM) has been treated as a mature technique to carry out the task of image orientation and 3D reconstruction. However, it is an ongoing challenge to obtain correct reconstruction results from image sets consisting of problematic match pairs. This paper investigated two types of problematic match pairs, stemming from repetitive structures and very short baselines. We built a weighted view-graph based on all potential match pairs and propose a progressive SfM method (PRMP-PSfM) that iteratively prioritizes and refines its match pairs (or edges). The method has two main steps: initialization and expansion. Initialization is developed for reliable seed reconstruction. Specifically, we prioritize a subset of match pairs by the union of multiple independent minimum spanning trees and refine them by the idea of cycle consistency inference (CCI), which aims to infer incorrect edges by analyzing the geometric consistency over cycles of the view-graph. The seed reconstruction is progressively expanded by iteratively adding new minimum spanning trees and refining the corresponding match pairs, and the expansion terminates when a certain completeness of the block is achieved. Results from evaluations on several public datasets demonstrate that PRMP-PSfM can successfully accomplish the image orientation task for datasets with repetitive structures and very short baselines and can obtain better or similar accuracy of reconstruction results compared to several state-of-the-art incremental and hierarchical SfM methods.

Keywords: structure from motion; match pair; cycle consistency inference; repetitive structure; very short baseline

Citation: Xiao, T.; Yan, Q.; Ma, W.; Deng, F. Progressive Structure from Motion by Iteratively Prioritizing and Refining Match Pairs. *Remote Sens.* **2021**, *13*, 2340. https://doi.org/10.3390/rs13122340

Academic Editors: Ayman F. Habib, Wanshou Jiang, San Jiang and Xiongwu Xiao

Received: 3 May 2021
Accepted: 5 June 2021
Published: 15 June 2021

Publisher's Note: MDPI stays neutral with regard to jurisdictional claims in published maps and institutional affiliations.

1. Introduction

Structure from motion (SfM) can automatically reconstruct sparse 3D points and estimate camera poses (also known as image orientation) from a set of 2D images, and has been extensively employed in photogrammetry [1,2]. Most feature-based SfM methods contain modules such as for feature extraction and matching [3], geometric verification [4–6], view-graph construction of match pairs [7–9], initial camera pose estimation [10–12], triangulation [13], and bundle adjustment [2,13,14]. Vertices in a view-graph denote images, and edges denote match pairs, which indicate a set of mutually overlapped image pairs. Taking the view-graph as an input, subsequent modules are used for image orientation tasks. According to the strategy of utilizing the view-graph, SfM can be categorized as incremental, hierarchical, or global.

Incremental SfM [10,12,15] typically starts with an initial reconstruction of a match pair or triplet and sequentially adds new images to the block. As the block grows, bundle adjustment is repeatedly conducted to refine the reconstruction results, which makes it time-consuming. While the efficacy of incremental methods has been widely demonstrated, they may have limitations in the presence of outliers in match pairs. First, the two-view reconstruction of the initial match pair is crucial, because the robustness and accuracy of the final reconstruction result relies heavily on it. Second, block growing is sensitive to the order of added new images. If a block grows in a wrong way, e.g., visually drifts of the newly oriented images arise, the whole reconstruction may be incorrectly estimated.

This can occur, for example, due to the incorrect match pair of a repetitive structure, which typically causes ambiguities when finding the correct order. Third, triangulation to estimate 3D points suffers from limited robustness on match pairs stemming from very short baselines. Based on incremental methods, hierarchical SfM [16–19] creates atomic reconstructions in a divide-and-conquer manner and combines them hierarchically. While efficient due to the parallelization of dealing with atomic reconstructions, it is sensitive to the method by which these reconstructions grow. Global SfM [11,20,21] simultaneously considers all images to obtain consistent image orientation results. It is of high efficiency but is sensitive to outliers in match pairs. Recent developments of SfM focus on large-scale image sets, such as internet photo collections [21–23] and UAV image sets [18,24]. They typically consider moderate match pairs and pay less attention to those due to repetitive structures [8,25,26] and very short baselines [27–29]. These problematic match pairs can degrade SfM reconstruction results, or even lead to failure. We propose an SfM method to obtain correct reconstruction results for image sets with problematic match pairs stemming from repetitive structures and very short baselines.

Before detailing our contributions, we shortly review the state of the art of those related methods. Many strategies have been tried to cope with problematic match pairs [8,26–31]. It has been suggested to find a reliable subset of match pairs before executing image orientation [8,9,28]. This strategy is known as refining match pairs (or view-graph filtering [9]), which is equivalent to obtaining a robust subgraph from the original view-graph. RANSAC was used to delete inconsistent edges, randomly sampling spanning trees, generating cycles by walking two edges in the tree and one edge in the remaining set, deleting edges that lead to large discrepancies on rotation over cycles, and keeping the solution with the largest number of edges [20]. A Bayesian framework was designed to infer incorrect edges based on the inconsistency over cycles, which we call cycle consistency inference (CCI) [32]. Consistency was checked by chaining relative transformations to find erroneous image triplets and eliminating all match pairs among these at once [33]. These last three schemes considered only inconsistency using rotation that might not be able to find unreliable match pairs if image sets were captured by nearly pure translation motion. Verification was performed on both rotation and translation for every triplet in the view-graph, eliminating edges that could not pass [34]. 3D relative translations of all match pairs were projected in multiple 1D directions, eliminating match pairs whose relative translations stood out in the majority of directions [21]. However, the paper's authors acknowledged that the this method fails to deal with match pairs due to repetitive structures [21,26]. Criteria were designed to indicate the probability of a match pair due to repetitive structure and a very short baseline [26]. The criterion on the repetitive structure was based on the assumption that match pairs overlap by a nearly constant amount, making it less general. The problem of the repetitive structure was addressed by prioritizing edges in one so-called verified maximum spanning tree and extending it to a sufficiently redundant view-graph [9].

Two strategies can be employed to improve the time efficiency and robustness of incremental SfM in the manipulation of the view-graph. First, view-graph partition and merging are typically used in hierarchical SfM. The idea was to divide the view-graph into a number of overlapping sub-graphs, each solved independently to create atomic reconstructions, and hierarchically merging these to obtain a complete reconstruction. To ensure accurate reconstruction, match pairs shared in different sub-graphs should be sufficiently redundant for reliable merging [24]. Thus, the quality of these shared match pairs has a significant influence on the merging process, and outliers might cause large drifts. A graph-based method for building reliable overlapping relations of images [29] was proposed to improve a previous hierarchical merging approach [16] in the presence of very short baselines and wide baselines (Wide baselines typically occur when terrestrial and UAV images are connected, and can lead to inaccurate SfM reconstruction results [29]). Second, the progressive scheme on the view-graph allows SfM to be carried out by iteratively prioritizing a subset of edges or match pairs, a strategy known as prioritizing match pairs. One maximum spanning tree was extracted from the view-graph to select the match pairs with the highest weights, which were used for an initial reconstruction, and expanded until no new images were added to the block among three consecutive iterations, implying that the block is completely built [35], alleviating the negative influence of some outliers in match pairs by carefully selecting the edges of the view-graph. However, a single tree is insufficiently robust if outliers are contained [9]; thus, it is dangerous to directly feed them into SfM pipelines, especially with a repetitive structure and very short baselines. A skeletal subset—not the full view-graph—was utilized to improve efficiency by up to an order of magnitude or more, with little or no loss in accuracy [36].

Based on an incremental method [10], termed COLMAP, we propose a progressive SfM pipeline (PRMP-PSfM) to obtain robust and accurate SfM reconstruction results. The remainder of this paper is organized as follows. Section 2 introduces COLMAP and discusses its limitations in the presence of a repetitive structure and very short baselines. Section 3 presents the proposed method, including initialization, which focuses on accurate seed reconstruction, and expansion, which yields a complete reconstruction result by progressively adding match pairs and corresponding images. Section 4 demonstrates the performance of our method on various image datasets. Some important components and settings of our method are discussed in Section 5. Section 6 concludes our work.

2. Incremental SfM Pipeline

2.1. COLMAP

Figure 1 depicts the workflow of the incremental SfM method (COLMAP), which is the basis of PRMP-PSfM. Given a set of images, feature extraction and matching (e.g., SIFT and nearest neighbor ratio matching [3]) are carried out to obtain all possible relations of images and corresponding conjugate points. These image pairs are verified by a two-view epipolar geometric constraint, i.e., geometric verification, which usually employs a five- [5] or eight-point method [4] with RANSAC [6] to yield all potential match pairs. They can be represented by a view-graph, where red vertices indicate images and gray edges indicate corresponding match pairs. Two-view reconstruction is initially built in the following estimation pipeline, and the next-best view selection and image registration orientate candidate images. Tracks are updated by concatenating the correspondences of match pairs, and track triangulation generates new 3D points to expand the reconstruction, which is refined by bundle adjustment. Outlier filtering eliminates inaccurate 3D points with large reprojection errors and images whose camera poses cause refinement to fail. The procedures of block growing, including next-best view selection, image registration, track triangulation, bundle adjustment, and outlier filtering, are repeated to obtain the final reconstruction result; red triangles in Figure 1 denote the estimated camera poses (i.e., image orientation parameters).

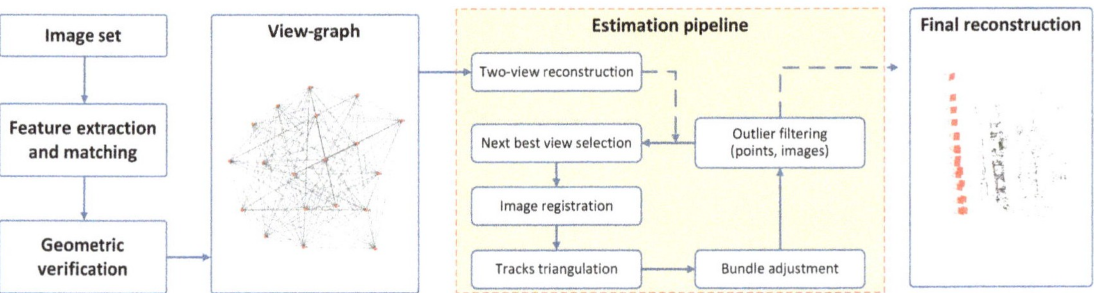

Figure 1. Workflow of COLMAP [10].

2.2. The Negative Influence of Problematic Match Pairs

Although many outliers in match pairs can be eliminated by RANSAC, some problematic match pairs still exist in the case of a repetitive structure or very short baselines.

We take COLMAP (Figure 1) as an example of a conventional incremental SfM method and study its limitations in these instances.

2.2.1. Repetitive Structure

A repetitive structure in a human-made environment often yields ambiguities in pairwise image matching, e.g., wrong match pairs that observe similar but different objects, such as rows of windows and similar building facades. Many suspicious correspondences can be generated due to very similar descriptors, and some can pass epipolar geometric verification. Consequently, a repetitive structure typically results in wrong match pairs [26]. Figure 2 shows an example of a dataset with a repetitive structure. Images were captured along a closed loop of one cup whose front and back sides are symmetric. Figure 2a shows a correct match pair whose images view the same area, and Figure 2b shows a wrong match pair due to a repetitive structure. We represent all potential match pairs by a view-graph (Figure 2c), and insert it in COLMAP to obtain the SfM reconstruction result (Figure 2d). To make a comparison, we manually select the correct match pairs, yielding the filtered view-graph in Figure 2e, and generate a more reasonable reconstruction in Figure 2f. Investigating these two reconstruction results, the one from the original view-graph is a folded scene with large drifts of camera poses. This implies that wrong match pairs due to a repetitive structure can indeed have a negative impact on the reconstruction result. COLMAP adds candidate images with more visible 3D points and a more uniform distribution of correspondences, reflecting the inherent assumption that the match pairs are correctly overlapped. However, this assumption is sometimes invalid due to ambiguities caused by a repetitive structure. Once the block grows in a wrong way, it yields an incorrect reconstruction result, such as the folded scene in Figure 2d.

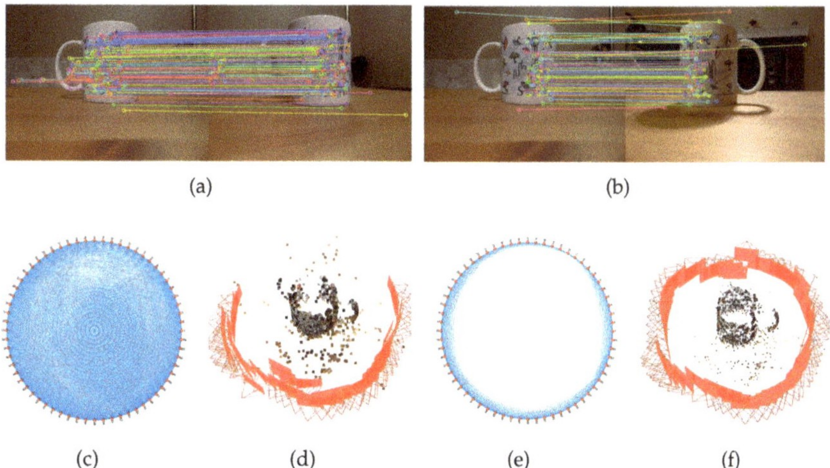

(a)   (b)

(c)   (d)   (e)   (f)

Figure 2. Example of a repetitive structure: (**a**) a correct match pair, and (**b**) a wrong match pair due to repetitive structure, where colored lines show correspondences; (**c**) original view-graph estimated by COLMAP; (**d**) reconstruction result from inputting (**c**) in [10], where red triangles describe camera poses; (**e**) filtered view-graph, and (**f**) corresponding reconstruction result, where edges in (**e**) are correct match pairs manually selected from (**c**).

2.2.2. Very Short Baselines

Very short baselines arise when the distance between images is insufficient or extremely close to pure rotation motion. Figure 3 shows two images (blue and red points) and one 3D point (a black point), which is the intersection point of two view rays (blue and red arrows). When we keep one image (blue) and the 3D point (back) fixed, and move the another image (red) along a circle (dotted line) with the center of the blue point with a constant radius, different cases of two-view intersection can be observed. If the radius is very small, i.e., with a very short baseline between these two images, it leads to a small intersection angle. Such poor intersection geometry typically results in an ill-posed problem of estimating coordinated 3D points [13,26,27].

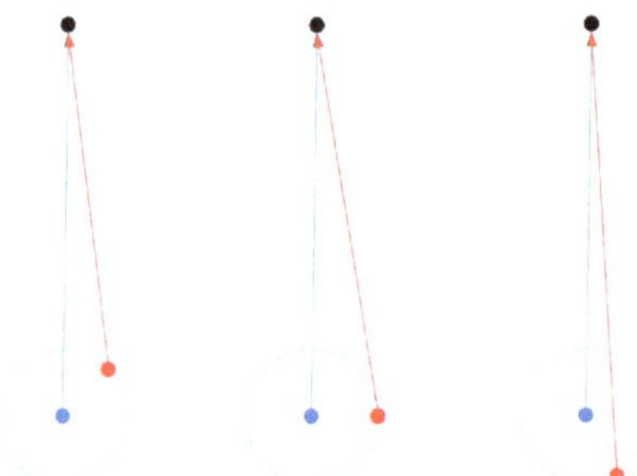

Figure 3. Different cases of two-view intersection with a constant baseline. Blue, red and black points indicate two images and one 3D point, respectively. Blue and red arrows indicate two view rays. The radius of circle (dotted line) indicates a constant baseline between two images

To illustrate the influence on SfM results, we tested a dataset with very short baselines in COLMAP; corresponding reconstruction results are shown in Figure 4. The reference reconstruction result (Figure 4a) was obtained by only using correct match pairs. When both correct match pairs and ones with very short baselines were input, the reconstruction result becomes worse, as can be seen in Figure 4b–f. In comparison to the reference, it can be seen that COLMAP can obtain a good two-view reconstruction (Figure 4b), but shows increasing drifts of camera poses as the block grows. As Figure 4c shows, some inaccurate 3D points (blue ellipses) are generated in the expanded reconstruction, mainly due to corresponding match pairs with very short baselines. The final reconstruction result (Figure 4f) suffers from obvious visual drifts due to error accumulation.

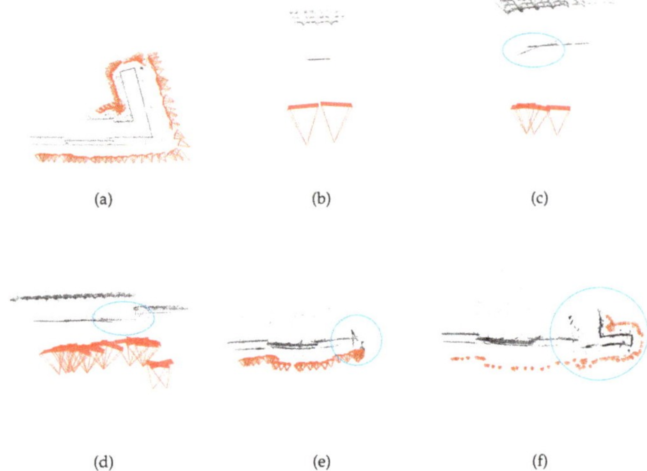

Figure 4. Example of very short baselines: (**a**) the reference reconstruction result; (**b**) initial two-view reconstruction; (**c–f**) reconstruction results as the block grows, where blue ellipses indicate areas where 3D points are incorrectly estimated.

3. Method

3.1. Overview

We present PRMP-PSfM, whose workflow is shown in Figure 5. The input is a set of images, following the procedures of COLMAP (Figure 1) to obtain all potential match pairs, from which we construct a weighted view-graph in Section 3.2. Weights of edges indicate the costs of match pairs; the smaller the weight, the more the possibility that a match pair is correct. The image orientation pipeline includes initialization and expansion. In initialization, we generate a seed view-graph comprising multiple independent minimum spanning trees (MSTs), containing subsets of match pairs with smaller costs, and apply outlier elimination using CCI for a filtered seed view-graph. We insert this robust view-graph in COLMAP to obtain an accurate reconstruction result, a procedure called seed reconstruction. Images are sometimes excluded from seed reconstruction due to the filtering procedure; expansion is designed to achieve a more complete reconstruction. Completeness is checked to decide whether the expansion is necessary. If this condition is met by the seed reconstruction, then we set it as the final reconstruction result. Otherwise, seed reconstruction is incomplete and expansion is carried out. New MSTs are progressively added to the filtered seed view-graph to realize a denser graph, i.e., an expanded view-graph, which is filtered by outlier elimination using CCI before applying incremental SfM. The procedure is repeated until completeness is achieved.

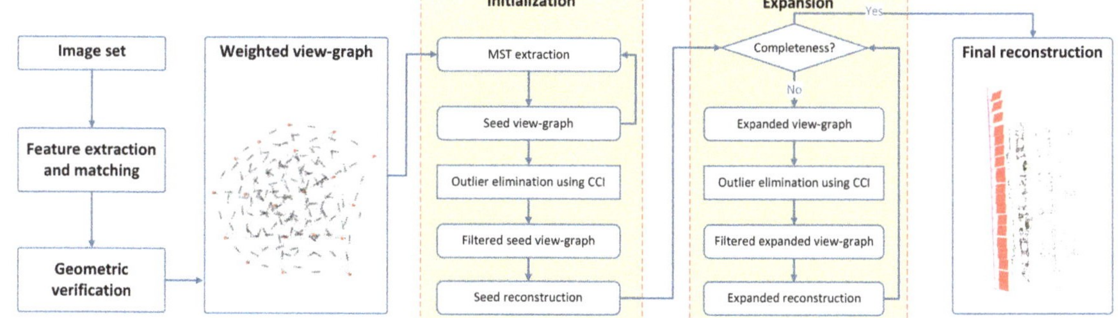

Figure 5. Workflow of PRMP-PSfM.

3.2. Construction of Weighted View-Graph

The undirected view-graph $G = (V, E)$ is often used to represent relations between images. Vertices V indicate images, and edges E indicate match pairs; $e_{ij} \in E$ means two images of vertices $\{v_i, v_j\} \in V$ are successfully matched after geometric verification. To prioritize the edges by MST, we need to construct a weighted view-graph $G = (V, E, W)$, where W is a set of scalar values indicating the costs of edges. The smaller the cost, the more likely it is that the edge is correct. It has been suggested to calculate the cost for each edge by Equation (1) [37], where M is the number of feature correspondences, and Θ is the mean intersection angle of all correspondences; they are normalized to [0, 1] and balanced by a factor μ, which is set by default to 0.1. After the costs of edges were calculated, one MST was extracted to select the most reliable match pairs. Only tracks generated from these selected match pairs were used in the proposed method, which yielded a robust method to avoid abundant or outlier tracks [37]. More feature correspondences of one match pair can generally produce a more robust estimate [8,9,37], which is also reasonable in the presence of repetitive structures because after geometric verification, the feature correspondences of a correctly overlapping match pair outnumber those of a match pair with a repetitive structure [9,35]. For very short baselines, which typically result in very small intersection angles, Θ is a suitable criterion to indicate their lengths. Hence, we generate the weight view-graph as [37]

$$W = \frac{1}{M} + \mu \times \frac{1}{\Theta} \tag{1}$$

3.3. Outlier Elimination Using CCI

We introduce a general outlier detection method, CCI [32], by analyzing the geometric consistencies over cycles. Given a view-graph (Figure 6a) and the estimated relative transformations $\{T_e\}$ of all potential match pairs, the exhaustive cycles of the length of all image triplets C are extracted (Figure 6b), and the deviation or inconsistency of each cycle is computed using a non-negative function $d(T_c)$, $c \in C$, where T_c is the chained transformation along the cycle. If one cycle is ideally consistent, then $d(T_c)$ should equal zero, while noise or outliers in match pairs can lead to a nonzero value. If $d(T_c)$ exceeds a threshold, then at least one problematic match pair exists in the cycle. Based on this idea, a Bayesian inference framework was proposed [32]. Define the following:

$P(d(T_c) \mid x_c = 1)$: probability of deviation $d(T_c)$ for a cycle, assuming all its edges are correct;

$P(d(T_c) \mid x_c = 0)$: probability of deviation $d(T_c)$ for a cycle, assuming at least one edge in the cycle is incorrect;

$P(x_e)$: prior probability for indicating the quality of an edge.

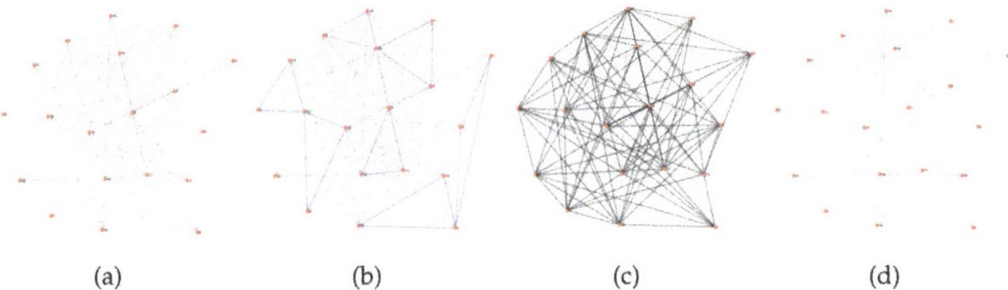

(a)   (b)   (c)   (d)

Figure 6. Illustrative process of outlier elimination using CCI to filter incorrect match pairs: (**a**) input original view-graph, where red vertices indicate images and gray edges are match pairs; (**b**) procedure of cycle extraction on view-graph, where only some sample cycles are highlighted by blue lines; (**c**) CCI infers incorrect edges, which are highlighted by black lines; and (**d**) filtered view-graph after eliminating incorrect edges.

Latent binary variables, x_e, x_c, are introduced for each edge and cycle, respectively, where $x_e = 0$ indicates an incorrect edge, $x_e = 1$ indicates a correct edge, and after chaining all the edges over a cycle, $x_c = \min_{e \in c} x_e$. Hence, $x_c = 1$ indicates all edges of the cycle are correct, and $x_c = 0$ if at least one incorrect edge exists. $x_e \in \{0, 1\}$ should be assigned to all edges to maximize the joint probability function:

$$\prod_{c \in C} P(x_{e \in c} \mid d(T_c)) \propto \prod_{e} P(x_e) \prod_{c \in C} P(d(T_c) \mid x_c) \tag{2}$$

The inference problem can be represented by factor graphs and solved by loopy belief propagation [38]. Once the inference is finished, we determined a set of incorrect edges (dark lines in Figure 6c), i.e., $\{x_e = 0\}$, and eliminate them to form a filtered view-graph, as seen in Figure 6d.

We elaborate on the calculation of the probability mode of cycles $P(d(T_c) \mid x_c = 1)$. For image sets with known intrinsic parameters, we obtained the estimated relative orientations of match pairs by decomposing the essential matrix. These relative orientations have five degrees of freedom, including relative rotations and translations. For a cycle formed by three images i, j, k, when the estimated relative rotation matrices of image pairs R_{ij}, R_{ik}, R_{jk} are known, we denote the chained rotation over the cycle as $c_R = R_{ij} R_{jk} R_{ik}^\top$, and the difference on rotation as $d_{\angle}(c_R) = \arccos(\frac{\text{trace}(c_R)-1}{2})$. Given the estimated relative translation vectors t_{ij}, t_{ik}, t_{jk}, the chained translation angle is $c_T = \theta_i + \theta_j + \theta_k$, where angles θ_i, θ_j, θ_k for the images are, respectively, calculated, i.e., $\theta_i = \arccos(\frac{t_{ij}^{\top} t_{ik}}{\|t_{ij}\|\|t_{ik}\|})$. We calculated the difference on translation $d_{\angle}(c_T) = |c_T - 180°|$, where $180°$ is the sum of the interior intersection angles of this triangle (cycle of length three). We set the deviation as the larger of the two differences in Equation (3). We fit the inlier portions $\{d(T_c) \le \sqrt{|c|}\varepsilon\}$ as an exponential distribution and empirically set ε to 2 degrees. $F(d(T_c); \lambda)$ is the cumulative distribution function, where λ is the parameter of the exponential function, which is adaptively estimated from the inlier data. As inconsistent cycles may have small $d(T_c)$, e.g., when the error of one incorrect edge is offset by another, we limit the maximum value of $P(d(T_c) \mid x_c = 1)$ in Equation (4) to 0.9 instead of to 1:

$$d(T_c) = \max\left[d_{\angle}(c_R), d_{\angle}(c_T)\right] \tag{3}$$

$$P(d(T_c) \mid x_c = 1) = \begin{cases} 0.9[1 - F(d(T_c); \lambda)], & d(T_c) \le \sqrt{|c|}\varepsilon, \\ 0, & d(T_c) > \sqrt{|c|}\varepsilon. \end{cases} \tag{4}$$

3.4. Initialization

For good seed reconstruction, the input view-graph should be as accurate as possible. Most incremental methods [10,33] employ the original view-graph. As discussed in Section 2.2, problematic match pairs must be eliminated because they have a negative influence on the reconstruction results of the incremental SfM method. One MST can be used as the input view-graph [35]. However, it is difficult to guarantee that its edges are correct, especially for image sets with problematic match pairs. Images in one MST are only two-fold overlapping, and the redundancy of edges is insufficient for accurate seed reconstruction if a selected edge in the MST is not correct [9].

As Figure 5 shows, we generate the filtered seed view-graph considering both accuracy and redundancy of match pairs, as shown in Figure 7. Given the original view-graph (Figure 7a), a number of MSTs comprise the seed view-graph (Figure 7b), and a filtered one, the filtered seed view-graph (Figure 7c), is obtained by outlier elimination using CCI. Specifically, given the weighted view-graph $G = (V, E, W)$, the first MST is extracted [39], which contains all vertices of V and $|V| - 1$ edges with the smallest costs, and these edges are removed from G, yielding a new graph G'. Here, $|\cdot|$ counts the number of vertices. The second MST is extracted from G', and this ensures there are no repeated edges between two MSTs, i.e., so-called orthogonal MSTs [25,35]. The above processes are repeated several times, and these extracted orthogonal MSTs compose a seed view-graph G_{seed} with N_{seed} iterations. With a smaller N_{seed}, the graph contains more accurate and less redundant edges, whereas a larger N_{seed} makes for a denser graph that may contain some unreliable edges. We first compute the averaging degree for all vertices, $\Delta(G)$, which indicates the density of the original view-graph, and multiply by a factor α to determine N_{seed} for adaptability:

$$N_{seed} = \alpha \times \Delta(G)$$
$$\Delta(G) = \frac{\sum_{i=1}^{|V|} deg(v_i)}{|V|} \tag{5}$$

where $deg(v_i)$ is the degree of vertex $v_i \in V$, i.e., the number of edges that are incident to it. Hence, our method is less sensitive to the density of view-graphs. CCI is then used to detect and eliminate outliers on G_{seed} to yield a filtered seed view-graph G_{seed}^{f}. We employ the largest connected component (in graph theory, a connected component of an undirected graph is a subgraph in which any two vertices are connected to each other by paths, and which is connected to no additional vertices in the rest of the graph) of G_{seed}^{f} for seed reconstruction. Note that some edges are possibly filtered out, which might cause G_{seed}^{f} to be disconnected and lead to several individual reconstructions. Therefore, the biggest connected component is selected for seed reconstruction.

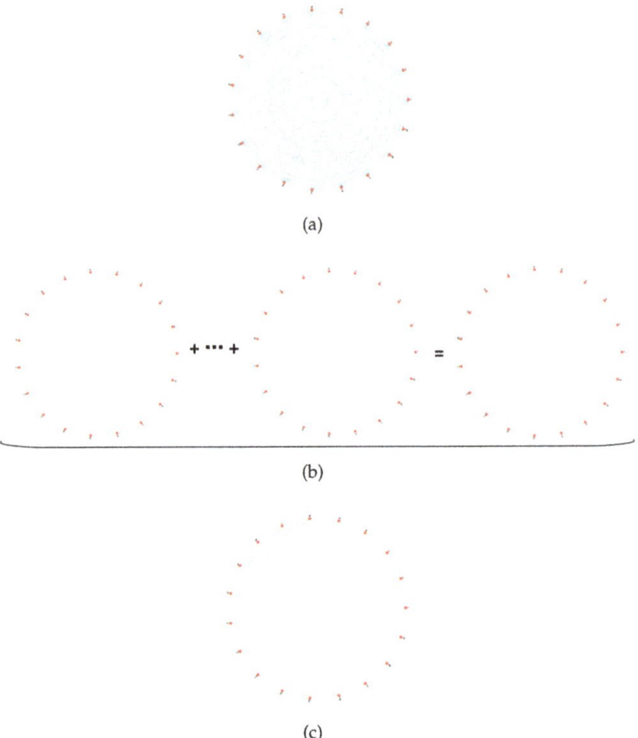

Figure 7. Illustrative process of generating the filtered seed view-graph: (**a**) original view-graph; (**b**) seed view-graphs obtained by uniting a number of MSTs; and (**c**) filtered seed view-graph after outlier elimination using CCI.

3.5. Expansion

Expansion aims to obtain a more complete reconstruction result. In general, when the block completely grows, complete reconstruction is reached; thus, regarding the condition of completeness, we suggest that all images in the input view-graph should be successfully orientated. Initialization uses the filtered seed view-graph with rather good match pairs to ensure accurate seed reconstruction. However, its completeness is not strictly guaranteed. Incomplete seed reconstruction might arise for two reasons: (1) the largest connected component of the filtered seed view-graph does not cover all images, yielding only part of the complete reconstruction result; (2) some unstable images are excluded by the procedure of outlier filtering (see Figure 1).

As Figure 5 shows, the condition of completeness is checked at the beginning of expansion. If the condition is reached, then the seed reconstruction is output as the final result. It is otherwise necessary to expand the seed reconstruction for completeness, and the filtered expanded view-graph and expanded reconstruction are similarly generated. A workflow is given in Figure 8. We progressively add new MSTs to the filtered seed view-graph to generate a denser graph. By conducting outlier elimination using CCI, we can obtain the filtered expanded view-graph. Analogously, we employ the largest connected component to add new edges and corresponding new images. Once these new images are orientated, the tracks are updated by concatenating the correspondences of match pairs of the current view-graph. Subsequent processes, as shown in Figure 1, are progressively conducted until the condition of completeness is reached.

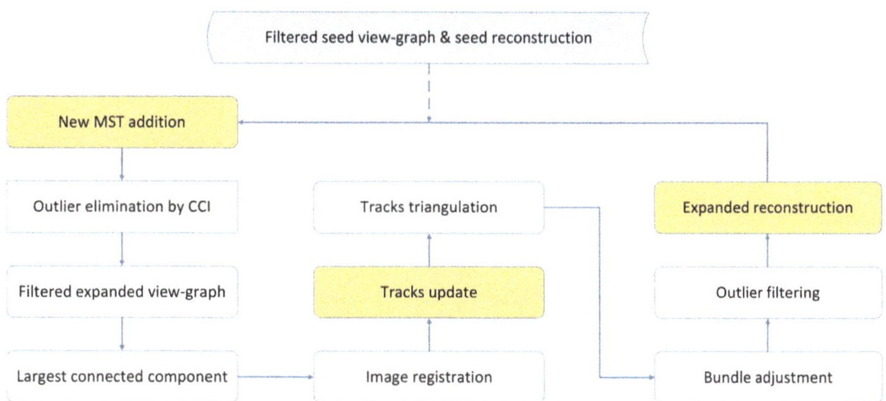

Figure 8. Workflow of expansion of seed reconstruction.

4. Results

We quantitatively and qualitatively evaluated the efficacy of PRMP-PSfM at generating accurate SfM reconstruction results. Experiments were conducted on datasets with different types of problematic match pairs, including repetitive structures and very short baselines. PRMP-PSfM was then compared to several state-of-the-art SfM pipelines, including three incremental methods (COLMAP [10], (COLMAP software was downloaded from https://github.com/colmap/colmap/releases, version 3.6-dev.2 [released 24 March 2019]. All experiments were conducted with the settings suggested by [10]) VisualSFM [15], (VisualSFM software was downloaded from http://ccwu.me/vsfm/, version V0.5.26 [accessed September 2020]. All experiments were conducted with the default settings) and OpenMVG [33] (Code was downloaded from https://github.com/openMVG/openMVG, version v1.5 [released 16 July 2019]. We used its incremental SfM pipeline. Details can be found at https://openmvg.readthedocs.io/en/latest/)) and two hierarchical methods (GraphSfM [19] (Code was downloaded from https://github.com/AIBluefisher/EGSfM. The version provided by the original author was implemented based on OpenMVG) and APE [29] (Note that we were unable to obtain the APE source code. We referred to the results of the original paper for comparison)). APE integrates a series of processes of dealing with problematic match pairs due to wide baselines and very short baselines. Some processes of these methods used in the experiments are listed in Table 1. Considering that they are all complex systems, the processes related to the view-graph are mainly summarized. We set $\alpha = 0.25$ for all experiments.

Table 1. Some properties of the methods used in our experiments: FE—feature extraction; FM—feature matching; MP—match pairs; BA—bundle adjustment; PRMP—prioritizing and refining match pairs; LM—Levenberg–Marquardt; PCG—preconditioned conjugate gradient; RBA—robust bundle adjustment. "Original" indicates that the original view-graph is taken as input and there is no specified process on its match pairs. "-" indicates that the corresponding items are unavailable.

Framework	FE	FM	MP	BA
PRMP-PSfM	SiftGPU [15]	Nearest neighbor ratio	PRMP	LM [14]
COLMAP	SiftGPU [15]	Nearest neighbor ratio	Original	LM [14]
VisualSFM	SiftGPU [15]	Preemptive feature matching [15]	Original	PCG [15]
OpenMVG	SIFT	Cascade hashing [40]	Original	LM [14]
GraphSfM	SIFT	Cascade hashing [40]	Original	LM [14]
APE	SiftGPU [15]	Wide baseline method [29]	Classification [29]	RBA [29]

4.1. Datasets

Table 2 lists the image datasets used in our experiments, consisting of five small public datasets (*Books*, *Cereal*, *Cup*, *Desk* and *Street* [8]), three middle-scale datasets *Indoor*, *Temple-of-Heaven (ToH)* [8], and *Redmond* [41]), three benchmark datasets (*B1*, *B2*, *B3* [26]), and one large image set (*Church* [29]). The "Type" column indicates two types of problematic match pairs: repetitive structure and very short baselines. To investigate the ability to cope with different problematic match pairs and demonstrate the performance of our method, eight image datasets with only repetitive structures were tested (Sections 4.2 and 4.3), followed by three benchmark datasets with both repetitive structures and very short baselines (Section 4.4), and one large-scale dataset with repetitive structures and very short baselines (Section 4.5).

Table 2. The description of image datasets used in our experiments.

Name	Images	Resolution	Type	Reference
Books	21			
Cereal	25			
Cup	64	1067×800	Repetitive structure	Yes
Desk	31			
Street	19			
Indoor	152	1200×800		
Redmond	148	3968×2232	Repetitive structure	No
ToH	341	4368×2912		
B1	182		Repetitive structure	
B2	215	3936×2624	Very short baselines	Yes
B3	342			
Church	1455	3264×2448 3648×2736 7360×4912	Repetitive structure Very short baselines	No

4.2. Performance on Five Small Datasets

We tested five small datasets with different degrees of repetitive structures; some sample images are shown in Figure 9. Since these datasets were rather small, we could manually establish the ground-truth view-graph by selecting a subset with the correct overlapping relations of match pairs from the original view-graph. We used the adjacency matrix to represent the view-graph (see Figure 10), the horizontal and vertical directions of which indicate image IDs, and dark pixels indicate that the corresponding match pairs are considered correct. Figure 10a corresponds to the original view-graphs generated after the default matching process [10], and Figure 10b to the ground-truth view-graphs. We can see that the original view-graphs had many wrong match pairs stemming from the repetitive structure.

We inserted the ground-truth view-graphs (Figure 10b) in the incremental SfM pipeline (COLMAP, see Figure 1) to obtain reference reconstruction results, as shown in Figure 11. To compare different SfM methods, we input the original view-graphs (Figure 10a) to the PRMP-PSfM, COLMAP and OpenMVG SfM pipelines to determine whether they were capable of dealing with images of a repetitive structure. The reconstruction results from these five small datasets are shown in Figure 11. Compared to the reference, PRMP-PSfM generated the best reconstruction results for all five datasets; COLMAP and OpenMVG only successfully reconstructed the *Desk* dataset and obtained various folded structures for the other four datasets. For repetitive structures, wrong overlapping relations of image pairs in the original view-graph can possibly make the next-best view selection (Figure 1) invalid. We propose an improved method to overcome this by manipulating the view-graph, and PRMP-PSfM could generate the best SfM reconstruction results on all five datasets (with repetitive structures).

Figure 9. Sample images of five small datasets with repetitive structures.

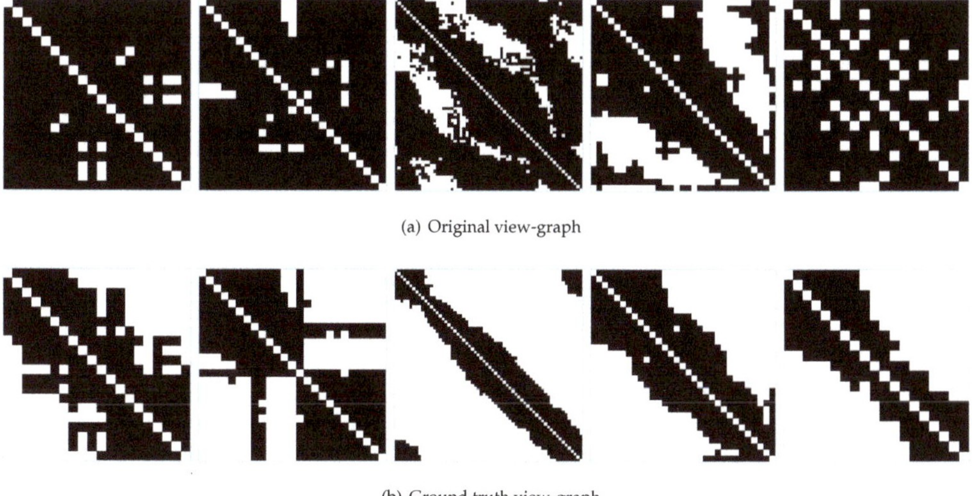

(a) Original view-graph

(b) Ground truth view-graph

Figure 10. Adjacency matrices of view-graph for five small datasets. From left to right: *Books, Cereal, Cup, Desk, Street*.

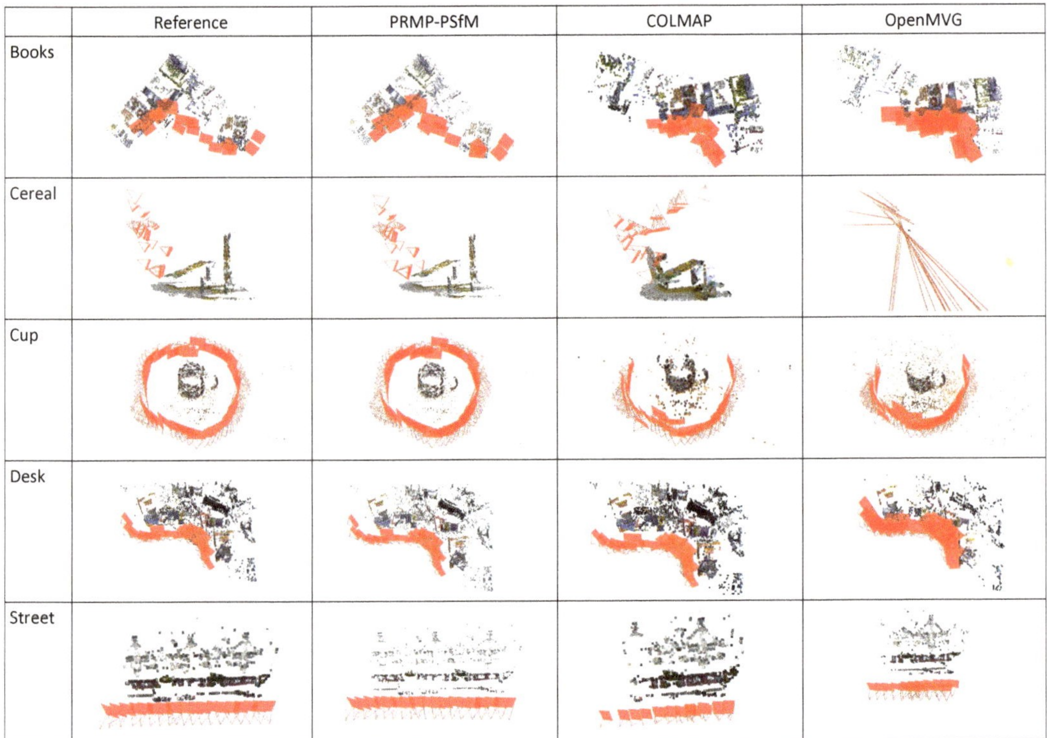

Figure 11. Reconstruction results from five small datasets.

4.3. Performance on Three Middle-Scale Datasets with Repetitive Structures

We reported on experiments on three middle-scale datasets with repetitive structures: *Indoor*, *Redmond* and *ToH*, whose sample images and reconstruction results are given in Figure 12a–c, respectively, [28]. The *Indoor* dataset was captured in an indoor scene, and its camera trajectory (red triangles) contained three single strips on three floors. For *Redmond*, the camera trajectory was nearly along a straight line when capturing a set of images in a row of some similar building facades. The scene of *ToH* contained nearly 360-degree symmetry, and images were captured with a closed loop.

Because there are no ground-truths of these three datasets, based on visualization results [28] (see Figure 12), we carried out qualitative evaluations of PRMP-PSfM, COLMAP, VisualSFM, OpenMVG, and GraphSfM, whose reconstruction results are shown in Figure 13. For *Indoor* and *Redmond*, the results of PRMP-PSfM and VisualSFM were visually similar to the visualization results, but COLMAP, OpenMVG, and GraphSfM failed with regard to correct and complete the camera poses. For *ToH*, only PRMP-PSfM and GraphSfM were able to close the loop, while COLMAP, VisualSFM, and OpenMVG could only solve parts of the whole scene. We generated the best results on these three datasets, further demonstrating the capability of PRMP-PSfM to deal with images with repetitive structures.

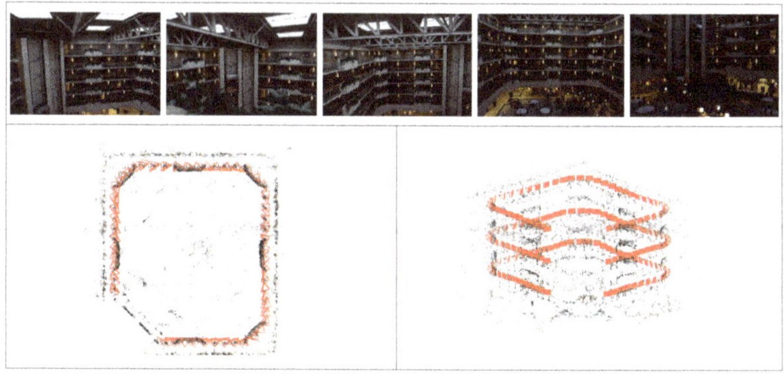

(a) Indoor

(b) Redmond

(c) ToH

Figure 12. Sample images and reconstruction results [28] of three middle-scale datasets with repetitive structures.

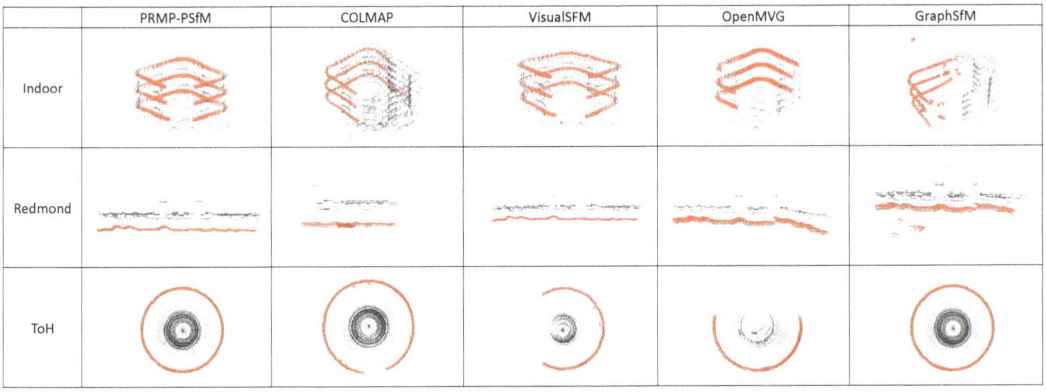

	PRMP-PSfM	COLMAP	VisualSFM	OpenMVG	GraphSfM
Indoor					
Redmond					
ToH					

Figure 13. Reconstruction results on three middle-scale datasets with repetitive structures.

4.4. Performance on Three Benchmark Datasets

The ground-truth for benchmark datasets *B1*, *B2* and *B3* for match pairs [26] means that the corresponding correct match pairs and problematic ones due to repetitive structures and very short baselines were manually found and labeled. We inserted these correct match pairs in COLMAP to obtain reference reconstruction results, which, along with sample images, are shown in Figure 14.

Figure 15 shows the final reconstruction results on these three benchmarks by the five SfM pipelines. Compared to the reference, only PRMP-PSfM and OpenMVG obtained similar results for all three datasets, while the other pipelines generated various visual drifts. For dataset *B1*, all pipelines except VisualSFM were able to obtain similar reconstruction results, which were generally identical to the reference. PRMP-PSfM, COLMAP, and OpenMVG were able to reconstruct dataset *B2*, but the result of VisualSFM showed large drift, and GraphSfM only recovered part of the scene. Dataset *B3* contained a closed loop around one building, and only PRMP-PSfM and OpenMVG could recover a complete reconstruction. To further demonstrate the performance of PRMP-PSfM, we quantitatively evaluated the reconstruction results that were qualitatively similar to the reference, i.e., those of *B1* and *B2* by PRMP-PSfM, COLMAP and OpenMVG, and those of *B3* by PRMP-PSfM and OpenMVG. We calculated the rotation and translation errors, which are listed in Table 3. It can be seen that our method obtained the highest accuracy on all three datasets. Although PRMP-PSfM and OpenMVG were able to obtain visually similar results, the numerical evaluation in Table 3 shows that reconstruction results of OpenMVG were less accurate than those of PRMP-PSfM. In particular, the results of max rotation error and max translation error indicate that some of the camera poses of OpenMVG were gross errors. In contrast, PRMP-PSfM was able to estimate the accurate rotation and translation parameters for all images, which shows its superiority at coping with problematic match pairs.

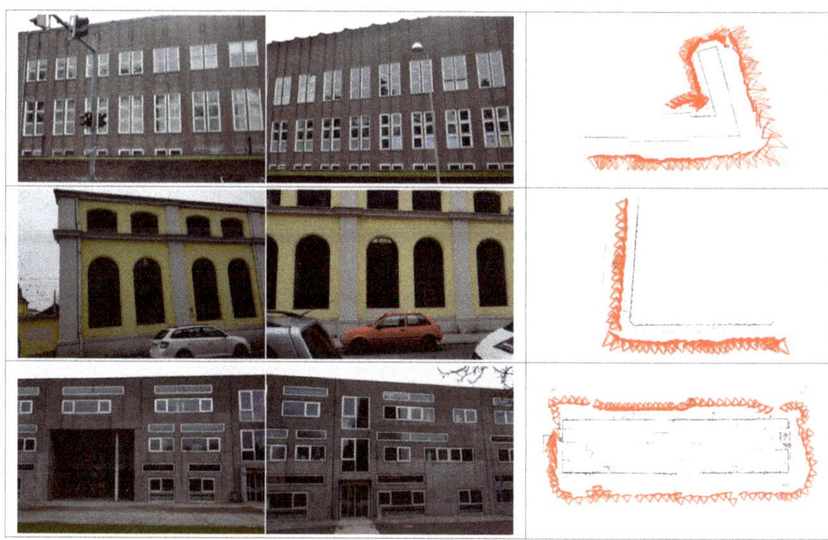

Figure 14. Sample images and correct reconstruction results of three benchmark datasets with both repetitive structures and very short baselines. From top to bottom: *B1*, *B2* and *B3*.

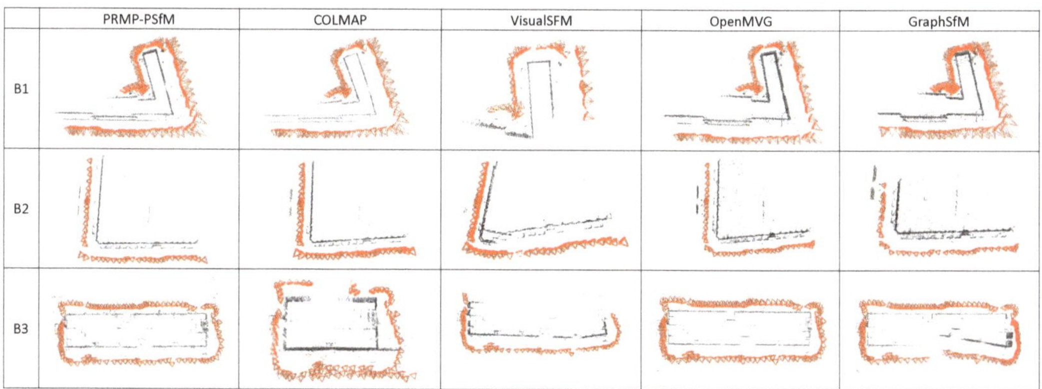

Figure 15. Reconstruction results of three benchmark datasets from different SfM pipelines.

Table 3. Quantitative evaluation for benchmark datasets *B1*, *B2* and *B3*. Rotation error is the angular difference from the reference, in degrees; translation error is the position difference from the reference in ground-truth units.

Dataset	Pipeline	Rotation Error				Translation Error ($\times 10^{-2}$)			
		Min	*Mean*	*Median*	*Max*	*Min*	*Mean*	*Median*	*Max*
B1	PRMP-PSfM	0.02	0.11	0.13	0.52	0.11	0.53	0.54	3.21
	COLMAP	0.17	1.47	1.27	2.21	1.25	8.93	9.78	18.4
	OpenMVG	0.18	1.7	1.63	3.74	1.55	10.02	10.40	18.66
B2	PRMP-PSfM	0.03	0.08	0.07	0.32	0.08	0.55	0.49	3.27
	COLMAP	0.15	0.91	1.02	1.84	1.05	4.58	4.86	9.89
	OpenMVG	0.15	0.66	0.46	4.54	0.22	7.29	4.84	48.61
B3	PRMP-PSfM	0.02	0.10	0.09	0.46	0.06	0.74	0.66	2.89
	OpenMVG	0.06	0.39	0.40	0.89	0.58	3.02	2.55	88.61

4.5. Performance on a Large-Scale Dataset

We evaluated PRMP-PSfM on the large-scale *Church* dataset [29], which consisted of 1455 unordered images with repetitive structures and very short baselines. Sample match pairs and our reconstruction results are shown in Figure 16. We present the numerical evaluation of five incremental and one hierarchical SfM pipeline in Table 4. The results of COLMAP, OpenMVG, and GraphSfM were obtained with default settings, and those of APE and VisualSFM are cited from Michelini et al. [29]. In terms of completeness, all pipelines except VisualSFM were able to orientate more than 98.9% of images, while COLMAP gave the most complete result (up to 99.9% of all images were solved). PRMP-PSfM, OpenMVG, GraphSfM, and APE generated similar mean reprojection errors, while COLMAP obtained the largest, implying that COLMAP did not obtain a convergent result. There are various reasons for the big differences in the number of reconstructed 3D points of these pipelines, such as different settings for feature extraction and rules for selecting tracks. PRMP-PSfM and COLMAP deleted tracks only observed by two images, while OpenMVG and GraphSfM retained them. Hence, it can be concluded that PRMP-PSfM, OpenMVG, GraphSfM, and APE obtained comparable precision and completeness results.

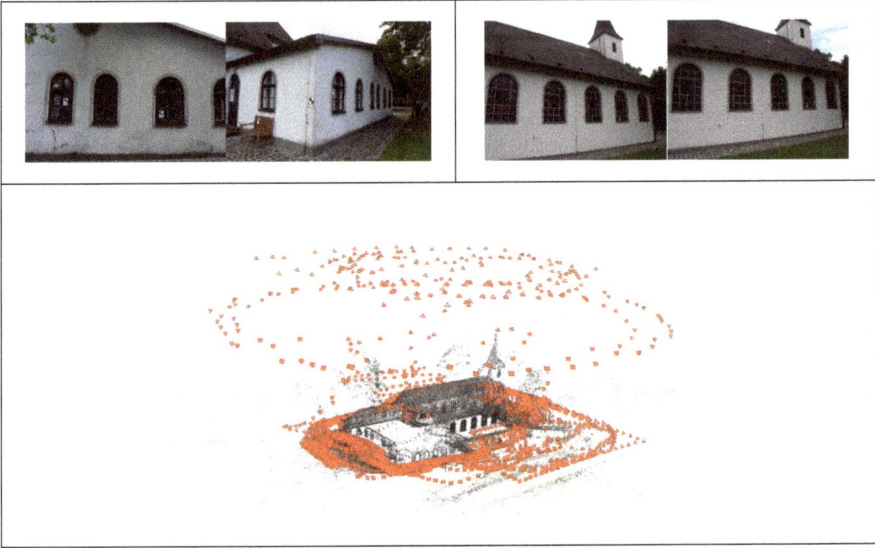

Figure 16. Sample images and our reconstruction result of the *Church* dataset. The top shows the sample match pairs, and on the left and right are match pairs due to repetitive structure and very short baselines, respectively. The bottom shows our SfM reconstruction result.

Table 4. Numerical comparison on *Church* dataset for different SfM pipelines. N_{img} is the number of orientated images in the final reconstruction results, and the % values refer to the number of orientated images compared with that of the input images, δ_0 is the mean projection error in pixels, N_p is the number of tie points, and "-" indicates that the corresponding items are unavailable.

Pipeline	N_{img} (%)	δ_0	N_p
PRMP-PSfM	1448 (99.5)	0.37	491,992
COLMAP	1454 (99.9)	1.09	549,957
VisualSFM	288(19.8)	0.74	14,295
OpenMVG	1452 (99.8)	0.54	1,687,694
GraphSfM	1439 (98.9)	0.51	2,762,371
APE	-	0.55	290,748

4.6. Performance of without Iteratively Refining Match Pairs

In PRMP-PSfM, the process of iteratively refining match pairs is implemented by repeatedly executing "outlier elimination using CCI" (Figure 5). To investigate how it influences SfM reconstruction results, we turned off that function. Figure 17 shows the reconstruction results for all datasets by PRMP-PSfM without iteratively refining match pairs, which we refer to as PSfM. Blue ellipses indicate visual drifts. The generated drifts in Books and Redmond were due to ambiguous tracks generated from match pairs with repetitive structures. The results of *Cereal* and *B1* contained large-scale drifts (see blue ellipses), which occurred at the beginning of the expansion. The reconstruction result of *B2* was negatively influenced by a repetitive structure and very short baselines.

We show a numerical comparison for the four datasets whose reconstruction results of PRMP-PSfM and PSfM were visually similar, i.e., *Cup*, *Desk*, *Street* and *B3*, in Table 5. Regarding the reprojection error, PRMP-PSfM showed better performance on *Desk* and *Street* and comparable results on *Cup* and *B3*. For errors on camera poses, we calculated the discrepancies between their results and the reference ones. PRMP-PSfM obtained higher accuracy than PSfM on both rotation and translation.

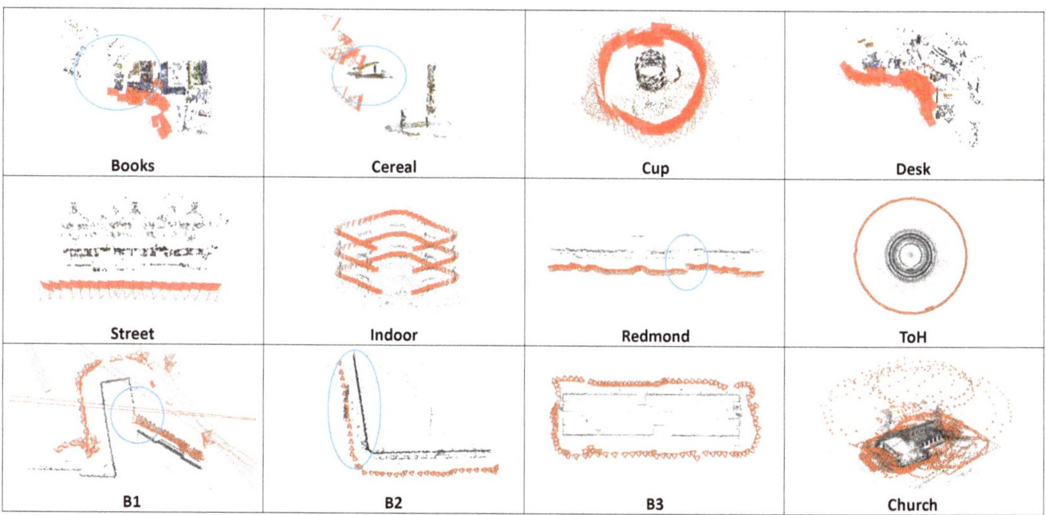

Figure 17. Reconstruction results for various datasets by PSfM, i.e., PRMP-PSfM without iteratively refining match pairs.

Table 5. Numerical comparison of reconstruction results between PRMP-PSfM and PSfM. δ_0 is the mean projection error, m_r is the mean rotation error in degrees and m_t is the mean translation error in ground-truth units $\times 10^{-1}$.

Dataset	PRMP-PSfM			PSfM		
	δ_0	m_r	m_t	δ_0	m_r	m_t
Cup	0.26	0.67	0.23	0.26	0.73	0.23
Desk	0.32	1.32	0.90	0.47	6.46	1.78
Street	0.18	0.78	0.11	0.35	1.16	0.29
B3	0.39	0.06	0.16	0.36	0.35	0.42

4.7. Settings of Parameter α

To investigate to what degree the key parameter α can influence the performance of PRMP-PSfM, we alternated the value of the free parameter α to qualitatively and quantitatively evaluate it on benchmark datasets *B1*, *B2* and *B3*, with results as shown in Figure 18 and Table 6. Figure 18 shows the seed reconstruction results (Section 3.4) obtained for different values of α. Table 6 shows the discrepancies between these and the reference results. As emphasized in Section 3.4, the objective of initialization is to obtain good seed reconstruction. As Table 6 shows, when setting $\alpha = 0.25$, our initialization step obtained the highest accuracy on *B1* and *B3*, and accuracy on *B2* comparable to that of $\alpha = 0.1$. We obtained the best result on *B2* with $\alpha = 0.1$, but less accurate results on *B1* and *B3*. Accuracy was lower for $\alpha = 0.40$ and $\alpha = 0.55$, because a denser view-graph leads to a higher possibility of including incorrect match pairs.

The case of $\alpha = 0.25$ was good enough for complete reconstruction results on *B1* and *B2*, revealing that PRMP-PSfM achieved the final reconstruction results by only applying the seed view-graph. Compared with other pipelines at handling the full view-graph, PRMP-PSfM has good potential to improve efficiency.

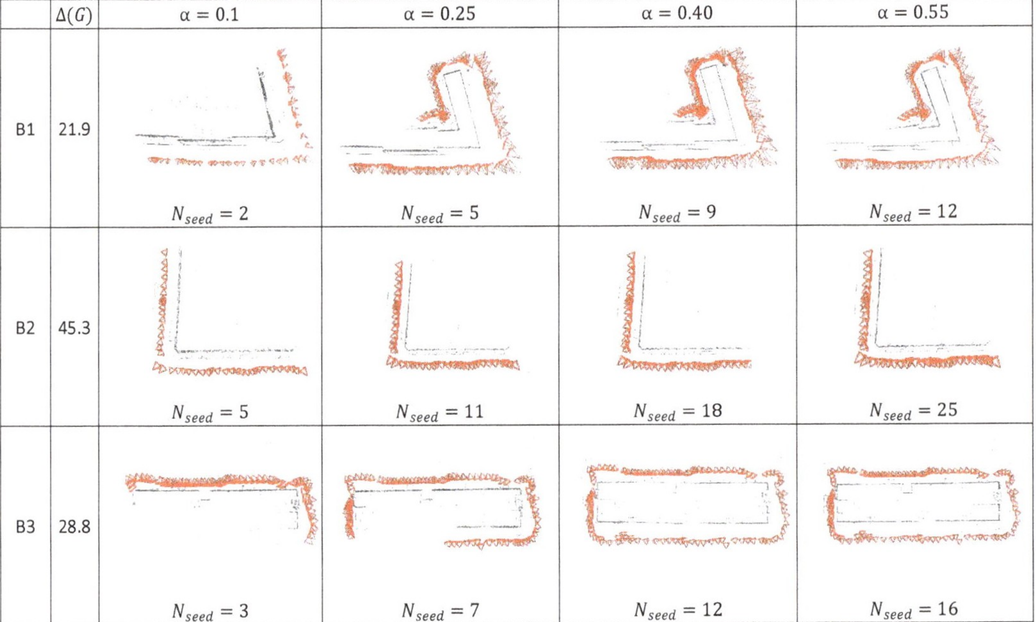

Figure 18. Seed reconstruction results for different values of α.

Table 6. Numerical comparisons of seed reconstruction results from different values of α. N_{img} is the number of orientated images in the final reconstruction results. m_r is the mean rotation error in degrees and m_t is the mean translation error in ground-truth units $\times 10^{-2}$.

Dataset	$\alpha = 0.1$			$\alpha = 0.25$			$\alpha = 0.40$			$\alpha = 0.55$		
	N_{img}	m_r	m_t	N_{img}	m_r	m_t	N_{img}	m_r	m_t	N_{img}	m_r	m_t
B1	109	0.93	7.82	182	0.11	0.53	182	0.18	1.24	182	1.47	8.89
B2	215	0.06	0.51	215	0.08	0.55	215	0.31	1.08	215	0.43	1.22
B3	182	0.26	1.26	275	0.22	1.07	342	0.35	1.44	342	0.47	1.93

5. Discussion

We investigate some characteristics of PRMP-PSfM and discuss the effect of iteratively prioritizing match pairs (Section 5.2) and the effect of iteratively refining match pairs (Section 5.1). A limitation on the condition of completeness is discussed in Section 5.3.

5.1. Effect of Iteratively Refining Match Pairs

In Section 4.6, without iteratively refining match pairs, our PRMP-PSfM becomes PSfM. From the visualization results in Figure 17, it can be seen that PSfM produced worse results than PRMP-PSfM, which could successfully reconstruct all these datasets (Figures 11, 13, 15 and 16). According to the numerical comparison presented in Table 5, PSfM obtained lower accuracy on SfM reconstruction. These results indicate that without iteratively refining match pairs, the outliers in match pairs degraded the SfM reconstruction results of PRMP-PSfM. In other words, iteratively refining match pairs can benefit PRMP-PSfM in regard to robustness and accuracy.

5.2. Effect of Iteratively Prioritizing Match Pairs

Iteratively prioritizing match pairs using MST yields the progressive scheme on view-graph, which aims to generate the filtered seed view-graph, and subsequently expands until the complete reconstruction is achieved. COLMAP directly applies the original view-graph, whereas PSfM employs the progressive scheme on the view-graph. Comparing the results of COLMAP (Figures 11, 13, 15 and 16) and PSfM (Figure 17), we can see that for datasets with a repetitive structure only, PSfM could successfully handle *Cup*, *Desk*, *Street*, *Indoor*, and *ToH*. In contrast, COLMAP could only obtain good results for *Desk*. This shows the advantages of the progressive scheme at handling images with repetitive structures. However, PSfM failed on *Books*, *Cereal*, and *Redmond* because the prioritized match pairs still exist some outliers that need to be eliminated. This demonstrates that only using the progressive scheme cannot perfectly prevent the negative influence of outliers in match pairs and implies that refining match pairs is necessary.

5.3. Limitation on Condition of Completeness

We discuss one limitation on the condition of completeness. For *Church* (Table 4), the condition of completeness that all images are successfully orientated could not be strictly met. For this dataset, the original view-graph obtained in PRMP-PSfM contained 1454 images, which was actually not 100% of all images. In the expansion period, some images causing the refinement of bundle adjustment to not converge were filtered. Hence, we set a relaxed condition of completeness, that 95% of images contained in the original view-graph were solved. Thus, the setting on the condition of completeness might need to be adjusted when dealing with various image sets, e.g., those containing both terrestrial and UAV images.

6. Conclusions

The SfM pipeline PRMP-PSfM was proposed for robust and accurate reconstruction from images with problematic match pairs due to repetitive structures and very short baselines; it can also be considered an improved incremental method. The limitations of conventional incremental methods in dealing with these problematic match pairs were discussed. All potential match pairs were first cast into a weighted view-graph, which could be manipulated to form a progressive scheme with initialization and expansion. In the initialization step, a subset of match pairs was prioritized using multiple MSTs. This was refined using an outlier elimination technique of consistency inference or CCI to generate the filtered seed view-graph for good seed reconstruction, which was expanded for more complete reconstruction by progressively adding new MSTs. As the reconstruction expanded, new match pairs were iteratively refined before carrying out the SfM process at each iteration. The above steps compose PRMP-PSfM, whose performance was demonstrated on datasets with repetitive structures and very short baselines. Experimental results showed that PRMP-PSfM could achieve better robustness and accuracy on reconstruction results than several state-of-the-art incremental and hierarchical SfM methods. Some cases may require the relaxation of the condition of completeness used in PRMP-PSfM. In the future, we hope to improve the condition of completeness and combine PRMR-PSfM with hierarchical SfM to efficiently deal with large-scale image sets.

Author Contributions: Conceptualization and methodology, T.X.; software and validation, T.X., Q.Y., W.M., F.D.; writing—original draft preparation, T.X.; project administration, F.D. All authors have read and agreed to the published version of the manuscript.

Funding: This research received no external funding.

Institutional Review Board Statement: Not applicable.

Informed Consent Statement: Not applicable.

Data Availability Statement: No new data were created or analyzed in this study. Data sharing is not applicable to this article.

Acknowledgments: We would like to thank Mario Michelini and Helmut Mayer, Universität der Bundeswehr München, for making the church dataset available to us. The first author would like to thank Xin Wang, Leibniz Universität Hannover, for the discussion on experimental design.

Conflicts of Interest: The authors declare no conflict of interest.

Abbreviations

Abbreviations
The following abbreviations are used in this manuscript:

SfM	Structure from motion
CCI	Cycle consistency inference
MST	Minimum spanning tree

References

1. Förstner, W.; Wrobel, B.P. *Photogrammetric Computer Vision*; Springer: Berlin/Heidelberg, Germany, 2016.
2. McGlone, C.; Mikhail, E.; Bethel, J. *Manual of Photogrammetry*, 5th ed.; American Society of Photogrammetry: Falls Church, VA, USA, 2004.
3. Lowe, D.G. Distinctive image features from scale-invariant keypoints. *Int. J. Comput. Vis.* **2004**, *60*, 91–110. [CrossRef]
4. Longuet-Higgins, H.C. A computer algorithm for reconstructing a scene from two projections. *Nature* **1981**, *293*, 133–135. [CrossRef]
5. Stewenius, H.; Engels, C.; Nistér, D. Recent developments on direct relative orientation. *ISPRS J. Photogramm. Remote Sens.* **2006**, *60*, 284–294. [CrossRef]
6. Fischler, M.A.; Bolles, R.C. Random sample consensus: A paradigm for model fitting with applications to image analysis and automated cartography. *Commun. ACM* **1981**, *24*, 381–395. [CrossRef]
7. Sweeney, C.; Sattler, T.; Hollerer, T.; Turk, M.; Pollefeys, M. Optimizing the viewing graph for structure-from-motion. In Proceedings of the IEEE International Conference on Computer Vision, Santiago, Chile, 7–13 December 2015; pp. 801–809.
8. Shen, T.; Zhu, S.; Fang, T.; Zhang, R.; Quan, L. Graph-based consistent matching for structure-from-motion. In *European Conference on Computer Vision*; Springer: Berlin/Heidelberg, Germany, 2016; pp. 139–155.
9. Cui, H.; Shi, T.; Zhang, J.; Xu, P.; Meng, Y.; Shen, S. View-graph construction framework for robust and efficient structure-from-motion. *Pattern Recognit.* **2020**, *114*, 107712. [CrossRef]

10. Schonberger, J.L.; Frahm, J.M. Structure-from-motion revisited. In Proceedings of the IEEE Conference on Computer Vision and Pattern Recognition, Las Vegas, NV, USA, 27–30 June 2016; pp. 4104–4113.

11. Wang, X.; Xiao, T.; Kasten, Y. A hybrid global structure from motion method for synchronously estimating global rotations and global translations. *ISPRS J. Photogramm. Remote Sens.* **2021**, *174*, 35–55. [CrossRef]

12. Snavely, N.; Seitz, S.M.; Szeliski, R. Modeling the World from Internet Photo Collections. *Int. J. Comput. Vis.* **2008**, *80*, 189–210. [CrossRef]

13. Hartley, R.; Zisserman, A. *Multiple View Geometry in Computer Vision*; Cambridge University Press: Cambridge, UK, 2003.

14. Triggs, B.; McLauchlan, P.F.; Hartley, R.I.; Fitzgibbon, A.W. Bundle adjustment—A modern synthesis. In *International Workshop on Vision Algorithms*; Springer: Berlin/Heidelberg, Germany, 1999; pp. 298–372.

15. Wu, C. Towards linear-time incremental structure from motion. In Proceedings of the 2013 International Conference on 3D Vision-3DV 2013, Seattle, WA, USA, 29 June–1 July 2013; IEEE: Piscataway, NJ, USA, 2013; pp. 127–134.

16. Mayer, H. Efficient hierarchical triplet merging for camera pose estimation. In *German Conference on Pattern Recognition*; Springer: Berlin/Heidelberg, Germany, 2014; pp. 399–409.

17. Toldo, R.; Gherardi, R.; Farenzena, M.; Fusiello, A. Hierarchical structure-and-motion recovery from uncalibrated images. *Comput. Vis. Image Underst.* **2015**, *140*, 127–143. [CrossRef]

18. Xie, X.; Yang, T.; Li, D.; Li, Z.; Zhang, Y. Hierarchical clustering-aligning framework based fast large-scale 3D reconstruction using aerial imagery. *Remote Sens.* **2019**, *11*, 315. [CrossRef]

19. Chen, Y.; Shen, S.; Chen, Y.; Wang, G. Graph-based parallel large scale structure from motion. *Pattern Recognit.* **2020**, *107*, 107537. [CrossRef]

20. Govindu, V.M. Robustness in motion averaging. In *Asian Conference on Computer Vision*; Springer: Berlin/Heidelberg, Germany, 2006; pp. 457–466.

21. Wilson, K.; Snavely, N. Robust global translations with 1dsfm. In *European Conference on Computer Vision*; Springer: Berlin/Heidelberg, Germany, 2014; pp. 61–75.

22. Agarwal, S.; Furukawa, Y.; Snavely, N.; Simon, I.; Curless, B.; Seitz, S.M.; Szeliski, R. Building rome in a day. *Commun. ACM* **2011**, *54*, 105–112. [CrossRef]

23. Wang, X.; Rottensteiner, F.; Heipke, C. Structure from motion for ordered and unordered image sets based on random kd forests and global pose estimation. *ISPRS J. Photogramm. Remote Sens.* **2019**, *147*, 19–41. [CrossRef]

24. Jiang, S.; Jiang, C.; Jiang, W. Efficient structure from motion for large-scale UAV images: A review and a comparison of SfM tools. *ISPRS J. Photogramm. Remote Sens.* **2020**, *167*, 230–251. [CrossRef]

25. Cui, H.; Shen, S.; Gao, W.; Liu, H.; Wang, Z. Efficient and robust large-scale structure-from-motion via track selection and camera prioritization. *ISPRS J. Photogramm. Remote Sens.* **2019**, *156*, 202–214. [CrossRef]

26. Wang, X.; Xiao, T.; Gruber, M.; Heipke, C. Robustifying relative orientations with respect to repetitive structures and very short baselines for global SfM. In Proceedings of the IEEE Conference on Computer Vision and Pattern Recognition Workshops, Long Beach, CA, USA, 16–17 June 2019.

27. Enqvist, O.; Kahl, F.; Olsson, C. Non-sequential structure from motion. In Proceedings of the 2011 IEEE International Conference on Computer Vision Workshops (ICCV Workshops), Barcelona, Spain, 6–13 November 2011; IEEE: Piscataway, NJ, USA, 2011; pp. 264–271.

28. Wang, X.; Heipke, C. An Improved Method of Refining Relative Orientation in Global Structure from Motion with a Focus on Repetitive Structure and Very Short Baselines. *Photogramm. Eng. Remote Sens.* **2020**, *86*, 299–315. [CrossRef]

29. Michelini, M.; Mayer, H. Structure from motion for complex image sets. *ISPRS J. Photogramm. Remote Sens.* **2020**, *166*, 140–152. [CrossRef]

30. Jiang, N.; Tan, P.; Cheong, L.F. Seeing double without confusion: Structure-from-motion in highly ambiguous scenes. In Proceedings of the 2012 IEEE Conference on Computer Vision and Pattern Recognition, Providence, RI, USA, 16–21 June 2012; IEEE: Piscataway, NJ, USA, 2012; pp. 1458–1465.

31. Heinly, J.; Dunn, E.; Frahm, J.M. Correcting for duplicate scene structure in sparse 3D reconstruction. In *European Conference on Computer Vision*; Springer: Berlin/Heidelberg, Germany, 2014; pp. 780–795.

32. Zach, C.; Klopschitz, M.; Pollefeys, M. Disambiguating visual relations using loop constraints. In Proceedings of the 2010 IEEE Computer Society Conference on Computer Vision and Pattern Recognition, San Francisco, CA, USA, 13–18 June 2010; IEEE: Piscataway, NJ, USA, 2010; pp. 1426–1433.

33. Moulon, P.; Monasse, P.; Perrot, R.; Marlet, R. Openmvg: Open multiple view geometry. In *International Workshop on Reproducible Research in Pattern Recognition*; Springer: Berlin/Heidelberg, Germany, 2016; pp. 60–74.

34. Jiang, N.; Cui, Z.; Tan, P. A global linear method for camera pose registration. In Proceedings of the IEEE International Conference on Computer Vision, Sydney, Australia, 1–8 December 2013; pp. 481–488.

35. Cui, H.; Shen, S.; Gao, W.; Wang, Z. Progressive large-scale structure-from-motion with orthogonal msts. In Proceedings of the 2018 International Conference on 3D Vision (3DV), Verona, Italy, 5–8 September 2018; IEEE: Piscataway, NJ, USA, 2018; pp. 79–88.

36. Snavely, N.; Seitz, S.M.; Szeliski, R. Skeletal graphs for efficient structure from motion. In Proceedings of the 2008 IEEE Conference on Computer Vision and Pattern Recognition, Anchorage, AK, USA, 24–26 June 2008; IEEE: Piscataway, NJ, USA, 2008; pp. 1–8.

37. Cui, Z.; Jiang, N.; Tang, C.; Tan, P. Linear Global Translation Estimation with Feature Tracks. *Proc. ECCV* **2014**, *3*, 61–75.

38. Kschischang, F.R.; Frey, B.J.; Loeliger, H.A. Factor graphs and the sum-product algorithm. *IEEE Trans. Inf. Theory* **2001**, *47*, 498–519. [CrossRef]
39. Prim, R.C. Shortest Connection Networks and Some Generalizations. *Bell Syst. Tech. J.* **1957**, *36*, 1389–1401. [CrossRef]
40. Cheng, J.; Leng, C.; Wu, J.; Cui, H.; Lu, H. Fast and accurate image matching with cascade hashing for 3d reconstruction. In Proceedings of the IEEE Conference on Computer Vision and Pattern Recognition, Columbus, OH, USA, 23–28 June 2014; pp. 1–8.
41. Cohen, A.; Zach, C.; Sinha, S.N.; Pollefeys, M. Discovering and exploiting 3d symmetries in structure from motion. In Proceedings of the 2012 IEEE Conference on Computer Vision and Pattern Recognition, Providence, RI, USA, 16–21 June 2012; IEEE: Piscataway, NJ, USA, 2012; pp. 1514–1521.

remote sensing

MDPI

Article

Accelerated Multi-View Stereo for 3D Reconstruction of Transmission Corridor with Fine-Scale Power Line

Wei Huang [1], San Jiang [2], Sheng He [1] and Wanshou Jiang [1,3,*]

[1] State Key Laboratory of Information Engineering in Surveying, Mapping and Remote Sensing, Wuhan University, 129 Luoyu Road, Wuhan 430079, China; hw1006@whu.edu.cn (W.H.); 2014301610342@whu.edu.cn (S.H.)

[2] School of Computer Science, China University of Geosciences, Wuhan 430074, China; jiangsan@cug.edu.cn

[3] Collaborative Innovation Center of Geospatial Technology, Wuhan University, 129 Luoyu Road, Wuhan 430079, China

* Correspondence: jws@whu.edu.cn; Tel.: +86-027-6877-8092 (ext. 8321)

Abstract: Fast reconstruction of power lines and corridors is a critical task in UAV (unmanned aerial vehicle)-based inspection of high-voltage transmission corridors. However, recent dense matching algorithms suffer the problem of low efficiency when processing large-scale high-resolution UAV images. This study proposes an efficient dense matching method for the 3D reconstruction of high-voltage transmission corridors with fine-scale power lines. First, an efficient random red-black checkerboard propagation is proposed, which utilizes the neighbor pixels with the most similar color to propagate plane parameters. To combine the pixel-wise view selection strategy adopted in Colmap with the efficient random red-black checkerboard propagation, the updating schedule for inferring visible probability is improved; second, strategies for decreasing the number of matching cost computations are proposed, which can reduce the unnecessary hypotheses for verification. The number of neighbor pixels necessary to propagate plane parameters is reduced with the increase of iterations, and the number of the combinations of depth and normal is reduced for the pixel with better matching cost in the plane refinement step; third, an efficient GPU (graphics processing unit)-based depth map fusion method is proposed, which employs a weight function based on the reprojection errors to fuse the depth map. Finally, experiments are conducted by using three UAV datasets, and the results indicate that the proposed method can maintain the completeness of power line reconstruction with high efficiency when compared to other PatchMatch-based methods. In addition, two benchmark datasets are used to verify that the proposed method can achieve a better F_1 score, 4–7 times faster than Colmap.

Keywords: power lines; UAV inspection; red-black propagation; depth map fusion; PatchMatch

Citation: Huang, W.; Jiang, S.; He, S.; Jiang, W. Accelerated Multi-View Stereo for 3D Reconstruction of Transmission Corridor with Fine-Scale Power Line. *Remote Sens.* **2021**, *13*, 4097. https://doi.org/10.3390/rs13204097

Academic Editor: Joaquín Martínez-Sánchez

Received: 10 September 2021
Accepted: 9 October 2021
Published: 13 October 2021

1. Introduction

In high-voltage transmission corridor scenarios, the power line is one of the key elements that should be regularly inspected by power production and maintenance departments. Recently, UAV photogrammetric systems equipped with optical cameras have been extensively used for data acquisition of transmission corridors, and a large number of high-resolution UAV images can be collected rapidly to achieve offsite visual inspection of power lines by using 3D point clouds of transmission corridors [1]. In the fields of photogrammetry and computer vision, 3D point clouds are usually generated through the combination of SfM (structure from motion) [2–6] for recovering camera poses and MVS (multi-view stereo) [7,8] for dense point clouds, which has been widely used in automatic driving [9], robot navigation [10], 3D visualization [11], DSM (digital surface model) generation [12], and vegetation encroachment detection [13]. In general, a majority of computational costs are consumed in MVS compared with SfM-based image orientation.

Thus, efficient reconstruction of 3D point clouds from UAV images is an urgent problem to be solved for the regular inspection of transmission corridors.

According to the work of [14], MVS methods can be divided into four groups: surface-evolution-based methods [15,16], voxel-based methods [17], patch-based methods [18,19], and depth-map-based methods [20–22]. As it is suitable for 3D dense matching of large-scale scenes, depth-map-based methods have been widely used, which can be verified from traditional dense matching algorithms, e.g., SGM (semi-global matching) [23] and PMVS (patch-based multi-view stereo) [18], to the recent PatchMatch-based methods. The PatchMatch algorithm was first proposed by [24], which can quickly find the nearest matching relationship with random searching and is successfully applied in the image interactive editing field. This method mainly includes three steps: random initialization, propagation, and searching, which are also the basic steps of the PatchMatch based dense matching methods. Barnes et al. [25] then expanded the algorithm in three aspects: k-proximity searching, multi-scale and multi-rotation searching, and matching with arbitrary descriptors and distances. The improved method is optimized and accelerated in parallel. Although these methods can be robustly applied in the field of image interactive editing, they cannot be directly used in the procedure of MVS because the matching correspondence between the two methods is only based on 2D similarity transformation [26]. Inspired by these two pioneering works, numerous PatchMatch-based dense matching methods emerge and achieve excellent performance in many fields.

According to the number of images used in the stereo matching, the PatchMatch based dense matching algorithms can be roughly divided into two categories: two-view-based stereo matching with disparities and multi-view-based stereo matching with depths. Bleyer et al. [27] was the first to apply PatchMatch to the two-view-based stereo-matching field. This method establishes the slanted support windows with disparities through two images, and the parameters of slanted support windows are estimated with the PatchMatch algorithm. Based on this work, Heise et al. [28] integrated the PatchMatch algorithm into a variational smoothing formulation with quadratic relaxation. With the estimated parameters of slanted support windows, the method can explicitly regularize the disparities and normal gradients. Besse et al. [29] analyzed the relationship between PatchMatch and belief propagation and proposed the PMBP algorithm, which combined PatchMatch with particle belief propagation optimization. It optimizes the disparity plane parameters by minimizing the global energy function in the continuous MRF (Markov random field), which improves the accuracy of disparities. However, due to the complicated matching cost calculation, its efficiency would slow down with the increase of the local support window size. Yu et al. [30] improved the PMBP algorithm in two aspects: cost aggregation and the propagation of PatchMatch, and proposed the SPM-BP method, which significantly improves the computational efficiency. All the above algorithms are matched at the pixel level, which is often inefficient and has poor performance on weak texture regions. Xu et al. [31] put forward the PM-PM method, which uses a unified variational formulation to combine object segmentation and stereo matching. The convex optimized multi-label Potts model is integrated into the PatchMatch techniques to generate disparity maps at object level while the efficiency and accuracy are maintained. Lu et al. [32] proposed PMF (PatchMatch filter) algorithm, which combined the random search procedure in PatchMatch with the edge-aware filtering technique. This method extends the random search procedure to the superpixels and takes the multi-scale consistency constraints into account. It can deal with the weak texture problem and improve efficiency. Similar to the strategy used in PMF, Tian et al. [33] proposed TDP-PM (tree dynamic programming-PatchMatch) method, which further combined the coarse-to-fine image pyramid matching strategy with global energy optimization based on continuous MRF. This method not only uses the local α-expansion-based tree dynamic programming, but also includes the PMF hierarchical strategy. UAV images often have a high overlap ratio and rich perspectives. For linear objects such as power lines, multi-view images are usually required to reconstruct the dense point clouds. Therefore, the two-view-based stereo matching is unsuitable for the UAV images in high-voltage power transmission lines.

In the PatchMatch-based multi-view stereo matching methods, Shen et al. [34] firstly extended the PatchMatch to multi-view stereo. The image orientation priors and the number of shared tie points computed by SfM (structure from motion) are applied to select neighbor images. The depth values of each pixel are then optimized by the lowest matching cost aggregation procedure with the support plane. Finally, the depth values are refined to improve the accuracy. Galliani et al. [20] presented a red-black symmetric checkerboard propagation mode to improve the efficiency of PatchMatch. The above two methods cannot handle the problem of occlusion, and the neighbor images are only selected based on the geometric information of the images. Zheng et al. [35] used the EM (expectation maximum) algorithm to achieve pixelwise visible image selection and established the visible probability of each pixel in the neighbor images through the graph model of HMM (hidden Markov field), which is solved jointly by EM optimization and the PatchMatch technique. Schonberger et al. [21] improved it by estimating the normal of the depth plane, fusing texture and geometric priors to perform pixelwise visible image selection, and using the multi-view geometric consistency to optimize the depth maps. Finally, the graph-based depth values and normal fusion strategy are proposed. This method achieves state-of-the-art results in accuracy, completeness, and efficiency, and the source codes are provided in Colmap as open source. Inspired by the work of [21], there are a variety of methods to improve the performance of PatchMatch mainly in two aspects: supporting weak texture regions [26,36] and taking into account the prior knowledge of planes [37,38]. To solve the problem of weak texture region matching, Romanoni et al. [36] modified the matching cost of photometric consistency to support the weak texture region depths estimation. The depth refinement and gaps filling strategies are then performed to eliminate the incorrect depth values and normal. Liao et al. [26] introduced a local consistency strategy with multi-scale constraints to alleviate the difficulty of weak texture region matching problems. In the methods with consideration of plane prior knowledge, Hou et al. [38] firstly segmented the image into superpixels and generated the candidate planes through the extended PatchMatch algorithm. The AMF (adaptive-manifold filter) is then applied to calculate and aggregate the matching cost. Finally, the BP is used to perform smoothing constraints. Xu et al. [37] integrated the plane assumptions into the PatchMatch framework with probabilistic graph models and formed a new way to aggregate the matching cost. However, with the plane assumptions or multi-scale constraints, the robustness and accuracy of weak texture regions can be improved. However, the application in other scenes is limited, which is unfavorable for the reconstruction of small objects or linear objects. Xu et al. [22] improved the PatchMatch method in the aspects of propagation mode, view selection and multi-scale constraints, and proposed the ACMH and ACMM methods. In terms of propagation, an efficient adaptive checkerboard mode is proposed, which is more efficient than the sequence propagation adopted in Colmap and the symmetric red-black checkerboard mode adopted in Gipuma [20]. The ACMM method ensures the efficiency and accuracy, and also supports the weak texture region matching.

Although the PatchMatch-based matching methods have been extensively explored, some issues still exist that should be addressed for the 3D reconstruction of transmission corridors. On the one hand, existing studies mainly focus on indoor and urban scenarios, and the dense matching results of UAV images in high-voltage power transmission lines need to be investigated. On the other hand, due to a large number of UAV images, existing methods are confronted with challenges of inefficiency, which cannot meet the demand of regular inspection of the high-voltage power transmission line. Thus, the main purpose of this paper is to improve the efficiency of dense matching while maintaining the completeness of the 3D point cloud of power lines under the framework of Colmap. First, with the assumption that the depth values of pixels with similar colors in the local region would be close, an efficient random red-black checkerboard propagation is proposed, which uses the most similar neighbor pixels to propagate the plane parameters. Furthermore, an improved strategy for the hidden variable state updating with HMM is proposed, which can make the random red-black checkerboard propagation adapt with the pixelwise view selection in Colmap. Second, two strategies for reducing the matching cost computation are adopted to improve efficiency. With the increase of iterations, the depth values converge gradually. The number of neighbor pixels propagated plane parameters to the current pixel is reduced in the later iterations. Considering that the depth error would be small with a lower matching cost, the number of combinations of depth values and normal in the refinement procedure is reduced for the pixels with low matching costs. Third, an efficient depth-map fusion method is proposed, which uses weight function based on the reprojection errors to fuse depths from multi-view images and is implemented under the GPU. Finally, three datasets of UAV images with high-voltage power transmission lines are used for analyzing the performance of power line reconstruction and efficiency. Two benchmark datasets are used in experiments for precision analysis.

The remainder of this paper is organized as follows. Section 2 describes the materials and methods. Three test sites of high-voltage power transmission lines and two benchmark datasets are introduced. Additionally, the framework of Colmap and three strategies for efficiency improvement are detailly described, including fast PatchMatch with random red-black checkerboard propagation, strategies for reducing matching cost calculation, and fast depth-map fusion with GPU acceleration. In Section 3, comprehensive experiments are presented and discussed with three UAV image datasets. In Section 4, the discussions about the accuracy analysis with two benchmark datasets are presented. Section 5 concludes the results of this study.

2. Materials and Methods

2.1. Study Sites and Test Data

To verify the applicability of the proposed method in the high-voltage power transmission line, three test sites of UAV images are selected for experimental analysis. The three test sites of UAV images are collected by means of a DJI Phantom 4 RTK UAV, including the voltage of 500 kV, 220 kV, and 110 kV in transmission lines, which used the rectangle closed-loop trajectory, the S-shaped strip trajectory, and the traditional multiple trajectories in the photogrammetry field, respectively. For the dense matching results of UAV images of the transmission line, this paper focuses on the analysis of the completeness of reconstructed power lines. To verify the precision of the proposed method, two benchmark datasets: the close-range outdoor dataset, Strecha [39], and the large-scale aerial dataset, Vaihigen [40], are selected to perform the experiments.

2.1.1. Test Sites of High-Voltage Power Transmission Lines

The three test sites of UAV images of high-voltage power transmission lines are shown in Figure 1. Test site 1 and test site 2 are both located in mountainous areas, which are mainly covered by vegetation; test site 3 is flat land includes roads and some part of a transformer substation. In test site 1, there are a total of 6 pylons and 5 spans of power lines which are 4-bundled conductors; in test site 2, there are 2 pylons and 1 span of power lines which are 2-bundled conductors; in test site 3, there are 2 pylons and one span of power lines which are 1-bundled conductors. The flight heights of the three test sites are 160 m, 80 m, and 65 m, respectively, which are relative to the location from where the UAV took off. Additionally, the GSD (ground resolution distance) of images in the three test sites are 4.70 cm, 2.72 cm, and 1.75 cm. The image numbers of the three test sites are 222, 191, and 103. The details of the three test sites are list in Table 1.

Table 1. Details of the three test sites.

Item Name	Test Site 1	Test Site 2	Test Site 3
Flight mode	rectangle	S-shaped	multiple trajectories
Flight height (m)	160	80	65

Table 1. *Cont.*

Item Name	Test Site 1	Test Site 2	Test Site 3
Voltage (kV)	500	220	110
Bundled conductors	4-bundled	2-bundled	1-bundled
Type of UAV		DJI Phantom 4 RTK	
Image size	5472 × 3078	4864 × 3648	5472 × 3078
Image number	222	191	103
GSD (cm)	4.70	2.72	1.75

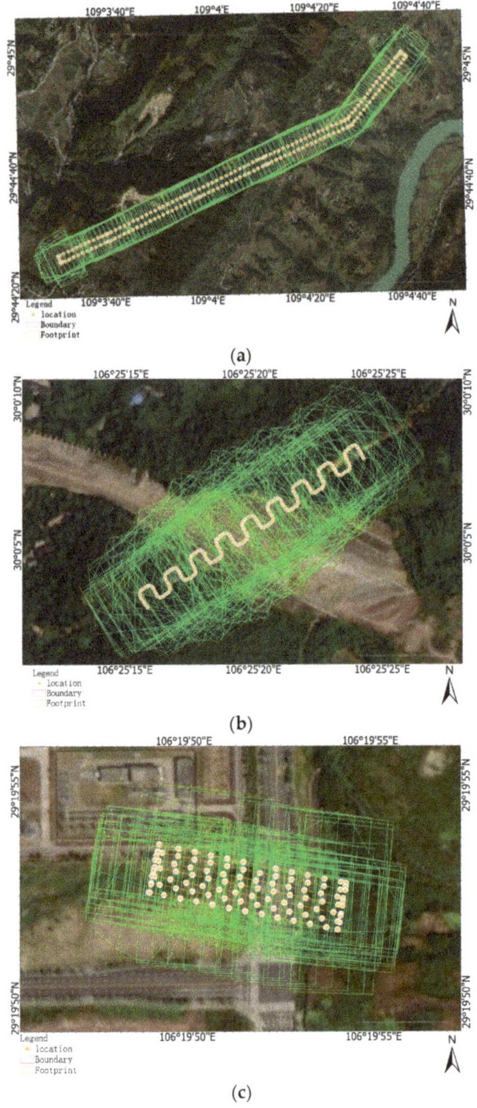

Figure 1. UAV images of the three test sites. (**a–c**) are the rectangle dataset in test site 1, S-shaped dataset in test site 2, and traditional multiple trajectories dataset in test site 3, respectively.

2.1.2. Benchmark Datasets

In the Strecha dataset, a Canon D60 camera was used for image collection, and the image resolution is 3072 × 2048 pixels. The ground truth meshes are provided with Fountain and Herzjesu in the Strecha dataset, which are acquired by Zoller+Forhlich IMAGER 5003. There are 11 and 8 images in Fountain and Herzjesu and the projection matrix of each image is provided. In the Vaihigen dataset, the Intergraph/ZI DMC is applied for the 20 pan-sharpened color infrared images collection. The forward and side overlaps are both 60%, the resolution of images is 7680 × 13,824 pixels, and the GSD is 8 cm. The ground truth airborne laser scanning (ALS) data is provided, which is collected by a Leica ALS50 system with 10 strips. Test site 1 and test site 3 in the Vaihigen dataset are selected for evaluating the precision of the reconstructed point clouds. The two test sites in the Vaihigen dataset are shown in Figure 2. Test site 1 is located in the center of the city and contains some historical buildings; test site 3 is located in the residential area with small detached houses and a few trees.

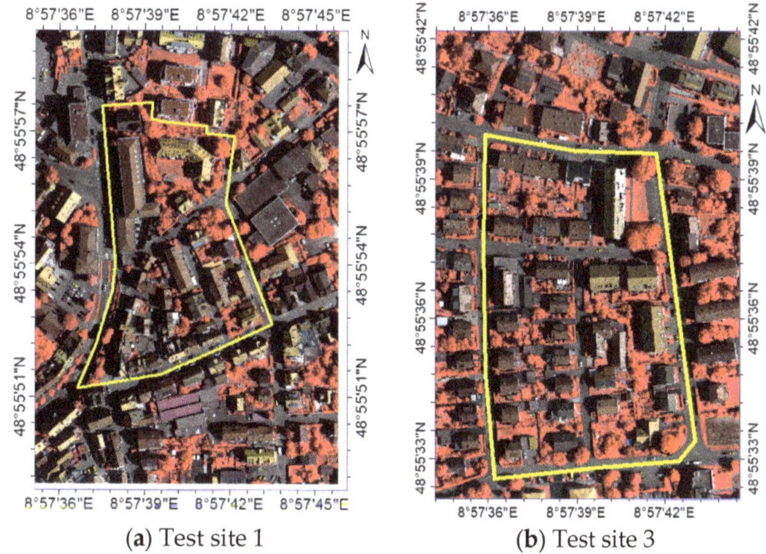

(**a**) Test site 1   (**b**) Test site 3

Figure 2. Two test sites in the Vaihigen dataset: (**a**,**b**) are test site 1 and test site 3, respectively.

2.2. Methodologies

In this section, the overview of PatchMatch-based dense matching is first introduced. The three aspects to improve the efficiency of dense matching for the 3D reconstruction of high transmission corridors are then presented: (1) fast PatchMatch with random red-black checkerboard propagation; (2) strategies for reducing matching cost calculation; and (3) fast depth-map fusion with GPU acceleration.

2.2.1. Overview of PatchMatch-Based Dense Matching

The Colmap framework proposed by [21] is improved based on [35]. Colmap can estimate the depth values and normal of the reference image at the same time. Additionally, the photometric and geometric priors are adopted to infer the pixelwise visibility probability from source images. The photometric and geometric consistency across multi-view images are used to optimize the depth and normal maps. The sequence propagation mode is applied in Colmap, which iteratively optimizes the depth values and normal in each row or column independently. For the convenience of the following discussion, this paper also uses l to describe the coordinates of the pixel in the image. Colmap estimates the depth d_l and normal n_l together with the binary visibility state variable $Z_l^j \in \{0,1\}$ for a pixel l of the reference image I^{ref} from the neighboring source images $I^{src} = \{I^j | j = 1...J\}$. In the binary visibility state variable $Z_l^j \in \{0,1\}$, 0 represents that the pixel l is occluded in the source image I^j while 1 means that pixel l is visible in the source image I^j. These parameters are inferred with a maximum a posterior (MAP) estimation, and the posterior probability is defined as:

$$
\begin{aligned}
P(Z, d, N \big| I) &= \frac{P(Z, d, N, I)}{P(I)} \\
&= \frac{1}{P(I)} \prod_{l=1}^{L} \prod_{j=1}^{J} [P(Z_{l,t}^j | Z_{l-1,t}^j, Z_{l,t-1}^j) P(I_l^j | Z_l^j, d_l, n_l) P(d_l, n_l | d_l^j, n_l^j)]
\end{aligned}
\tag{1}
$$

where L is the number of rows or columns in the reference image during the current iteration, $I = \left\{ I^{ref}, I^{src} \right\}$ and $N = \{n_l | l = 1...L\}$.

In formula (1), the first likelihood term $P(Z_{l,t}^j | Z_{l-1,t}^j, Z_{l,t-1}^j)$ enforces the smoothness constraint of the visibility probability calculation during the optimization process, which can ensure the smoothness both spatially and along with the successive iteration; the second likelihood term $P(I_l^j | Z_l^j, d_l, n_l)$ means that the photometric consistency between the pixels in the windows B_l centered on the pixel l in the reference image I^{ref} and the corresponding projection pixels in the non-occluded source image I^j. The bilaterally weighted NCC (normalized cross-correlation) cost function is applied to compute the photometric consistency, which can achieve better accuracy at the boundary of the occluded regions. In the cost aggregation procedure, the Monte Carlo sampling method is used to randomly sample the neighbor images in the sampling distribution function $P_l(j)$, and the matching costs of the image of which the probability $P_l(j)$ is bigger than the randomly generated probability are accumulated; the sampling distribution function $P_l(j)$ takes full consideration of the triangulation prior, resolution prior, incident prior, and visibility probability of images; the last likelihood term $P(d_l, n_l | d_l^j, n_l^j)$ represents the geometric consistency from multi-view images, which enforces the depth consistency and the accuracy of normal estimation.

Solving formula (1) directly is intractable. Analogous to [35], Colmap factorizes the real posterior $P(Z, d, N | I)$ as an approximation function $q(Z, d, N) = q(Z)q(d, N)$ and adapts the GEM (generalized expectation-maximization) algorithm to optimize. In the E-step, the parameters of depth and normal (d, N) are kept fixed, and the parameter Z is regarded as the hidden state variable of HMM. The function $q(Z_{l,t}^j)$ is estimated by the forward-backward message passing algorithm during each iteration, the formula is as follows:

$$
q(Z_{l,t}^j) = \frac{1}{A} \overrightarrow{m}(Z_{l,t}^j) \overleftarrow{m}(Z_{l,t}^j)
\tag{2}
$$

where $\overrightarrow{m}(Z_{l,t}^j)$ and $\overleftarrow{m}(Z_{l,t}^j)$ represent the recursively forward and backward message passing, respectively. The specific formulas of $\overrightarrow{m}(Z_{l,t}^j)$ and $\overleftarrow{m}(Z_{l,t}^j)$ are

$$
\overrightarrow{m}(Z_l^j) = P(I_l^j | Z_l^j, d_l, n_l) \sum_{Z_{l-1}^j} \overrightarrow{m}(Z_{l-1}^j) P(Z_{l,t}^j | Z_{l-1,t}^j, Z_{l,t-1}^j)
\tag{3}
$$

$$
\overleftarrow{m}(Z_l^j) = \sum_{Z_{l+1}^j} \overleftarrow{m}(Z_{l+1}^j) P(I_{l+1}^j | Z_{l+1}^j, d_{l+1}, n_{l+1}) P(Z_{l,t}^j | Z_{l+1,t}^j, Z_{l,t-1}^j)
\tag{4}
$$

where $\overrightarrow{m}(Z_0^j)$ and $\overleftarrow{m}(Z_{l+1}^j)$ are set to 0.5 as an uninformative prior. The variable $q(Z_{l,t}^j)$ together with the triangulation prior, resolution prior, and incident prior determines the Monte Carlo sampling distribution function $P_l(j)$. In the M-step, the function $q(Z)$ is kept fixed, and the parameters $q(d, N)$ are estimated through PatchMatch. Finally, by iterating the E-step and the M-step in the row- or column-wise propagation, the depth values, normal, and pixelwise visibility probability are estimated.

In the depth map filtering stage, the characteristics of photometric and geometric consistency, the number of images that are visible in the source images, the visibility probability, the triangulation angle, resolution, and the incident angle for a pixel in the reference image are considered. In the depth-map-fusion stage, the pixel in the reference image and the set of corresponding pixels in source images with photometric and geometric consistency are regarded as a directed graph. These corresponding pixels are the nodes of the graph and the directed edges of the graph point from the pixel in the reference image to the pixels in the source images. Colmap recursively finds all the pixels with photometric and geometric consistency, and then uses the media depth value and mean normal as the fused depth value and normal, respectively. Finally, all the pixels that participated in the fusion stage in the directed graph are removed and the steps above are repeated to fuse the next point until the directed graph is empty.

2.2.2. Fast PatchMatch with Random Red-Black Checkerboard Propagation

Galliani et al. [20] firstly introduced the symmetric red-black checkerboard propagation mode into the PatchMatch framework and proposed the Gipuma method, which makes full use of the parallel processing of GPU and improves the efficiency of Patch-Match. Xu et al. [22] further proposed the adaptive red-black checkerboard propagation mode to improve the efficiency of PatchMatch. The diffusion-like red-black checkerboard propagation scheme is proved to be more efficient than the sequence propagation scheme. The purpose of the paper is to improve the efficiency of the Colmap by adopting the diffusion-like propagation scheme while preserving the innovational pixelwise view selection strategy with HMM inference.

Through the analysis of the symmetry red-black pattern proposed by [20] and the adaptive red-black pattern proposed by [22], it can be discovered that these two propagation modes both use fixed neighbor positions to propagate the plane parameters. Gipuma employs the fixed positions of 8 neighbor points for propagation, while ACMH and ACMM expand the neighbor ranges and sample 8 points from specific patterns with the smallest matching cost for propagation, which fully takes into account the structural region information and makes the propagating range further and more effective. Different from the two propagation modes, this paper adopts the random red-black checkerboard pattern to propagate the plane parameters. A fixed number N_s of sampling points with different color patterns are randomly generated within the local window range centered at the current pixel. In the experiment, the N_s is set to 32 and the local window radius is set to the same with the matching window radius. Then 8 neighbor points with the most similar color to the current pixel are selected to propagate their plane parameters. The advantage of employing the randomly sampling neighbor pixels is that it can break through the limitation of fixed positions and the pixels with other color patterns in the local window have the opportunity to propagation their plane parameters.

Each thread unit in the GPU processes a single pixel instead of the entire row or column of pixels when the random red-black checkerboard propagation is applied in PatchMatch. The hidden state variable updating schedule of Colmap cannot be used directly in such propagation. To combine the pixelwise view selection strategy in Colmap with the random red-black checkerboard propagation, the updating schedule should be improved.

The GEM is applied in Colmap to approximate the solution of the function $q(Z, d, N) = q(Z)q(d, N)$, and in the E-step, the forward-backward message passing algorithm is used to update the hidden state variable $q(Z_{l,t}^j)$. In Colmap, the updating schedule with $q(Z_{l,t}^j)$ proposed by [35] is deeply integrated with the sequence propagation, as shown in Figure 3.

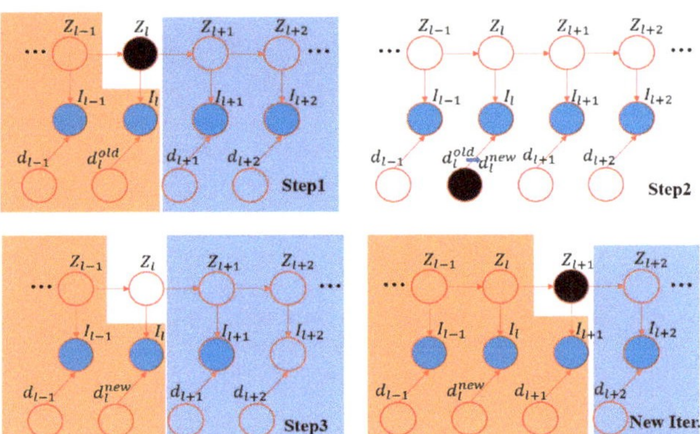

Figure 3. The visibility state variable updating schedule adopted in Colmap [35].

The main steps of the updating schedule in Colmap are: (a) traverse all the pixels of the row or column in the reference image from the end to the start, and visit the matching cost of each pixel in the source images, and compute the backward message $\overleftarrow{m}(Z_l^j)$ with the matching cost using formula (4); (b) traverse all the pixels of the row or column in the reference image from the start to the end, and visit the matching cost in the source images, and compute the forward message $\overrightarrow{m}(Z_l^j)$ with the matching cost using formula (3), and update the visibility probability $q(Z_{l,t}^j)$ with $\overleftarrow{m}(Z_l^j)$ and $\overrightarrow{m}(Z_l^j)$ using formula (2); (c) use the PatchMatch technique to compute the matching cost and update the parameters of depth and normal $q(d, N)$; and (d) visit the matching cost and the previous forward message of the current pixel in all the source images, recompute the forward message $\overrightarrow{m}(Z_l^j)$ with the matching cost using formula (3). Algorithm 1 shows the details below.

Algorithm 1 Updating schedule of sequence propagation in Colmap [35]

Input: All images, depth map, and normal map
       (randomly initialized or from previous propagation)
Output: Updated depth map and normal map, the visible probability for each pixel
       (j—image index,l—pixel index)
1: For $l = L$ to 1
2:    For $j = 1$ to J
3:       Compute backward message $\overleftarrow{m}(Z_l^j)$
4: For $l = 1$ to L
5:    For $j = 1$ to J
6:       Compute forward message $\overrightarrow{m}(Z_l^j)$
7:       Compute $q(Z_{l,t}^j)$
8:    Estimate $q(d, N)$ by PatchMatch
9:    For $j = 1$ to J
10:      Recompute forward message $\overrightarrow{m}(Z_l^j)$

This paper adopts the following steps to improve the updating schedule: (a) After the random initialization of depth values and normal in the reference image, the bilateral weighted NCC is adopted to calculate the matching cost with source images; (b) given the traversal direction, traverse the whole image row- or column-wise in direction, compute the backward message with the previous matching cost using formula (4); (c) traverse all the pixels of the row or column in the opposite direction of step (b), compute the forward message with the previous matching cost using formula (3) and update $q(Z_{l,t}^j)$ using formula (2); (d) propagate the plane parameters with red pattern pixels to the black pattern pixels, and update $q(d, N)$ of black pattern pixels by PatchMatch; (e) rotate the traversal direction 90° clockwise, repeat steps (b) and (c) to finish the updating of hidden variable $q(Z_{l,t}^{j})$; (f) propagate the plane parameters with black pattern pixels to the red pattern pixels, and update $q(d, N)$ of red pattern pixels by PatchMatch; (g) repeat the steps from (b) to (f), process the random red-black checkerboard propagation with PatchMatch until reaching the maximum iteration number. Algorithm 2 shows the details below.

Algorithm 2 Updating schedule of random red-black checkerboard propagation

---

Input: All images, depth map, and normal map
　　　　(randomly initialized or from previous propagation)
Output: Updated depth map and normal map, the visible probability for each pixel
　　　　(j—image index, l—pixel index)
1: procedure Update($q(Z)$,L)
2: For $l = L$ to 1
3:　　For $j = 1$ to J
4:　　　Compute backward message $\overleftarrow{m}(Z_l^j)$
5: For $l = 1$ to L
6:　　For $j = 1$ to J
7:　　　Compute forward message $\overrightarrow{m}(Z_l^j)$
8:　　　Compute $q(Z_{l,t}^j)$
9: end procedure
10: Initialize the traversal direction D
11: Execute Update($q(Z)$, L_{row}) in direction D
12: Update the $q(d, N)$ of back pattern pixels with the plane parameters of red pattern pixels by PatchMatch
13: Rotate the direction D 90° clockwise, execute Update($q(Z)$,L_{col}) in direction D
14: Update the $q(d, N)$ of red pattern pixels with the plane parameters of black pattern pixels by PatchMatch
15: Repeat steps from 11 to 14 until reaching the maximum iteration number

---

Algorithm 1 and Algorithm 2 show the updating schedule of sequence propagation in Colmap and the updating schedule of random red-black checkerboard propagation of the proposed method. The main difference is that: the E-step and M-step in Colmap are closely related to each other, but they are separated in the proposed method. In each iteration of the row or column of the referenced image, the hidden variables $q(Z_{l,t}^j)$ are inferred through a forward-backward message passing algorithm, and the hidden variables $q(Z_{l,t}^j)$ affect the pixelwise view selection during the matching cost aggregation in PatchMatch. The E-step and M-step are completed together in each iteration of the row or column. However, with the proposed updating schedule, the E-step and M-step are completed individually. The traversal unit is each row or column in E-step, while it is an independent pixel in M-step. The updating schedule in Colmap strictly follows the Markov assumption in HMM, that is, the current hidden state is only related to the previous hidden state. It can better approximate the hidden state function $q(Z)$, but it also limits the scalability of the propagation. To better adapt to the diffusion-like propagation, the proposed schedule completely separates the E-step and M-step. Although this schedule is not as rigorous as in Colmap, it is still an optional strategy with the consideration of the scalability of propagation and efficiency.

Figure 4 shows the two depth maps calculated with three different view selection and propagation strategies. The first row is the images from the South Building dataset [41]; and the second row is the images from the UAV dataset of a high-voltage power transmission line, which is located at the border of the corridor and has fewer neighboring images overlapped with them. From the first row, it can be seen that the depth map of "top-k-winners-take-all" view selection with symmetric red-black checkerboard propagation strategy has more incorrect depth values at the boundary of the occluded regions of the tree, while the depth map generated with the proposed strategy with random red-black checkerboard propagation is close to the result of Colmap. From the second row, it can be seen that the "top-k-winners-take-all" view selection with symmetric red-black checkerboard propagation strategy cannot infer the depth values in the region where there are few overlapped source images, while the proposed strategy and Colmap can infer the depth values better. To conclude, the depth maps with the proposed strategy are close to the results calculated by Colmap, which proves the effectiveness of the proposed strategy.

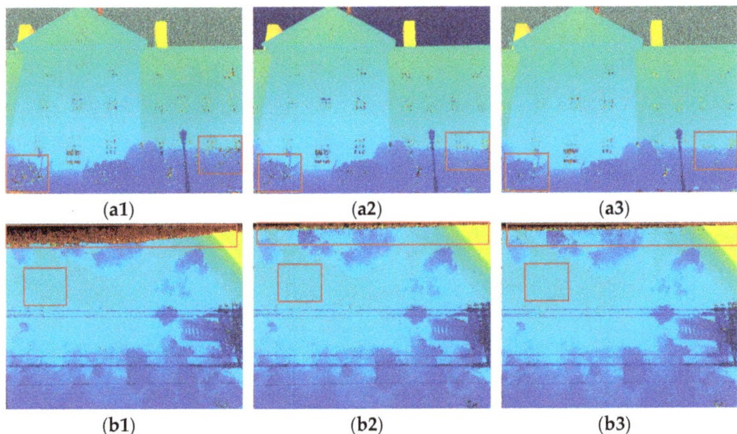

Figure 4. Comparison of the depth maps with three different view selection and propagation strategies. (**a1,b1**) are the results of the "top-k-winners-take-all" strategy with symmetric red-black propagation; (**a2,b2**) are the results of Colmap with sequence propagation; (**a3,b3**) are the results of the proposed strategy with random red-black propagation.

2.2.3. Strategies for Reducing Matching Cost Calculation

In Colmap, the most time-consuming processing step of PatchMatch is the matching cost calculation with bilateral weighted NCC. This paper adopts the following strategies to further improve efficiency by reducing the unnecessary calculation of bilateral weighted NCC.

Firstly, the number of neighbor pixels to propagate plane parameters is reduced with the increasing iteration number. From Figure 5, it can be seen that the accuracy of the reconstructed depth values gradually converges to being stable as the number of iterations increases. At about the third iteration, the convergence speed of the depth values increases significantly, which indicates that a large number of depth values and normal in the reference image are correctly estimated. Therefore, the 8 neighbor points with the most similar to the current pixel are employed to propagate the plane parameters in the early iterations. Additionally, the number of neighbor points can be reduced to improve the efficiency as the number of iterations increases. In this paper, when the number of iterations is greater than 3, the number of neighbor points is reduced to 4.

Secondly, the number of combinations with the depth and normal hypotheses is reduced in the plane refinement procedure according to the matching cost. The current best depth d_l and normal n_l obtained from the neighbor pixels propagation have the three states: neither of them, one of them, or both of them are close to the optimal depth d_l^* and normal n_l^* [21]. Therefore, Colmap generates a new depth d_l^{prp} and normal n_l^{prp} by perturbing the current best depth d_l and normal n_l within a small range. Additionally, to increase the chance of sampling the correct depth and normal, the random depth d_l^{rnd} and normal n_l^{rnd} are generated at the same time. Finally, these three hypotheses of depth values and normal are combined to obtain 6 new hypotheses, as shown in formula (5). These hypotheses are then applied to calculate the matching cost, and the depth and normal with the smallest matching cost are regarded as the optimal solution in the plane refinement.

$$\left\{ (d_l^{prp}, n_l), (d_l^{rnd}, n_l), (d_l, n_l^{rnd}), (d_l^{rnd}, n_l^{rnd}), (d_l^{prp}, n_l), (d_l, n_l^{prt}) \right\} \tag{5}$$

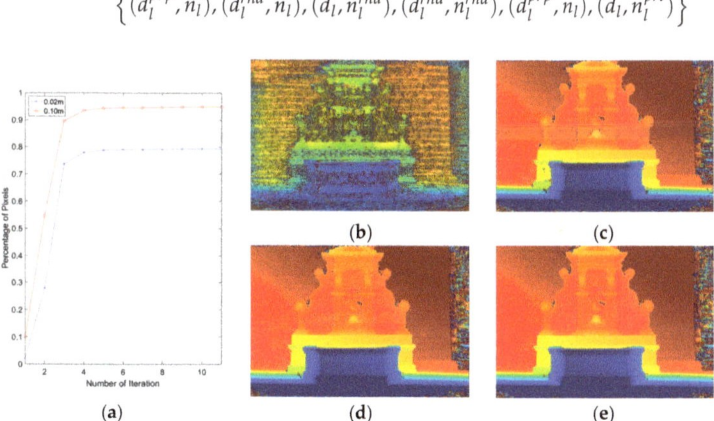

Figure 5. The relationship of accuracy and iteration number. (**a**) is the percentage of pixels with absolute errors below 2 cm and 10 cm. (**b**–**e**) are the depth maps with iterations of 1, 3, 5, and 7.

In practical application, it can be found that the current best depth d_i and normal n_l is close to the optimal depth d_l^* and normal n_l^* when the matching cost is relatively small. Therefore, the random depth d_l^{rnd} and normal n_l^{rnd} should be considered whether to join the new six hypotheses according to the current best matching cost. If the matching cost is less than a threshold of 0.5, only three new hypotheses are adopted in the plane refinement, as shown in formula (6); otherwise, the six hypotheses are still used as shown in formula (5). In this way, the number of matching cost computations can be reduced.

$$\left\{ (d_l^{prp}, n_l), (d_l^{prp}, n_l), (d_l, n_l^{prt}) \right\} \tag{6}$$

2.2.4. Fast Depth Map Fusion with GPU Acceleration

In the depth-map-fusion stage, Colmap uses a recursive method to fuse the depth values and normal which meet the condition of photometric and geometric consistency. However, this method faces the following problems: (1) firstly, using the recursive method to traverse through the directed graph is inefficient, which is not suitable for GPU parallel processing; (2) secondly, due to the incorrectly estimated depth value, this method may merge the 3D points of different pixels in the same image, which increases the number of iterations. Therefore, this paper proposed a fast depth-map-fusion method accelerated by the GPU, and the speed of fusing each depth map is stable.

The constraints of fusing depth maps similar to [20–22] are adopted in the proposed method. Firstly, the depths with photometric consistency are considered to be fused. For the depth estimated with PatchMatch, the number k_{cost} of matching costs with the source images that are less than a threshold $T_{ncc} = 0.5$ are counted. If $k_{cost} \geq 3$, the depth is regarded as stable. Secondly, the constraints with geometric consistency, such as reprojection error $\psi \leq 2$ and relative depth difference $\varepsilon \leq 0.01$, are adopted in the depth map fusion. Additionally, statistically the number k_{geo} in source images that satisfy the constraints. If $k_{geo} \geq 3$, these depths should be fused. It should be pointed out that the normal angle constraint is not adopted in the proposed method. The reason is that the estimated normal of linear objects such as power lines are not accurate. If the normal angles are considered in the depth map fusion stage, most reconstructed point clouds of power lines would be filtered out and become too sparse.

Once the depth values from the source images that satisfy all the constraints are clustered, the median location and mean normal are adopted as the fused depth value and normal in Colmap. However, it does not take into account the influence of the depth error. The reprojection errors from the source images can reflect the error to some extent. Therefore, the depth values are fused by the weighted reprojection errors in the proposed depth map fusion stage. The weighting method proposed by [42] is adopted in this paper, as shown in formulas (7) and (8), which is theoretically equivalent to the least-square solution.

$$D_{k+1}(x) = \frac{W_i(x)D_i(x) + w_{i+1}(x)d_{i+1}(x)}{W_i(x) + w_{i+i}(x)} \tag{7}$$

$$W_{i+1}(x) = W_i(x) + w_{i+1}(x) \tag{8}$$

where $D_i(x)$ and $W_i(x)$ are the depth value and weight accumulated for the i-th time, respectively. The weighting formula according to the reprojection errors is $w_i(x) = \frac{1}{1+\exp(\sum\limits_{i \in N} \cos t_j)}$, where is the reprojection error between the depth value x in the j-th image and the corresponding depth values in the neighbor images (the index of neighbor images is $i = 1, 2, ..., N$) which satisfy all the constraints.

Finally, the main steps of the depth map fusion in the GPU are as follows:

(a)   Initialize the global binary state values of all the depth maps as 0;

(b)   Load the reference image I^i, source image I^{neib}, the corresponding depth maps D, and normal maps N into the GPU;

(c)   Iterate all the pixels in the reference image I^i. For a current pixel p_j in the image I^i, if $k_{\cos t}$ of p_j is bigger than 3 and the state value of p_j is 0, then the depth value D_j^i of p_j is back-projected to the source images I^{neib} and find the depths cluster D_s^{neib} that satisfies all the geometric constraints; otherwise, process the next pixel p_{j+1} of the referenced image I^i;

(d)   Count the number V_{ij} in the depths cluster D_s^{neib} and the corresponding binary state values S^s. If $V_{ij} \geq 3$ and the values in S^s are 0, then use formulas (7)and (8) to fuse the depth D_j^i and D_s^{neib}, average the corresponding normal, and set all the corresponding binary state values to 1; otherwise, process the next pixel p_{j+1}. Algorithm 3 shows the details below.

Algorithm 3 Fast depth maps fusion with GPU

Input: All images I, depth maps D, and normal maps N
Output: Fused dense point cloud
        (i—image index,j—pixel index)
1: Initialize the global binary state values S of depth maps as 0
2: foreach image I^i in I
3:     Load the reference image I^i, the source images I^{neib},
      the depth maps D^i and D^{neib}, and the normal maps N^i and N^{neib} into GPU
4:     foreach pixel p_j in I^i
5:     if $k_{\cos t} \geq 3$ and the binary state value of D_j^i is 0
6:        Compute the depth cluster D_s^{neib} that satisfies all the geometry constraints
7:        Statistic the number V_{ij} in the depths cluster D_s^{neib}
8:        if $V_{ij} \geq 3$ and all the binary state values S^s of D_s^{neib} is 0
9:          Set the binary state values of D_j^i and D_s^{neib} as 1
10:        Use formula (7) and (8) to fuse the depth D_j^i and D_s^{neib},
11:        Average the corresponding normal
12:        endif
13:     endif
14: end foreach
15: end foreach

---

3. Results

The PatchMatch-based methods: Colmap [21], Gipuma [20], and ACMH [22], are selected for the comparative analysis of precision and efficiency. All the methods are implemented in the GPU and their codes are open source. All the experiments are conducted on eight Intel Core i7-7700 CPUs with Nvidia GeForce GTX 1080 graphic card, 32GB RAM, and 64-bit Windows 10 OS.

3.1. Analysis of the Power Line Reconstruction

In this experiment with the three datasets of high-voltage power transmission lines, the image size is set to half the width and height of the image: in test site 1 and test site 3 the image size is set to 2736 × 1824 pixels, in test site 2 the image size is set to 2432 × 1824 pixels. The matching widows are all set to 15 × 15 pixels, the step size is 1 pixel, and the number of iterations is set to 6. Since the rectangle closed-loop trajectory is adopted in test site 1, the maximum number of views selected for PatchMatch is set to 10 to ensure that the side-overlapping images can be selected to reconstruct more stable power lines, while it is set to 5 in test site 2 and test site 3. Only the photometric consistency matching cost is applied in Colmap and ACMH without the geometric consistency since the geometric consistency matching cost is not conducive for reconstructing small objects such as power lines. In the depth-map fusion, the normal angle constraint is not taken into consideration with all the methods since the normal of power lines estimated by PatchMatch is not accurate. The minimum number k_{geo} of images satisfying the geometry constraints are set to 3 for all four methods to ensure the reconstructed point cloud with less noise. Additionally, the left parameters are maintained at default. Since the median filter is applied for the depth maps in ACMH, it would filter out most of the power lines. In the experiments, the median filter is not adopted in ACMH. The depth-map fusion program provided in Gipuma only processes one depth map fusion with source images and the fused points in previous fusion can still be used in the next depth map fusion procedure, which leads to the final fused point clouds being redundant. Additionally, the depth map fusion constraints used in Gipuma are similar to ACMH, so the fusion program provided by ACMH is used for fusing the depth maps generated with Gipuma in the experiments.

Firstly, the depth maps for the image in test site 1 generated by the four methods are selected for comparative analysis, as shown in Figure 6. It can be seen that there are more noisy speckles in the depth map generated by Gipuma because only the "top-k-winners-take-all" strategy is adopted without visible view selection and the matching cost is a weighted combination of the absolute color and gradient differences, which is not as robust as weighted bilateral NCC adopted in Colmap, ACMH, and the proposed method. The heuristic multi-hypothesis joint view selection adopted in ACMH uses the neighbor best matching cost to infer the visibility, which is sensitive to the threshold. In the vegetation coverage area, the matching costs between different images are different due to the perspective change. This visible view selection method would fail to select the right visible image and lead to large speckles in the depth map. Unlike Gipuma and ACMH, Colmap uses the HMM to infer the pixelwise visible probability in the source images, which is more robust. Therefore, the depth map generated by Colmap has less noise and higher completeness. The HMM inference strategy in Colmap is improved to adapt to the random red-black checkerboard propagation in this paper. Although the inference strategy is not as rigorous as Colmap, it can be seen that the depth map generated by the proposed method is still better than Gipuma and ACMH, and is slightly worse than Colmap in some local details. Through the comparison of the depth maps, it can be found that the updating strategy of HMM adopted in the proposed method is still suitable for the UAV images of the high-voltage power transmission line.

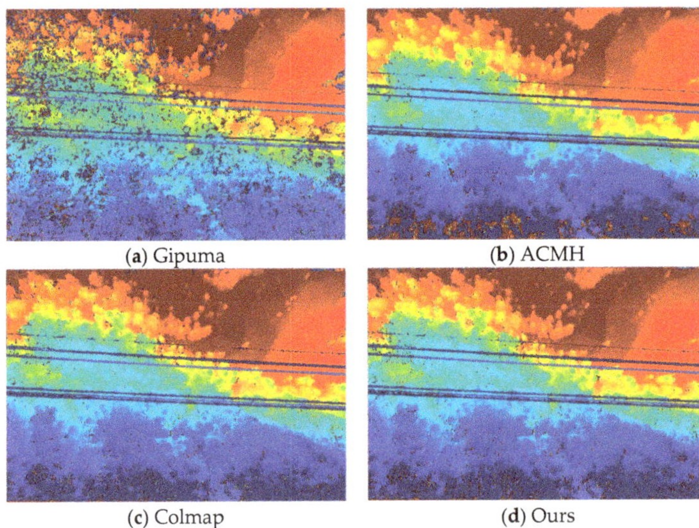

(**a**) Gipuma   (**b**) ACMH

(**c**) Colmap   (**d**) Ours

Figure 6. Depth maps comparison with different methods. (**a**–**d**) are generated by Gipuma, ACMH, Colmap, and the proposed method, respectively.

Secondly, this paper focuses on the comparative analysis of the reconstructed point clouds of power lines in the UAV images of the three test sites with the four different methods, as shown in Figure 7, Figure 8, and Figure 9, respectively. In test site 1, three spans' point clouds of power lines are selected for visual comparison. From Figure 7, it can be seen that the power lines reconstruction result by Gipuma is the worst, and only a few power lines can be reconstructed in each bundled conductor. The ACMH can reconstruct the point clouds of power lines on both sides of the spans, but the point clouds in the middle of the spans are sparse and incomplete. Colmap and the proposed method can reconstruct more complete point clouds in each span, and the results of power lines are significantly better than those reconstructed by Gipuma and ACMH.

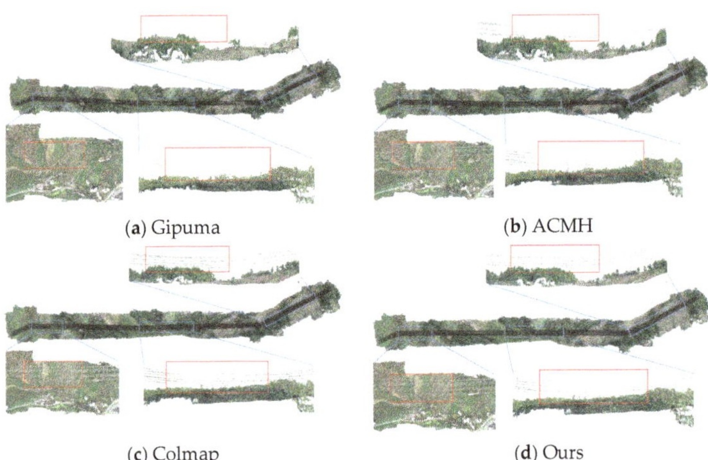

(**a**) Gipuma   (**b**) ACMH

(**c**) Colmap   (**d**) Ours

Figure 7. The comparison of reconstructed point clouds of power lines in test site 1. (**a**–**d**) are generated by Gipuma, ACMH, Colmap, and the proposed method, respectively.

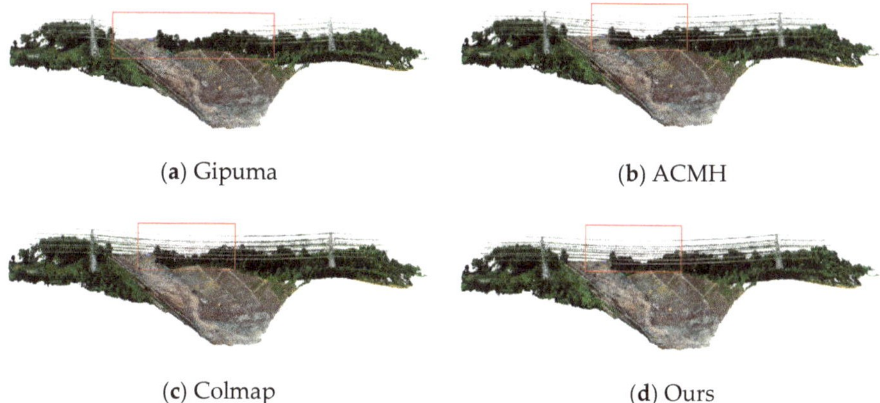

(**a**) Gipuma   (**b**) ACMH

(**c**) Colmap   (**d**) Ours

Figure 8. The comparison of reconstructed point clouds of power lines in test site 2. (**a–d**) are generated by Gipuma, ACMH, Colmap, and the proposed method, respectively.

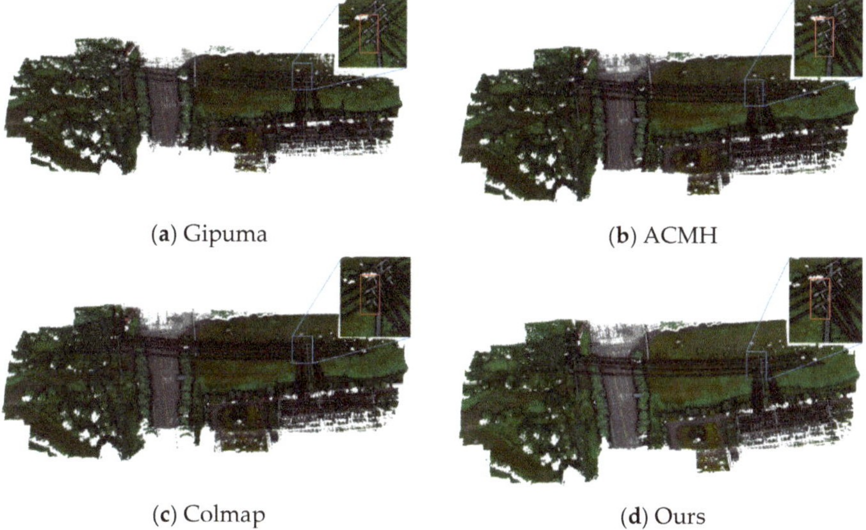

(**a**) Gipuma   (**b**) ACMH

(**c**) Colmap   (**d**) Ours

Figure 9. The comparison of reconstructed point clouds of power lines in test site 3. (**a–d**) are generated by Gipuma, ACMH, Colmap, and the proposed method, respectively.

The two sides of test side 2 are hillsides, while the middle part is low, with a large height difference between both sides. Figure 8 is the results of the reconstructed power lines with different methods. From Figure 8, it can be seen that Gipuma can only reconstruct a few of the power lines on both sides of the span; while the ACMH can reconstruct relatively more complete point clouds of power lines than Gipuma, but in the middle region, parts of power line cannot be reconstructed; Colmap and the proposed method reconstruct power lines more completely than ACMH and Gipuma.

The terrain in test site 3 is relatively flat, including part of the transformer substation and roads. Figure 9 shows the comparison of the reconstructed power lines with different methods in test site 3. It can be seen that the Gipuma can only reconstruct part of the power lines, and there are many breaks at the uppermost power lines; ACMH, Colmap, and the proposed method can reconstruct power lines more completely than Gipuma. In addition, from the reconstructed jumper lines marked as blue rectangles in Figure 9, it can be seen that Gipuma failed to reconstruct the jumper lines; while ACMH can only reconstruct part of the jumper lines. Similarly, Colmap and the proposed method can reconstruct jumper lines more completely than Gipuma and ACMH.

Since the rectangle closed-loop trajectory is applied for UAV image collection, the intersection angle between adjacent images on the same side is small. If the images on the same side are selected for dense matching, the depth error of power lines would be augmented, which would be regarded as noise and removed in depth-map fusion. Therefore, the pixelwise visible image selection is extremely important for power line reconstruction with the rectangle closed-loop trajectory. In addition, because the power lines are suspended and the background in the image with different perspectives is different, the matching cost function directly affects the reconstruction result of power lines. The "top-k-winner-take-all" strategy is applied in Gipuma without robust pixelwise view selection, and the matching costs are the weight combinations of the color and gradient difference, which lead to poor performance on the reconstruction of power lines. Unlike Gipuma, the weighted bilateral NCC matching cost function is adopted in ACMH, Colmap, and the proposed method. Therefore, the main factors that affect the completeness of power lines are the view selection and the propagation mode. ACMH performs poorly in the completeness of power line reconstruction in test site 1 because it only uses pixels with the smallest matching cost in the fixed neighbor positions to select the visible image without taking into account the influence of the intersection angle. In addition, these pixels with sorted smallest matching costs are used to propagate the plane parameters. However, the matching cost of power lines is usually greater than that of the pixels of the ground. In this case, the pixels selected to propagate their plane parameters are located in the background of power lines, which leads to the low efficiency of propagation and the convergence speed of power lines is very slow in limited iterations. The structures of UAV images in test site 2 and test site 3 are stable, the propagation modes become the main factor that affects the completeness of reconstructed power lines. Due to the large terrain undulations and the large height difference between the terrain and power lines in test site 2, ACMH updates the depth and normal of power lines through the neighbor pixels with the smallest matching cost, which has poor propagation efficiency. The sequence propagation is adopted in Colmap, and the propagation direction in each iteration is changed to realize the depth and normal updating with the four neighbor pixels, which has high effectiveness for power line reconstruction. Random red-black propagation is applied in the proposed method, and the depth and normal are updated through the neighbor pixels with the most similar color, which can also ensure the effectiveness of the propagation for power lines. Compared with the results of Colmap, the proposed method has little difference from Colmap in the completeness of the reconstructed power lines.

3.2. Analysis of the Performance of Efficiency

In this experiment, the three datasets of the UAV images in the high-voltage power transmission line are selected to analyze the runtime performance. All the parameters are maintained the same as those in Section 3.1 and all the methods are run on the same platform. Figure 10 is a comparison chart of the total runtime of dense matching and depth map fusion with the four methods, and Table 2 lists the detailed runtime with the three high-voltage power transmission line datasets. Through comparative analysis, it can be seen that Colmap has the slowest runtime due to the sequence propagation, while Gipuma, ACMH, and the proposed method use diffusion-like propagation, which is more efficient and convenient for GPU parallel processing. However, bisection refinement is employed in Gipuma, which is time-consuming to generate more hypotheses to verify. ACMH directly accesses the color values from the texture memory in the GPU for matching cost computation, which does not make full use of the advantage of the shared memory technique of GPU. This paper fully combines the advantages of the above methods and adopts the random red-black checkerboard propagation and shared memory technique in

GPU to improve efficiency. Moreover, two strategies for reducing the number of matching cost computations are adopted in the proposed method. Therefore, the runtime of the proposed method of PatchMatch is about 3–5 times faster than Colmap. With regards to depth-map fusion, it can be found that Colmap is the slowest, and the runtime is about 1/3 of the dense matching. ACMH and the proposed depth map fusion methods are more efficient than Colmap. The total runtime of the PatchMatch and depth-map fusion of the proposed method is about 4–7 times faster than Colmap.

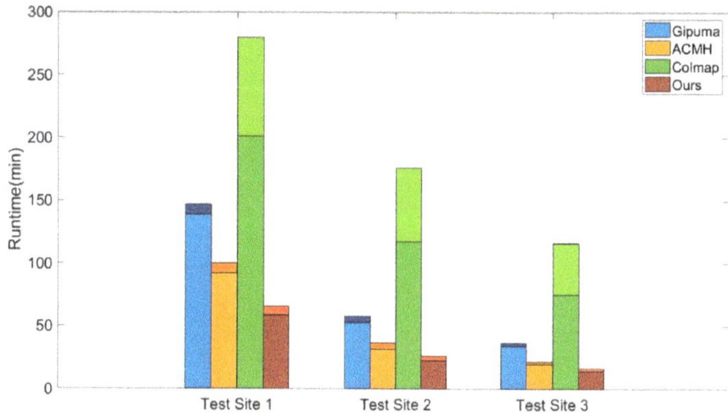

Figure 10. The comparison of total runtimes of dense matching and depth map fusion with different methods.

Table 2. Details of the runtime of dense matching and depth map fusion with different methods for the three datasets in the high-voltage power transmission line.

Methods		Test Site 1/(Min)	Test Site 2/(Min)	Test Site 3/(Min)
	PatchMatch	139.04	52.74	34.12
Gipuma	Depth fusion	7.75	4.88	2.38
	Total	146.79	57.62	36.50
	PatchMatch	92.39	31.82	19.56
ACMH	Depth fusion	7.75	4.88	2.38
	Total	100.14	36.70	21.94
	PatchMatch	201.03	117.28	75.04
Colmap	Depth fusion	78.54	58.62	41.07
	Total	279.57	175.90	116.11
	PatchMatch	58.55	22.62	14.09
Ours	Depth fusion	7.07	3.86	2.35
	Total	65.62	26.48	16.44

4. Discussion

In this section, the analysis of the precision with the proposed method is discussed. The Strecha dataset and Vaihigen dataset are applied to verify the precision of the proposed method. The two benchmark datasets both provide the parameters of image orientations, the intrinsic parameters of the camera, and the ground truth mesh or point cloud. The accuracy, completeness, and F_1 score [43] are adopted for precision analysis.

In the experiment with the Strecha dataset, the image size is set 1563 × 1024 pixels and the maximum number of views selected for PatchMatch is set 10. In this experiment with the Vaihigen dataset, the image size is set 3889 × 7000 pixels, and the maximum number of views selected for PatchMatch is set 5. The remaining parameters are consistent with the experiment of the UAV images in the high-voltage power transmission line.

Figure 11 shows the comparison of reconstructed results of different methods in the Fountain and Herzjesu datasets. It can be seen that ACMH can match more point clouds on both sides of the Fountain dataset and at the gates of the Herzjesu dataset. This is mainly because there are coplanar areas in the two datasets, and the adaptive red-black checkerboard propagation adopted in ACMH can propagate the depth and normal in a larger range, which is more efficient in the coplanar areas and can improve the completeness in the low-textured areas. Gipuma performs worse than other methods in these two datasets. The proposed method can match slightly more point clouds on both sides of the Fountain dataset than Colmap, indicating that the random red-black checkerboard propagation adopted in the proposed method is better than the sequence propagation of Colmap in the coplanar regions but still worse than the adaptive red-black propagation of ACMH.

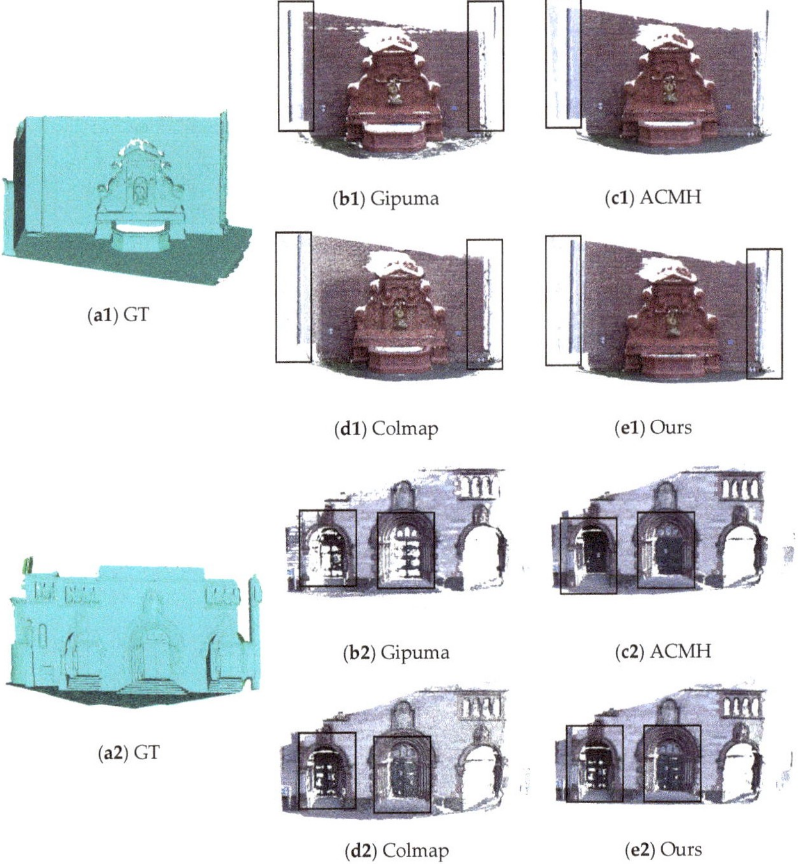

Figure 11. The comparison of reconstructed point clouds with the Strecha dataset. (**a1,a2**) are the ground truth meshes of the Fountain and Herzjesu datasets, respectively. (**b1–e1**) and (**b2–e2**) are generated by Gipuma, ACMH, Colmap, and the proposed method with the Fountain and Herzjesu datasets, respectively.

In addition, this paper quantitatively analyses the precision of point clouds reconstructed by the four methods in the Strecha dataset, in which the accuracy, completeness, and F_1 score of point clouds are used. The vertexes of the meshes in the Fountain and Herzjesu datasets are used as ground truth point clouds for the precision analysis. Table 3 shows the accuracy, completeness, and F_1 score with the evaluation threshold of 2 cm and 10 cm in percentage. It can be seen that Gipuma achieves the highest accuracy of the two datasets with the 2 cm and 10 cm thresholds because the bisection refinement is applied in Gipuma to obtain more accurate depth values. However, Gipuma performs worse in terms of completeness and F_1 score. ACMH has the highest completeness and F_1 score in the Fountain dataset with 2 cm and 10 cm, and in the Herzjesu dataset with a 2 cm threshold, indicating that ACMH has advantages in the coplanar regions. The F_1 score of the proposed method is higher than that of Colmap, which verifies that the random red-black propagation can improve the propagation efficiency.

Table 3. The precision comparison of point clouds with different methods in the Strecha dataset. The accuracy (A), completeness (C), and F_1 score (in %) are evaluated with 2 cm and 10 cm thresholds.

Datasets	Methods	2 cm/(%)			10 cm/(%)		
		A	C	F_1	A	C	F_1
Fountain	Gipuma	84.47	41.39	55.56	97.54	54.33	69.79
	ACMH	74.83	48.95	59.19	95.83	60.32	74.04
	Colmap	75.08	44.72	56.05	95.18	58.71	72.62
	Ours	74.61	47.20	57.82	95.26	59.44	73.20
Herzjesu	Gipuma	78.42	29.53	42.90	96.43	47.78	63.90
	ACMH	74.08	39.51	51.53	94.52	54.75	69.34
	Colmap	67.28	32.84	44.13	92.26	55.20	69.07
	Ours	73.16	37.78	49.83	94.28	54.73	69.25

Figure 12 shows the ground truth point clouds and the results of reconstructed point clouds in test site 1 and test site 3 of the Vaihigen with different methods. It can be seen that Gipuma has the worst completeness with a large number of holes in both test sites. ACMH, Colmap, and the proposed method all have poor performance in the road regions because the roads are weakly textured with fewer color changes and they are difficult to match with bilateral weighted NCC. It can also be found that the completeness of ACMH in test site 1 marked with a black rectangle is worse than Colmap and the proposed method. The reconstructed point clouds of Colmap in test site 3 marked with a black rectangle are better than the other three methods.

Similarly, the accuracy, completeness, and F_1 score are used for quantitative evaluation with the thresholds of 0.2 m and 0.5 m, as shown in Table 4. It can be seen that Gipuma has the highest accuracy in test site 1 with the evaluation threshold of 0.5 m, but has the lowest F_1 score in both test sites similar to the Strecha dataset. ACMH has the highest accuracy and F_1 score in both test sites with the evaluation threshold of 0.2 m; the proposed method achieves the highest F_1 score in both test sites with an evaluation threshold of 0.5 m. It can also be found that the F_1 scores in both test sites of the Vaihigen dataset of the proposed method are better than Colmap.

Table 4. The precision comparison of point clouds with different methods in test site 1 and test site 3 of the Vaihigen dataset. The accuracy (A), completeness (C), and F_1 score (in %) are evaluated with 0.2 m and 0.5 m thresholds.

Datasets	Methods	0.2 m (%)			0.5 m(%)		
		A	C		A	C	F_1
Test site 1	Gipuma	37.11	23.35	28.66	71.68	50.32	59.13
	ACMH	39.91	41.10	40.50	68.35	65.23	66.76
	Colmap	32.64	26.59	29.31	55.80	66.32	60.61
	Ours	38.62	42.11	40.29	66.29	67.58	66.93
Test site 3	Gipuma	36.47	16.27	22.50	73.85	41.43	53.08
	ACMH	40.92	38.88	39.87	74.93	65.99	70.17
	Colmap	36.26	26.92	30.90	64.41	71.18	67.63
	Ours	37.04	37.40	37.22	72.08	69.78	70.91

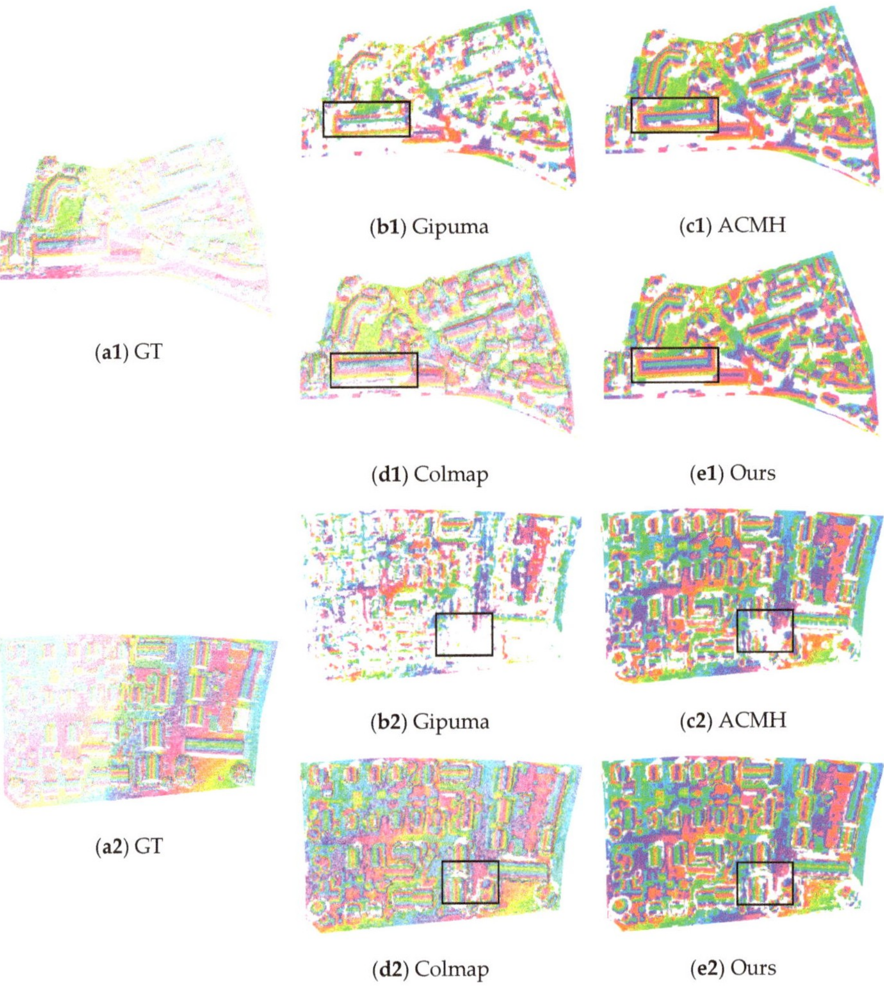

Figure 12. The comparison of reconstructed point clouds with test site 1 and test site 3 in the Vaihigen dataset. (**a1**,**a2**) are the ground truth point clouds of test site 1 and test site 3, respectively. (**b1**–**e1**) and (**b2**–**e2**) are generated by Gipuma, ACMH, Colmap, and the proposed method with test site 1 and test site 3, respectively.

5. Conclusions

An improved fast PatchMatch method for the UAV images of high-voltage power transmission lines is proposed based on Colmap, which can greatly improve efficiency while ensuring the completeness of the reconstruction of power lines. This paper employs the following three aspects to improve the efficiency of Colmap. Firstly, a new random red-black checkerboard propagation is proposed. By randomly sampling the neighbor pixels with different color patterns, the pixels with the most similar color to the current pixel are selected to propagate the plane parameters, which is more conducive to the reconstruction of power lines compared with the adaptive red-black propagation in ACMH. To combine the pixelwise view selection strategy in Colmap with the efficient random red-black checkerboard propagation, the updating schedule of hidden variables adopted in Colmap is improved. Secondly, strategies for reducing the number of matching cost computations are adopted. The number of neighbor pixels for the plane parameters propagation is reduced with the increasing of iteration number; the number of combinations with the depth and normal hypotheses is reduced in the plane refinement procedure according to the matching cost. Finally, an efficient depth map fusion is implemented in the GPU, which uses the weighted function based on reprojection error to fuse depth values. Through these strategies, the efficiencies of dense matching and depth-map fusion are greatly improved.

The experiments with UAV images of high-voltage power transmission lines from three test sites show that the proposed method can reconstruct more complete point clouds of power lines than Gipuma and ACMH, and the reconstructed power lines are more similar to Colmap. With the analysis of runtime performance, the proposed method achieves 4–7 times faster than that of Colmap. Experiments of the precision analysis with two benchmark datasets, Strecha and Vaihigen, demonstrate that the score of the proposed method is higher than Colmap. Comprehensive experiments indicate that the proposed method has promising application for high-voltage power transmission lines.

Author Contributions: Conceptualization, W.J. and W.H.; methodology, W.H. and W.J.; software, W.H. and S.J.; validation, W.J., S.J. and W.H.; formal analysis, W.J. and S.J.; investigation, W.H.; resources, W.J.; data curation, S.J.; writing—original draft preparation, W.H.; writing—review and editing, W.J., S.J. and S.H.; visualization, W.H. and S.H.; supervision, W.J.; project administration, W.J.; funding acquisition, W.J. and S.J. All authors have read and agreed to the published version of the manuscript.

Funding: This research was supported by funds from the "National Natural Science Foundation of China" (grant No. 42001413) and the "China High-Resolution Earth Observation System Based Demonstration System of Remote Sensing Application for Urban Refined Management (Phase II)" (grant No. 06-Y30F04-9001-20/22).

Institutional Review Board Statement: Not applicable.

Informed Consent Statement: Not applicable.

Data Availability Statement: Data available in a publicly accessible repository that does not issue DOIs Publicly available datasets were analyzed in this study. Strecha and Vaihigen Datasets can be found here: [https://documents.epfl.ch/groups/c/cv/cvlab-unit/www/data/multiview/denseMVS.html, accessed on 13 October 2021; https://www2.isprs.org/commissions/comm2/wg4/benchmark/2d-sem-label-vaihingen/, accessed on 13 October 2021]. The three test site datasets of UAV images in high transmission lines are available on request from the corresponding author.

Acknowledgments: The authors thank the researchers of Colmap, Gipuma and ACMH for providing the codes. The authors would like to express their gratitude to the editors and the anonymous reviewers for their professional and help comments, which are of great help to the improvement of this study.

Conflicts of Interest: The authors declare no conflict of interest.

References

1. Jiang, S.; Jiang, W.; Huang, W.; Yang, L. UAV-Based Oblique Photogrammetry for Outdoor Data Acquisition and Offsite Visual Inspection of Transmission Line. *Remote Sens.* **2017**, *9*, 278. [CrossRef]
2. Agarwal, S.; Furukawa, Y.; Snavely, N.; Simon, I.; Curless, B.; Seitz, S.; Szeliski, R. Building Rome in a Day. *Commun. ACM.* **2011**, *54*, 105–112. [CrossRef]
3. Schönberger, J.L.; Frahm, J.-M. Structure-from-Motion Revisited. In Proceedings of the IEEE Conference on Computer Vision and Pattern Recognition (CVPR), Las Vegas, NV, USA, 27–30 June 2016; 2016; pp. 4104–4113. [CrossRef]
4. Jiang, S.; Jiang, W. Efficient SfM for Oblique UAV Images: From Match Pair Selection to Geometrical Verification. *Remote Sens.* **2018**, *10*, 1246. [CrossRef]
5. Jiang, S.; Jiang, W. Efficient Structure from Motion for Oblique UAV Images Based on Maximal Spanning Tree Expansion. *ISPRS J. Photogramm. Remote Sens.* **2017**, *132*, 140–161. [CrossRef]
6. Jiang, S.; Jiang, C.; Jiang, W. Efficient Structure from Motion for Large-scale UAV Images: A Review and a Comparison of SfM Tools. *ISPRS J. Photogramm. Remote Sens.* **2020**, *167*, 230–251. [CrossRef]
7. Stentoumis, C.; Grammatikopoulos, L.; Kalisperakis, I.; Karras, G. On Accurate Dense Stereo-matching Using a Local Adaptive Multi-cost Approach. *ISPRS J. Photogramm. Remote Sens.* **2014**, *91*, 29–49. [CrossRef]

8. Furukawa, Y.; Hernandez, C. Multi-View Stereo: A Tutorial. *Found. Trends Comput. Graph. Vis.* **2015**, *9*, 1–148. [CrossRef]
9. Chen, C.; Seff, A.; Kornhauser, A.; Xiao, J. Deepdriving: Learning Affordance for Direct Perception in Autonomous Driving. In Proceedings of the IEEE International Conference on Computer Vision (ICCV), Las Condes, Chile, 11–18 December 2015; pp. 2722–2730. [CrossRef]
10. Haque, A.U.; Nejadpak, A. Obstacle Avoidance Using Stereo Camera. *arXiv* **2017**, arXiv:1705.04114.
11. Günen, M.; Beşdok, P.; Besdok, E. Use of Potree and Cesium Platforms for Presentation of Point Clouds. In Proceedings of the International Symposium on Innovative Approaches in Scientific Studies, Antalya, Turkey, 19 April 2018; p. 58.
12. Rhee, S.; Kim, T. Automated DSM Extraction from UAV Images and Performance Analysis. In Proceedings of the International Conference on Unmanned Aerial Vehicles in Geomatics, Toronto, ON, Canada, 30 August–2 September 2015; pp. 351–354. [CrossRef]
13. Zhang, Y.; Yuan, X.; Li, W.; Chen, S. Automatic Power Line Inspection Using UAV Images. *Remote Sens.* **2017**, *9*, 824. [CrossRef]
14. Seitz, S.M.; Curless, B.; Diebel, J.; Scharstein, D.; Szeliski, R. A Comparison and Evaluation of Multi-View Stereo Reconstruction Algorithms. In Proceedings of the 2006 IEEE Computer Society Conference on Computer Vision and Pattern Recognition, New York, NY, USA , 17–22 June 2006; 2006; pp. 519–528. [CrossRef]
15. Vu, H.H.; Keriven, R.; Labatut, P.; Pons, J.P. Towards High-resolution Large-scale Multi-view Stereo. In Proceedings of the IEEE Conference on Computer Vision and Pattern Recognition, Miami, FL, USA, 20–25 June 2009; pp. 1430–1437. [CrossRef]
16. Cremers, D.; Kolev, K. Multiview Stereo and Silhouette Consistency via Convex Functionals over Convex Domains. *IEEE Trans. Pattern Anal. Mach. Intell.* **2011**, *33*, 1161–1174. [CrossRef]
17. Sinha, S.N.; Mordohai, P.; Pollefeys, M. Multi-View Stereo via Graph Cuts on the Dual of an Adaptive Tetrahedral Mesh. In Proceedings of the IEEE International Conference on Computer Vision, Rio de Janeiro, Brazil, 14–21 October 2007; pp. 1–8. [CrossRef]
18. Furukawa, Y.; Ponce, J. Accurate, Dense, and Robust Multiview Stereopsis. *IEEE Trans. Pattern Anal. Mach. Intell.* **2010**, *32*, 1362–1376. [CrossRef]
19. Goesele, M.; Snavely, N.; Curless, B.; Hoppe, H.; Seitz, S.M. Multi-View Stereo for Community Photo Collections. In Proceedings of the IEEE 11th International Conference on Computer Vision, Rio de Janeiro, Brazil, 14–21 October 2007; pp. 1–8. [CrossRef]
20. Galliani, S.; Lasinger, K.; Schindler, K. Massively Parallel Multiview Stereopsis by Surface Normal Diffusion. In Proceedings of the IEEE International Conference on Computer Vision, Las Condes, Chile, 11–18 December 2015; pp. 873–881. [CrossRef]
21. Schönberger, J.L.; Zheng, E.; Pollefeys, M.; Frahm, J.M. Pixelwise View Selection for Unstructured Multi-View Stereo. In Proceedings of the European Conference on Computer Vision; Springer: Cham, Switzerland, 2016; pp. 501–518. [CrossRef]
22. Xu, Q.; Tao, W. Multi-Scale Geometric Consistency Guided Multi-View Stereo. In Proceedings of the IEEE/CVF Conference on Computer Vision and Pattern Recognition, Long Beach, CA, USA, 15–21 June 2019; pp. 5483–5492. [CrossRef]
23. Hirschmüller, H. Stereo Processing by Semiglobal Matching and Mutual Information. *IEEE Trans. Pattern Anal. Mach. Intell.* **2008**, *30*, 328–341. [CrossRef]
24. Barnes, C.; Shechtman, E.; Finkelstein, A.; Dan, B.G. PatchMatch: A Randomized Correspondence Algorithm for Structural Image Editing. *ACM Trans. Graph.* **2009**, *28*, 24. [CrossRef]
25. Barnes, C.; Shechtman, E.; Dan, B.G.; Finkelstein, A. The Generalized PatchMatch Correspondence Algorithm. In Proceedings of the European Conference on Computer Vision, Heraklion, Greece, 5–11 September 2010; pp. 29–43. [CrossRef]
26. Liao, J.; Fu, Y.; Yan, Q.; Xiao, C. Pyramid Multi-View Stereo with Local Consistency. *Comput. Graph. Forum.* **2019**, *38*, 335–346. [CrossRef]
27. Bleyer, M.; Rhemann, C.; Rother, C. PatchMatch Stereo—Stereo Matching with Slanted Support Windows. In Proceedings of the British Machine Vision Conference, Dundee, UK, 29 August–2 September 2011; pp. 1–11.
28. Heise, P.; Klose, S.; Jensen, B.; Knoll, A. PM-Huber: PatchMatch with Huber Regularization for Stereo Matching. In Proceedings of the IEEE International Conference on Computer Vision, Columbus, OH, USA, 23–28 June 2014; pp. 2360–2367. [CrossRef]
29. Besse, F.; Rother, C.; Fitzgibbon, A.; Kautz, J. PMBP: PatchMatch Belief Propagation for Correspondence Field Estimation. *Int. J. Comput. Vis.* **2014**, *110*, 2–13. [CrossRef]
30. Yu, L.; Min, D.; Brown, M.S.; Do, M.N.; Lu, J. SPM-BP: Sped-Up PatchMatch Belief Propagation for Continuous MRFs. In Proceedings of the IEEE International Conference on Computer Vision, Santiago, Chile, 7–13 December 2015; pp. 4006–4014. [CrossRef]
31. Xu, S.; Zhang, F.; He, X.; Shen, X.; Zhang, X. PM-PM: PatchMatch with Potts Model for Object Segmentation and Stereo Matching. *IEEE Trans. Image Process.* **2015**, *24*, 2182–2196. [CrossRef]
32. Lu, J.; Yu, L.; Yang, H.; Min, D.; Eng, W.; Do, M.N. PatchMatch Filter: Edge-Aware Filtering Meets Randomized Search for Visual Correspondence. *IEEE Trans. Pattern Anal. Mach. Intell.* **2017**, *39*, 1866–1879. [CrossRef]
33. Tian, M.; Yang, B.; Chen, C.; Huang, R.; Huo, L. HPM-TDP: An efficient hierarchical PatchMatch depth estimation approach using tree dynamic programming. *ISPRS J. Photogramm. Remote Sens.* **2019**, *155*, 37–57. [CrossRef]
34. Shen, S. Accurate Multiple View 3D Reconstruction Using Patch-Based Stereo for Large-Scale Scenes. *IEEE Trans. Image Process.* **2013**, *22*, 1901–1914. [CrossRef]
35. Zheng, E.; Dunn, E.; Jojic, V.; Frahm, J.M. PatchMatch Based Joint View Selection and Depthmap Estimation. In Proceedings of the IEEE Conference on Computer Vision and Pattern Recognition, Columbus, OH, USA, 23–28 June 2014; pp. 1510–1517. [CrossRef]

36. Romanoni, A.; Matteucci, M. TAPA-MVS: Textureless-Aware PAtchMatch Multi-View Stereo. In Proceedings of the 2019 IEEE/CVF International Conference on Computer Vision, Seoul, Korea, 27 October–2 November 2019; pp. 10413–10422. [CrossRef]
37. Xu, Q.; Tao, W. Planar Prior Assisted PatchMatch Multi-View Stereo. In Proceedings of the AAAI Conference on Artificial Intelligence, New York, NY, USA, 7–12 February 2020; pp. 12516–12523. [CrossRef]
38. Hou, Y.; Peng, J.; Hu, Z.; Tao, P.; Shan, J. Planarity Constrained Multi-view Depth Map Reconstruction for Urban Scenes. *ISPRS J. Photogramm. Remote Sens.* **2018**, *139*, 133–145. [CrossRef]
39. Strecha, C.; Hansen, W.V.; Gool, L.V.; Fua, P.; Thoennessen, U. On Benchmarking Camera Calibration and Multi-View Stereo for High Resolution Imagery. In Proceedings of the IEEE Conference on Computer Vision and Pattern Recognition, Anchorage, AK, USA, 23-28 June 2008; pp. 1–8. [CrossRef]
40. Cramer, M. The DGPF-test on digital airborne camera evaluation overview and test design. *Photogramm. Fernerkund. Geoinf.* **2010**, 73–82. [CrossRef]
41. Hane, C.; Zach, C.; Cohen, A.; Angst, R.; Pollefeys, M. Joint 3D Scene Reconstruction and Class Segmentation. In Proceedings of the IEEE Conference on Computer Vision and Pattern Recognition, Portland, OR, USA, 23–28 June 2013; pp. 97–104. [CrossRef]
42. Curless, B.; Levoy, M. A Volumetric Method for Building Complex Models from Range Images. In Proceedings of the 23rd Annual Conference on Computer Graphics and Interactive Techniques, New Orleans, LA, USA, 4–9 August 1996; pp. 303–312.
43. Knapitsch, A.; Park, J.; Zhou, Q.-Y.; Koltun, V. Tanks and Temples: Benchmarking Large-scale Scene Reconstruction. *ACM Trans. Graph.* **2017**, *36*, 1–13. [CrossRef]

Article

DP-MVS: Detail Preserving Multi-View Surface Reconstruction of Large-Scale Scenes

Liyang Zhou [1,†], Zhuang Zhang [1,†], Hanqing Jiang [1,†], Han Sun [1], Hujun Bao [2] and Guofeng Zhang [2,*]

[1] SenseTime Research, Beijing 100080, China; zhouliyang@sensetime.com (L.Z.); zhangzhuang@sensetime.com (Z.Z.); jianghanqing@sensetime.com (H.J.); sunhan@sensetime.com (H.S.)
[2] State Key Lab of CAD&CG, Zhejiang University, Hangzhou 310058, China; baohujun@zju.edu.cn
* Correspondence: zhangguofeng@zju.edu.cn
† These authors contributed equally to this work.

Abstract: This paper presents an accurate and robust dense 3D reconstruction system for detail preserving surface modeling of large-scale scenes from multi-view images, which we named DP-MVS. Our system performs high-quality large-scale dense reconstruction, which preserves geometric details for thin structures, especially for linear objects. Our framework begins with a sparse reconstruction carried out by an incremental Structure-from-Motion. Based on the reconstructed sparse map, a novel detail preserving PatchMatch approach is applied for depth estimation of each image view. The estimated depth maps of multiple views are then fused to a dense point cloud in a memory-efficient way, followed by a detail-aware surface meshing method to extract the final surface mesh of the captured scene. Experiments on ETH3D benchmark show that the proposed method outperforms other state-of-the-art methods on F1-score, with the running time more than 4 times faster. More experiments on large-scale photo collections demonstrate the effectiveness of the proposed framework for large-scale scene reconstruction in terms of accuracy, completeness, memory saving, and time efficiency.

Keywords: multi-view reconstruction; detail preserving; depth estimation; surface meshing

Citation: Zhou, L.; Zhang, Z.; Jiang, H.; Sun, H.; Bao, H.; Zhang, G. DP-MVS: Detail Preserving Multi-View Surface Reconstruction of Large-Scale Scenes. *Remote Sens.* **2021**, *13*, 4569. https://doi.org/10.3390/rs13224569

Academic Editor: George Karras

Received: 3 September 2021
Accepted: 8 November 2021
Published: 13 November 2021

Publisher's Note: MDPI stays neutral with regard to jurisdictional claims in published maps and institutional affiliations.

1. Introduction

Multi-view stereo (MVS) reconstruction of large-scale scenes is a research topic of vital importance in computer vision and photogrammetry. With the popularization of digital cameras and unmanned aerial vehicles (UAV), it is becoming more and more convenient to capture large numbers of high resolution photos of the real scenes, which makes it more feasible to reconstruct 3D digitalized models of the scenes from the captured high-quality images. With the development of smart cities and digital twin, 3D reconstruction of large-scale scenes has attracted more attentions due to its usefulness in providing digitalized content for various applications such as urban visualization, 3D navigation, geographic mapping, and model vectorization. However, these applications usually require reconstruction of high-quality dense surface models. Specifically, 3D visualization and navigation demand realistically textured 3D surface models with complete structures and few artifacts, while geographic mapping and model vectorization depends on highly accurate dense point clouds or models with geometric details as reliable 3D priors, which are great challenges to multi-view reconstruction.

Over the past few years, significant progresses have been made in MVS, especially in the reconstruction of aerial scenes. However, most existing state-of-the-art (SOTA) methods lack sufficient details in their reconstruction results, or take huge time to achieve high reconstruction accuracy. Besides, it consumes a lot of memory to fuse high resolution depth maps to dense point cloud. Learning-based multi-view depth estimation schemes do not perform so well as traditional methods in generalization and scene detail recovery, and usually have difficulties in handling high-resolution images. In this paper, we propose a novel MVS framework for detail preserving reconstruction of dense surface model from multiple images captured by a digital camera or UAV, which we named DP-MVS. Our DP-MVS framework is designed for large-scale scene reconstruction which takes accuracy, robustness and efficiency into account, to ensure that the reconstruction is carried out in a time-and-memory-efficient way to recover accurate geometric structures with fine details. The key contributions of our system can be summarized as:

- We propose a detail preserving PatchMatch approach to ensure an accurate dense depth map estimation with geometric details for each image view.
- Considering that high resolution depth map fusion is usually memory consuming, we propose a memory-efficient depth map fusion approach for handling extremely high resolution depth map fusion, to ensure accurate point cloud reconstruction of large-scale scenes without out-of-memory issues.
- We propose a novel detail-aware Delaunay meshing to preserve fine surface details for complicated scene structures.

Experiments with quantitative and qualitative evaluations demonstrate the effectiveness and efficiency of our DP-MVS method by achieving SOTA performance on large-scale image collections captured by digital cameras or UAVs.

2. Related Work

According to the taxonomy given in [1], existing multi-view reconstruction approaches can be generally divided into four categories: voxel based methods, surface evolution based methods, feature point growing based methods, and depth-map merging based methods.

2.1. Voxel Based Methods

The voxel based methods compute a cost function on a 3D volume within a bounding box of the object. Seitz et al. [2] proposed a voxel coloring framework that identifies the voxels with high photo-consistency across multiple image views in the 3D volume space of the scene. Vogiatzis et al. [3] use graph-cut optimization to compute a photo-consistent surface that encloses the largest possible volume. These methods are limited in reconstruction accuracy and space by the voxel grid resolution. Sinha et al. [4] proposed to use photo-consistency to guide the adaptive subdivision of the 3D volume to generate a multi-resolution volumetric mesh that is densely tesselated around the possible surface, which breaks through the voxel resolution limitation to some extent. However, large-scale scenes are difficult for this method due to its high computational and memory costs. Besides, these methods are only suitable for compact objects with a tight enclosing bounding box.

2.2. Surface Evolution Based Methods

The surface evolution based methods iteratively evolve from an initial surface guess to minimize the photo-consistency measurement. Faugeras and Keriven [5] deduce a set of PDEs from a variational principle to deform an initial set of surfaces toward the objects to be detected. Hernández et al. [6] proposed a deformable model framework which fuses texture and silhouette driven forces for the final object surface evolution, based on an initial surface that should be close enough to the objective one. Hiep et al. [7] use a minimum s-t cut based global optimization to generate a initial visibility consistent mesh from dense point cloud, and then capture small geometric details with a variational mesh refinement approach. Li et al. [8] use an adaptive resolution control to classify the initial mesh into significant and insignificant regions, and accelerate the stereo refinement by culling out and simplifying most insignificant regions, while still refining and subdividing the significant regions to a SOTA level of geometry details. Romanoni and Matteucci [9] proposed a model-based camera selection method to increase the quality of pairwise camera selection, and an occlusion-aware masking to improve the model refinement robustness by avoiding the influence of occlusions on photometric error computation. A common drawback of these methods is the requirement of a reliable initial surface which is usually difficult for outdoor scenes. Cremers et al. [10] formulate multi-view reconstruction as a convex functional minimization problem that does not rely on initialization, with the exact silhouette consistency imposed as convex constraints which restrict the domain of feasible functions. However, this method uses voxel representation for reconstruction space, and is therefore unsuitable for large-scale scenes.

2.3. Feature Growing Based Methods

The feature point growing based methods first reconstruct 3D feature points from regions with textures, and then expand these feature points to textureless areas. Lhuillier et al. [11] proposed a quasi-dense approach to acquire 3D surface model, which expands the sparse feature points by resampling quasi-dense points from the quasi-dense disparity maps generated by match propagation. Goesele et al. [12] proposed a method to handle challenging Internet photo collections using per-view and per-pixel image selection for stereo matching, with a region growing process to expand the reconstructed SIFT features [13]. Based on these methods, Furukawa et al. [14] presented the SOTA MVS method called Patch-based MVS (PMVS) that first reconstructs a set of sparse matched keypoints, and then iteratively expands these keypoints till visibility constraints are invoked to filter away noisy matches. Based on PMVS, Wu et al. [15] proposed Tensor-based MVS (TMVS) for quasi-dense 3D reconstruction which combines the complementary advantages of photo-consistency, visibility and geometric consistency enforcement in MVS under a 3D tensor framework. These feature point growing methods attempt reconstructing a global 3D model using all the input images, and therefore will suffer from the scalability problem with a large number of images. Although this problem can be alleviated by dividing the input images into clusters with small overlaps like [16], its computational complexity still remains to be a problem for large-scale scenes.

2.4. Depth-Map Merging Based Methods

The depth-map merging based methods first estimate a depth map for each view and then merge all the depth maps together into a single model by taking visibility into account. Strecha et al. [17] jointly model depth and visibility as a hidden Markov Random Field (MRF), and use an EM-algorithm to alternate between estimation of visibility/depth and optimization of model parameters, without merging the depth maps to a final model. Goesele et al. [18] compute depth maps using a window-based voting approach with good matches and then merge them with volumetric integration. Merrell et al. [19] proposed a real-time 3D reconstruction pipeline, which utilizes visibility-based and confidence-based fusion for multi-view depth map fusion to online large-scale 3D model. Zach et al. [20] presented a method for range image integration by globally minimizing an energy functional consisting of a total variation (TV) regularization force and an L^1 data fidelity term. Kuhn et al. [21] use a learning-based TV prior to estimate uncertainties for depth map fusion. Liu et al. [22] produced high quality MVS reconstruction using continuous depth maps generated by variational optical flow, which requires visual hull as an initialization. However, these methods use volumetric representation for depth map fusion or rely on an initial model, and are therefore limited in scalability. Some other methods fuse the estimated depth map to point cloud, and focus on estimating confidence or uncertainty constraint to guide the depth map fusion process, which turns out to be more suitable for large-scale scenes. For example, a confidence-based MVS method in [23] developed a self-supervised deep learning method to predict the spatial confidence for multiple depth maps. Bradley et al. [24] proposed to use a robust binocular scaled-window stereo matching technique, followed by adaptive filtering of the merged point clouds, and efficient high-quality mesh generation. Campbell et al. [25] use multiple depth hypotheses with a spatial consistency constraint to extract the true depth for each pixel in a discrete MRF framework, while Schönberger et al. [26] perform MVS with pixelwise view selection for depth and normal estimation and fusion. Shen [27] computes the depth map for each image using PatchMatch, and fuses multiple depth maps by enforcing depth consistency at neighboring views, which is similar to [28], with the difference that Tola et al. used DAISY features [29] to produce depth maps. Li et al. [30] also generate depth maps using DAISY, and applied two stages of bundle adjustment to optimize the positions and normals of 3D points. However, these methods usually require complex computation for high-quality depth map estimation. To expand the reconstruction scale to a larger extent at a lower computational cost, Xue et al. [31] proposed a novel multi-view 3D dense matching method for large-scale aerial images using a divide-and-conquer scheme, and Mostegel et al. [32] innovatively proposed to prioritize the depth map computation of MVS by confidence prediction to efficiently obtain compact 3D point clouds with high quality and completeness. Wei et al. [33] proposed a novel selective joint bilateral upsampling and depth propagation strategy for high-resolution unstructured MVS. Wang et al. [34] proposed a mesh-guided MVS method with pyramid architecture, which uses the surface mesh obtained from coarse-scale images to guide the reconstruction process. However, these methods do not consider too much about how to preserve the true geometric details in depth map estimation and fusion stages. Some learning-based multi-view stereo reconstruction approaches such as [35–40] have achieved significant improvements on various benchmarks, but the robustness and generalization of these methods are still limited for natural scenes compared to the traditional methods. To better tackle practical problems such as dense reconstruction of textureless regions, some recent works try to combine learning methods with traditional MVS methods to improve generalization. For example, Yang et al. [41] use a light-weight depth refinement network to improve the noisy depths of textureless regions produced by multi-view semi-global matching (SGM). Yang and Jiang [42] combine deep learning algorithms with traditional methods to extract and match feature points from light pattern augmented images to improve a practical 3D reconstruction method for weakly textured scenes. Stathopoulou et al. [43] tackle the textureless problem by leveraging semantic priors into a PatchMatch-based MVS in order to increase confidence and better support depth and normal map estimation on weakly textured areas. However, even with these combination of traditional and learning algorithms, visual reconstruction of large textureless areas commonly present in urban scenarios of building facades or indoor scenes still remains to be a challenge. Some recent works such as [44,45] focus on novel path planning methods for the high-quality aerial 3D reconstruction of urban areas. Pepe et al. [46] apply SfM-MVS approach to airborne images captured by nadir and oblique cameras to build 2.5D map and 3D models for urban scenes. These works make efforts to improve the global reconstruction completeness and scalability for large-scale urban scenes, but pay less attention to the local reconstruction geometric details or textureless challenge.

Depth map estimation is vitally important for a high-quality MVS reconstruction. Recently, PatchMatch stereo methods [26,47–50] have shown great power in depth map estimation with their fast global search for the best matches in other images, with different kinds of propagation schemes developed or improved. For example, Schönberger et al. [26], Zheng et al. [48] both use sequential propagation scheme, while [49,51] both utilize checkerboard-based propagation to further reduce runtime. ACMM [50] extends the work of [51] by introducing a coarse-to-fine scheme for better handling textureless areas. Assuming the textureless areas are piecewise planar, ACMP [52] extends ACMM by contributing a novel multi-view matching cost aggregation which takes both photometric consistency and planar compatibility into consideration, and TAPA-MVS [53] proposed novel PatchMatch hypotheses to expand reliable depth estimates to neighboring textureless regions. Furthermore, Schönberger et al. [26], Xu and Tao [50] both use a forward/backward reprojection error as an additional error term for PatchMatch. MARMVS [54] additionally select the optimal patch scale for each pixel to reduce matching ambiguities. However, these methods focus on speeding up computation and handling textureless regions, but seldom have any strategy for geometric detail preserving, which is exactly the main focus of our method.

3. Materials and Methods

3.1. System Overview

Suppose a large-scale scene is captured by multiple RGB images with digital cameras mounted on terrestrial or UAV platforms, denoted as $\{\mathbf{I}_i | i = 1, 2, \cdots, N\}$, where N is the number of input images. Our dense 3D reconstruction system is applied for the inputed multi-view images to robustly reconstruct an accurate surface model of the captured scene. We now outline the steps of the proposed multi-view reconstruction framework, as shown in Figure 1.

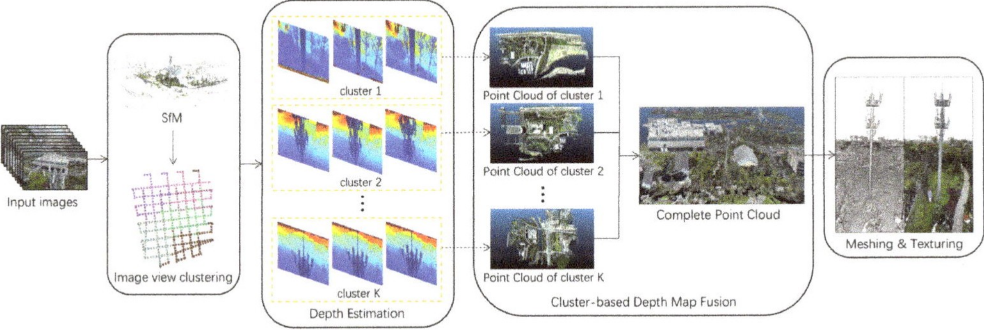

Figure 1. System framework of DP-MVS, which consists of six modules including SfM, image view clustering, detail preserving depth map estimation, cluster-based depth map fusion, detail-aware surface meshing, and multi-view texture mapping.

Our DP-MVS framework first reconstructs a sparse map for the input multi-view images using an incremental Structure-from-Motion (SfM) framework similar to Schönberger and Frahm [55]. Then, the image views are divided into a number of clusters according to covisibility relationship based on the reconstructed sparse map. For each cluster, a novel detail preserving PatchMatch approach is applied to estimate a dense depth map $\mathbf{D}_i$ for each image view i. Then, the depth maps in each cluster are fused to a noise-free dense point cloud. After that, point clouds of all the clusters are merged into a final point cloud denoted by $\mathbf{P}$. Finally, a detail-aware Delaunay triangulation is used to extract the final surface mesh of the captured scene from the merged point cloud, which is represented as $\mathbf{S}$. The main steps of our framework will be described in detail in the following subsections.

3.2. Detail Preserving Depth Map Estimation

Our method adopts a novel PatchMatch based stereo method for accurate depth map estimation with detail preservation. A well-known PatchMatch scheme is sequential propagation, which alternatively performs upward/downward propagation during odd iterations and reverse propagation during even iterations. Because only neighborhood pixels are referred during one propagation, this scheme is more sensitive to textureless regions. Furthermore, sequential propagation can only be parallelized at the row or column level, which cannot fully utilize the strength of modern multi-core GPU. Another PatchMatch scheme is checkerboard-based propagation, in which the reference image is partitioned into a checkerboard pattern of "red" and "black" pixels. Propagation and optimization is performed for all "red" pixels in odd iterations and all "black" pixels in even iterations, and is therefore more suitable for parallelized handling of high-resolution images. The standard checkerboard propagation scheme was firstly introduced by [49], which consumes a lot of time to calculate normalized cross correlation (NCC) of multiple sample points from multiple views at each single propagation. ACMM [50] improves the scheme by introducing a multi-hypothesis joint view selection strategy. The strategy is more suitable for depth estimation of planar structures, where the samples with high confidence could be propagated readily along smooth surface. However, if there is a foreground object with structure thinner than the sampling window size, the hypotheses are very likely to be sampled to the background regions, which might force the foreground depth to shift to the background position. Inspired by [26], we propose a detail preserving PatchMatch method based on the diffusion-like propagation scheme, which ensures both high accuracy and completeness of the estimated depth map, especially for accurate reconstruction of detailed structures.

Figure 2 shows the comparison results of depth estimation by ACMM and our proposed method for the experimental case "B5 Tower", with both sampling strategies illustrated to show the difference. Take the pixel highlighted in Figure 2a as an example. Most sample points of ACMM lie in the background regions, which mistakenly wipes its depth to background level, as illustrated in Figure 2b. To better solve this problem, we change 4 V-shaped areas of ACMM to oblique long strip areas to obtain more even distribution of hypotheses. This improved sampling strategy is more favorable to the recovery of thin objects than ACMM, considering it increases the probability of sampling foreground thin object regions. In addition, we observe that the sequential-based propagation strategy is helpful to detail recovery. An important reason is that it only propagates neighboring depths and a reliable hypothesis could be spread further along horizontal and vertical directions. Inspired by this observation, we further add the four-neighboring hypotheses. Thus, there are 12 hypotheses totally, which increases the time complexity of a single propagation by half. In order to improve computational efficiency, the four-neighboring hypotheses are sorted in descending according to their NCC cost and the top-K ones are selected as the final hypotheses, with $K = 2$ in our experiments. We exhibit the generated depth maps and normal maps of case "B5 Tower" estimated by ACMM and our strategy in Figure 2c,d. Here, we use the same multi-scale framework as ACMM for our strategy as a fair comparison. As can be seen in the red rectangles, the depth map estimated by our proposed method contains richer geometric details, with more accurate normal map, especially for thin structures. Actually, our dense matching method can reconstruct thin structures with at least 2 pixels width, the corresponding Ground Sample Distance (GSD) is 2.4 cm at flying altitude of 30 m.

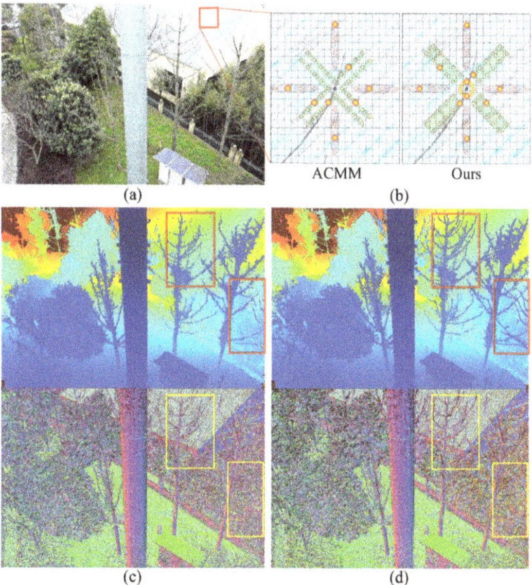

Figure 2. Illustration of our PatchMatch sampling strategy: (**a**) A source image view of case "B5 Tower". (**b**) The ACMM [50] sampling strategy of a pixel in the red rectangle of (**a**), and our improved sampling strategy. (**c**) The estimated depth map and normal map by ACMM. (**d**) Our estimated depth map and normal map, which contain better details highlighted in the red and yellow rectangles.

In the refinement step, for each pixel p of the current image view, we generate a perturbed depth and normal hypothesis (d_p^{pert}, n_p^{pert}) by perturbing the current depth and normal estimation (d_p, n_p). A random depth and normal hypothesis is also generated denoted as (d_p^{rand}, n_p^{rand}). These newly generated depths and normals are combined with the current one to yield 6 additional candidate depth and normal pairs (d_p, n_p), (d_p^{pert}, n_p^{pert}), (d_p^{pert}, n_p), (d_p, n_p^{pert}), (d_p^{rand}, n_p), (d_p, n_p^{rand}). During each iteration, for each pixel, we choose the depth and normal estimation with the best NCC cost from the set of candidate depths and normals, to further refine its current depth and normal estimation. Usually, 3~5 iterations are sufficient for the depth maps and normal maps to converge.

We compare our proposed scheme with SOTA methods including ACMM, Open-MVS [56] and COLMAP [26,55] in four cases "ZJU Fan", "B5 Tower", "B5 West" and "B5 Wire" captured by UAV with resolution 4864×3648. For fair comparisons, we use our SfM results as the input to these methods. Thus for COLMAP, we actually only use its MVS module [26]. We perform depth estimation by choosing 8 reference images for the current image, except OpenMVS, for which we use the default setting. The depth map results are given together with their corresponding point clouds by projecting the depth values forward to 3D space to more directly show the 3D geometry of the depth maps. As shown in the highlighted rectangle regions of the depth maps and the point clouds in Figure 3, with the proposed sampling strategy, our depth estimation method performs better than the checkerboard propagation scheme of ACMM and the sequential propagation scheme of OpenMVS and COLMAP in thin structure depth recovery. Additionally, as can be seen in Figure 3, the produced depth maps are more complete and less noisy compared with OpenMVS and COLMAP, which validates the proposed method.

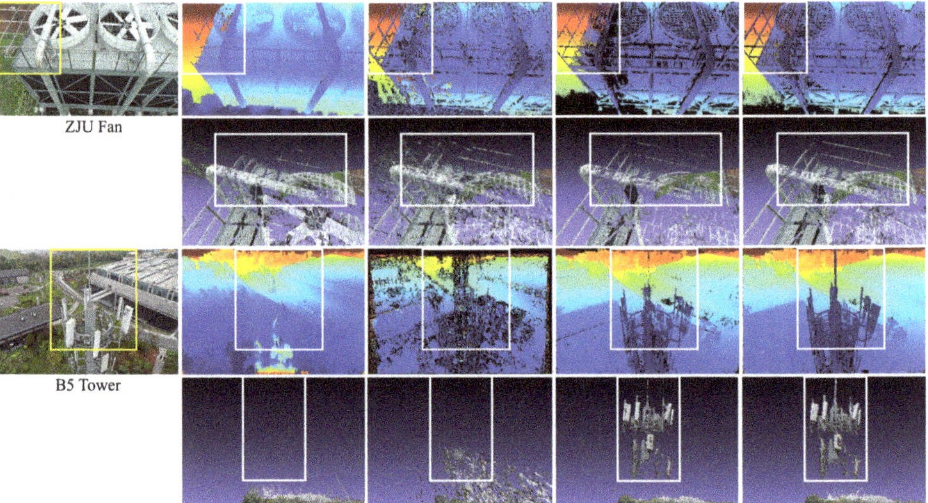

Figure 3. *Cont.*

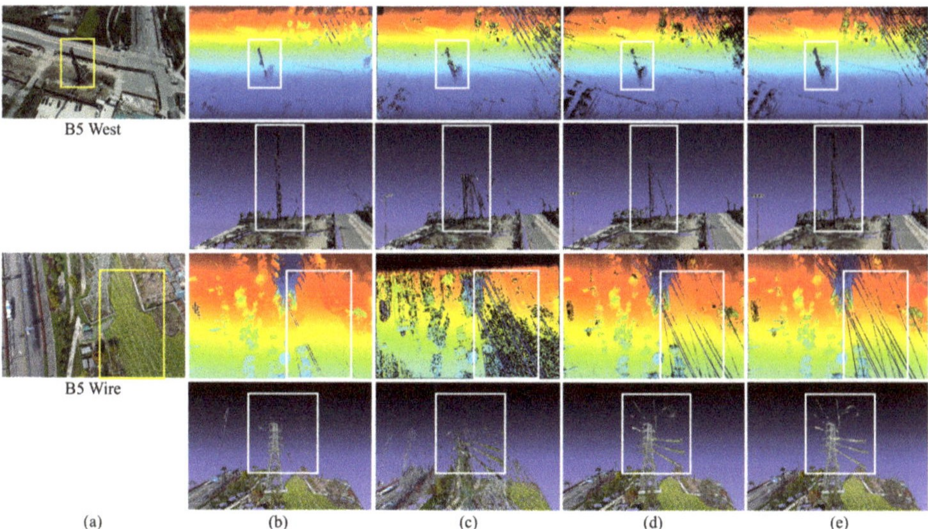

Figure 3. (**a**) A source image view for each case of "ZJU Fan", "B5 Tower", "B5 West" and "B5 Wire". (**b**) The depth map results of ACMM [50] and their corresponding point clouds by projecting the depths forward to 3D space. (**c**) The depth maps and point clouds of OpenMVS [56]. (**d**) The depth maps and point clouds of COLMAP [26]. (**e**) Our results of depth maps and point clouds, which turn out to be the best in both details and noisylessness.

3.3. Memory-Efficient Depth Map Fusion

We adopt the the graph-based framework proposed by [26] for depth map fusion, in which the consistent pixels are connected according to the geometry and depth consistency from multi-view images recursively. This method requires loading all the depth maps and normal maps into memory in advance. Therefore, for large-scale scenes of high-resolution images, out-of-memory problem will be the bottleneck. To solve this problem, we divide the scene into multiple clusters and fuse the depth maps of each cluster to an individual point cloud separately. Finally, all the clusters are merged into a complete point cloud.

Theoretically, the memory complexity of N image views with resolution $W \times H$ is $O(N \times W \times H)$, since the main memory bottleneck is to load all the depth maps and normal maps in advance. Therefore, to avoid out-of-memory, the image views should be evenly divided into a few clusters, so that all the depth maps inside one cluster could be loaded into a single computer at once without out-of-memory risk. We adopt K-means algorithm to perform the partition. Specifically, we first estimate a cluster number K, based on the total image view number N divided by the maximum number of image views supported by each cluster denoted as n, that is:

$$K = \lceil N/n \rceil. \tag{1}$$

We set $n = 120$ for image resolution 4864×3648 in our experiments. Then, we initialize K seed image views and iteratively classify all the other image views based on their distances to the cluster centers and covisibility scores. Therefore, for each image view $\mathbf{I}_i$, we define the distance criterion to the kth cluster for K-means as:

$$\begin{aligned}
\mathcal{D}_i^k &= (C_i - \hat{C}_k)(1 - \hat{S}) \\
\hat{S} &= \frac{max_{j \in \mathcal{C}(k)} \mathcal{M}_{ij}}{max_{j=1,\cdots,N} \mathcal{M}_{ij}},
\end{aligned} \tag{2}$$

where $k = 1, \cdots, K$. Here $\mathcal{D}_i^k$ is a newly defined distance between $\mathbf{I}_i$ and the kth cluster, which measures both the Euclidian distance and the covisibility between them. C_i is the camera center of $\mathbf{I}_i$. $\hat{C}_k$ is the barycenter of the camera centers of all the image views contained in the kth cluster. $\hat{S}$ is the normalized covisibility score. $\mathcal{M}_{ij}$ is the number of SIFT feature correspondences between image views i and j, and $\mathcal{C}(k)$ is the set of image views in the kth cluster. Intuitively, an image view with closer distance and stronger connectivity to a cluster will be prioritized into the cluster. In order to ensure the time efficiency of fusion, the number of images contained in each cluster should be almost the same. We first choose the cluster k' with farthest distance to all the other clusters. If it has less than n number of views, we push image views from other clusters into it in the ascending order of $\mathcal{D}_i^{k'}$ until the number of views reaches n, otherwise the redundant images are popped out in the descending order of $\mathcal{D}_i^{k'}$, and pushed into the cluster k with the minimal $\mathcal{D}_i^k$ until the number of views reaches n. Meanwhile, the fused point clouds of neighboring clusters might be inconsistent at the boundary. The reason is that the partition makes the connected pixels broken into different parts, which results in a slight boundary difference from the original fused point cloud. Thus, we add additional connected images to ensure there are sufficient overlapping regions between neighboring clusters, which increases the point cloud redundancy to a certain extent. To eliminate redundancy, we merge those points from adjacent clusters if their projections on the overlapping images are the same, and the projection depth error and normal error are below certain thresholds, which are set to 1% and $10°$ respectively in the experiments. In this way, we ensure that the final point cloud is almost the same as the result without image view clustering.

We show the image view clustering result of case "B5 West" with totally 513 images of resolution 4864×3648 and $n = 120$ in Figure 4a,b, which runs on our server platform with 500 GB memory. We set $K = 5$ according to Equation (1). The fused point clouds with and without image view clustering are also given in Figure 4c,d to show the effectiveness of our memory-efficient depth map fusion, which saves the memory cost from 87 GB without clustering, to 28 GB with 5 clusters for the case "B5 West", but brings almost no difference to the quality of the final fused point cloud, with the point cloud redundancy increased only by 10% of the total fused point cloud size.

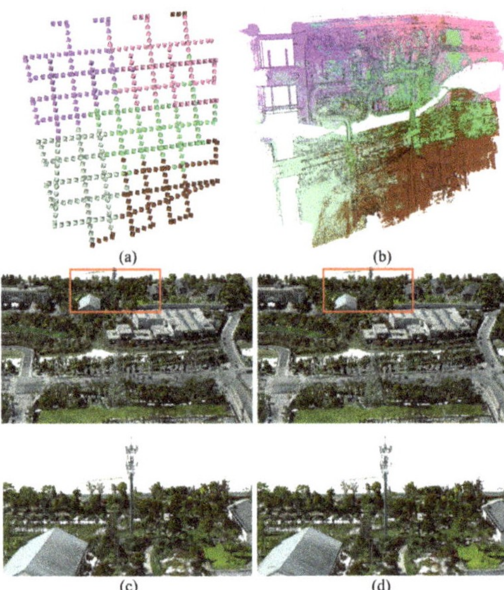

Figure 4. (**a**) Camera views of case "B5 West" divided into 5 clusters. (**b**) The final fused point cloud with points from different clusters visualized in different colors. (**c**) The fused point cloud without clustering. (**d**) The clustering-based point cloud fusion, with magnified rectangle regions to show almost no difference from (**c**).

3.4. Detail-Aware Surface Meshing

After obtaining a dense point cloud fused with multiple depth maps, we can reconstruct a surface mesh from this point cloud using Delaunay triangulation. Conventional Delaunay triangulation usually has difficulty in reconstructing geometric details such as thin structures or rough surfaces, due to its sensitivity to the noisy points that easily drones out surface details. One straightforward idea to handle thin structures is to extract 3D curves from multi-view image edges, and generate mesh from the tetrahedra topologized by both points and curves, which is used in [57]. However, since detailed structures like rough surfaces cannot be represented by curves, these methods are designed to better handle line structures specially. In this subsection, we propose a more general visibility-based Delaunay triangulation method for meshing dense point cloud, which pays more attention to the reconstruction of thin structures and rough surfaces by improving the visibility constraint to further eliminate the impact of noisy points, and using point density as a new density constraint to better preserve detailed geometry.

Each point P in the dense point cloud $\mathbf{P}$ contains the set of image views from which it has been triangulated and visible. 3D Delaunay triangulation is applied to these points to build tetrahedra $\mathbf{T}$. Then, the tetrahedra are labeled inside or outside the surface through an energy minimization framework, with the labeling denoted by $\mathbf{L}$. We follow the previous methods [58,59] by using MRF approach to solve the tetrahedron binary labeling problem. Here, we use a graph-cuts framework similar to [58] to set up the graph of tetrahedra. Denote the directed graph as $\mathbf{G} = (\mathbf{T}, \mathbf{F})$, where each node $\tau \in \mathbf{T}$ is a tetrahedron, and each edge $f \in \mathbf{F}$ is the triangular facet shared by two neighboring tetrahedra. For each tetrahedron $\tau \in \mathbf{T}$, $\mathbf{L}(\tau) \in \{inner, outer\}$. The energy function for tetrahedron labeling problem is defined as:

$$E(\mathbf{T}, \mathbf{F}, \mathbf{L}) = \sum_{\tau \in \mathbf{T}} E_{data}(\tau, \mathbf{L}(\tau)) + \sum_{f \in \mathbf{F}} E_{smooth}(f, \mathbf{L}(\tau), \mathbf{L}(\tau')), \tag{3}$$

where E_{data} is the data term for tetrahedron τ, and E_{smooth} is the smooth term for facet f which is shared by two neighboring tetrahedra (τ, τ'). All the data terms and smooth terms are initialized to 0. As shown in Figure 5a, for each point $P \in \mathbf{P}$ and one of its visible image view $\mathbf{I}_i$, a line of sight shoots from the camera center C_i of $\mathbf{I}_i$ to P, and intersects with a number of facets in $\mathbf{F}$. For the tetrahedron τ_v that C_i lies in, we consider it more likely to be outside the surface, and penalize its data term for *inner* case. For the tetrahedron τ_p that contains P and intersects with the extended line of sight, we consider it inside the surface, and penalize its data term for *outer* case. For each facet f_i intersected with the line of sight, we consider it less likely to be shared by two tetrahedra with different labels, and penalize its smooth term for the case of different labels. Therefore, for each shooting of line of sight, we follow the strategies discussed above to accumulate the data terms and smooth terms as follows:

$$\begin{aligned}
E_{data}(\tau_v, inner) &+ = \alpha_v \\
E_{data}(\tau_p, outer) &+ = \alpha_v \\
E_{smooth}(f_i, inner, outer) &+ = \omega_d(f_i)\omega_v(f_i)\omega_q(f_i)\alpha_v \\
E_{smooth}(f_i, outer, inner) &+ = \omega_d(f_i)\omega_v(f_i)\omega_q(f_i)\alpha_v,
\end{aligned} \tag{4}$$

where α_v is a constant parameter, which we set to 1 in our experiments. ω_d, ω_v and ω_q are the density weight, visibility weight and quality weight respectively. Labatut et al. [58] only use visibility weight and quality weight, which has limitation in preserving geometric details. In comparison, we propose this novel density weight to enforce the accuracy of rough surface geometry, considering the output surface should be closer to where the point cloud has denser spacing, while the sparse points are more likely to be outliers. The density weight is computed for each facet $f \in \mathbf{F}$ by:

$$\omega_d(f) = 1 - \lambda_d \exp\left(-\frac{V(f)^2}{\sigma_d{}^2}\right), \tag{5}$$

where $V(f)$ is the total edge length of f divided by the total number visible image views for vertices of f, which encourage the facet to be denser in spacing and have more sufficient visible images to be reliable. The value of σ_d is set according to the distribution of $\{V(f)|f \in \mathbf{F}\}$, which we set to be $1/4$-order minimum of $\{V(f)\}$. λ_d controls the the influence of density weight, which is set to 0.8 in our experiments. Affected by this weight, the surface will tend to appear in denser point regions with more visible image view supports, which is helpful to preserving rough surface details.

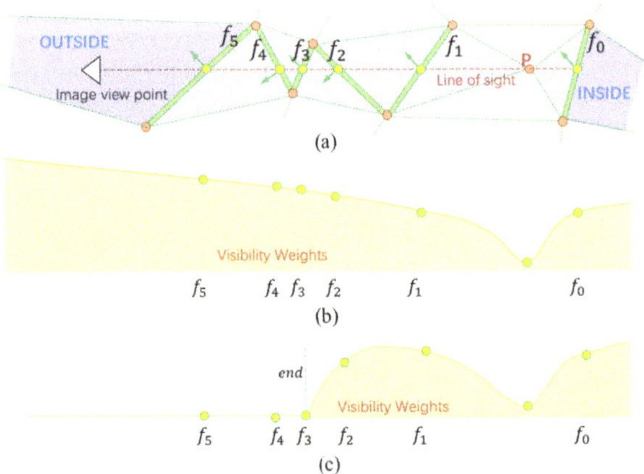

Figure 5. Illustration of our visibility weight strategy: (**a**) A line of sight from the camera center of an image view traverses a sequence of tetrahedra to a 3D point to accumulate the visibility weighted smooth terms. (**b**) The visibility weights proposed by [58]. (**c**) The visibility weights by our strategy, which are stopped by the end facet.

As in [58], visibility weight is used to penalize the visibility conflicts of dense points. Labatut et al. define the visibility weight as:

$$\omega_v(f) = 1 - \exp\left(-\frac{D(f)^2}{\sigma_v^2}\right),\tag{6}$$

where $D(f)$ is the distance between the intersection of f with the line of sight and the point P, which penalizes the facet far from P to appear in the final surface. However, considering that noisy points may introduce the incorrect accumulations of visibility weights along the line of sight, which might lead to the loss of thin structures. To better handle the influence of noisy points, we propose an intersection stop mechanism for smooth term accumulation. Denoting the distance between C_i and P as $\hat{D}$, when a facet f_j intersected with the line of sight satisfies the two conditions which are $V(f_j) > \sigma_d$ and $D(f) > (1 - S(P))\hat{D}$, this facet will be the end of the intersection process, and the facets left to be intersected will be ignored. Here $S(P)$ is the score of uncertainty of P computed in Section 3.3. In this way, the incorrect accumulation caused by noisy points will be relieved by reliable facets with sufficient visible image views and dense vertex spacing, to better reconstruct details with thin structure.

As defined in [58], the quality weight $\omega_q(f) = (1 - \min(\cos\phi, \cos\varphi))$, where ϕ and φ are the angles between f_j and the circumspheres of the two neighboring tetrahedra τ and τ' respectively, which ensures the global surface quality by giving stronger smoothness connection between tetrahedra of better shape.

The defined energy function is finally solved by applying s-t cut on the graph to determine the binary labels, and the surface **S** is extracted from the labeled tetrahedra by collecting the triangular facets between two neighboring tetrahedra with different labels as the final mesh surface. Experiments of cases "B5 Tower" and "ZJU Fan" in Figure 6 show that our detail-aware surface meshing approach can reconstruct more accurate surface mesh than the approach of [58], with more complete details preserved, such as the thin structures of tower antennas and fan railings, and the rough surface details of the fans, by using our improved visibility-based Delaunay triangulation.

After surface mesh of the scene being reconstructed, we can use the input multi-view images with poses to perform texture mapping for the reconstructed surface mesh. We follow the approach in [60] to perform a multi-view texture mapping to get a final textured 3D model.

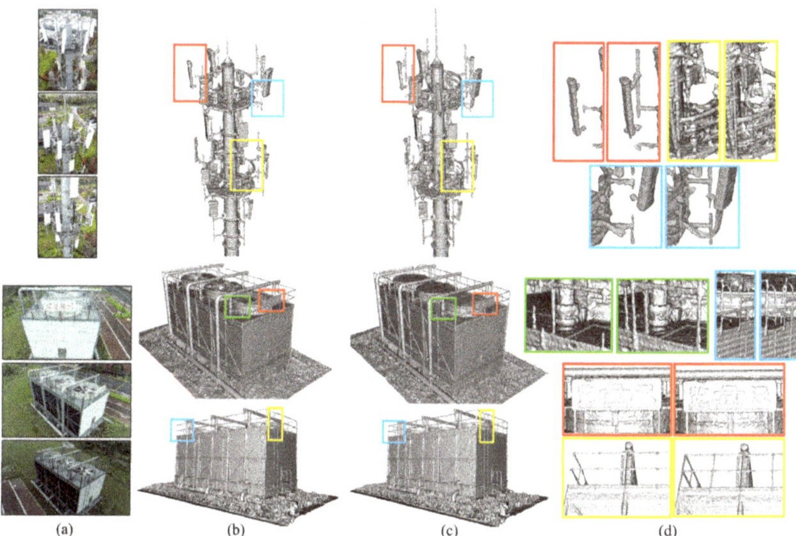

Figure 6. (**a**) Three representative image views for each case of "B5 Tower" and "ZJU Fan". (**b**) The reconstructed surface models of the two cases by [58]. (**c**) The surface models generated by our meshing approach. (**d**) Comparisons of the details in the rectangles of (**b**,**c**), which shows the effectiveness of our detail-aware surface meshing.

4. Results

In this section, we exhibit quantitative and qualitative comparisons of our DP-MVS framework with other SOTA methods on several experimental cases. We also report the time consumption on the stages of depth estimation, fusion and meshing of different methods to show the runtime efficiency of our method. All the cases were captured by DJI PHANTOM 4 RTK UAV, except for the case "Qiaoxi Street" that was recorded by a Huawei Mate 30 mobile phone. The image resolution is 4864 × 3648 for DJI PHANTOM 4 RTK and 2736 × 3648 for Huawei Mate 30.

4.1. Qualitative Evaluation

We first give the qualitative comparisons of our surface reconstruction method with other SOTA methods implemented by third party source libraries including OpenMVS [56] and COLMAP [26]. For fair comparisons of surface reconstruction, we run OpenMVS and COLMAP based on our SfM input. Figure 7 shows the reconstruction results of all the methods on 5 cases "B5 West", "B5 Tower", "B5 Wire", "Qiaoxi Street", "ZJU Fan", each of which contains some thin objects or detailed structures. For fairness to other methods, we turned off the mesh optimization process when experimenting with OpenMVS. From the experimental results we can see that our DP-MVS approach performs better than the other methods in the finally generated 3D models, especially in those regions which contain rough surface structures and thin structures, which validates the effectiveness of our DP-MVS method. As shown in the rectangle regions of Figure 7, the geometric details of the rough surface structures of the buildings, and the thin structures of the tower antennas and fan railings are better reconstructed by our detail preserving depth map estimation and detail-aware surface meshing, compared to other SOTA methods.

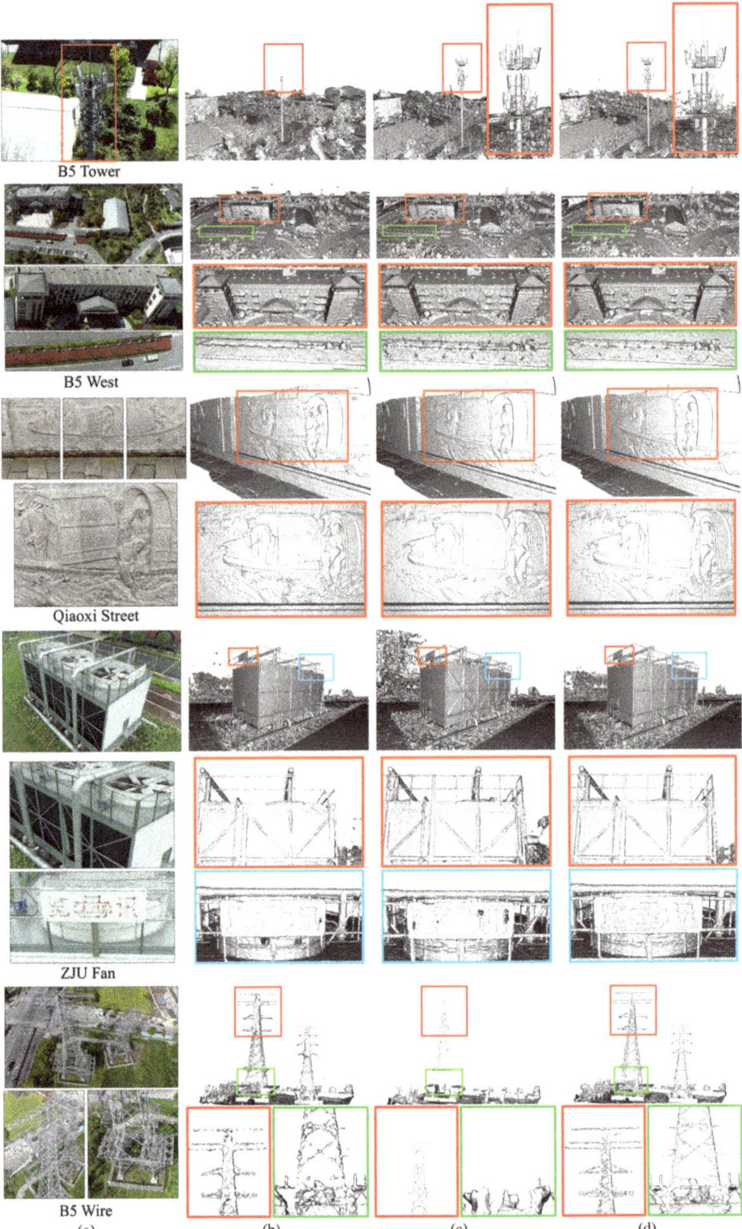

Figure 7. Comparison of our DP-MVS pipeline with other SOTA methods on 5 cases: "B5 Tower", "B5 West", "Qiaoxi Street", "ZJU Fan" and "B5 Wire". (**a**) Some representative source images of each case. (**b**) The reconstructed surface models of OpenMVS [56]. (**c**) The reconstruction results of COLMAP [26]. (**d**) The reconstructed 3D models by our DP-MVS. Some detailed structures are highlighted in the rectangles to show the effectiveness of our reconstruction pipeline.

We also qualitatively compare our surface reconstruction method with the third party software RealityCapture v1.2 by Capturing Reality (www.capturingreality.com accessed on 3 September 2021) on the cases "ZJU Fan" and "B5 West", as shown in Figure 8. From the reconstructed models, we can see that RealityCapture loses geometric details especially in thin structures. In comparison, our DP-MVS performs better in both reconstruction completeness and geometric details, as can be seen in the highlighted rectangle regions.

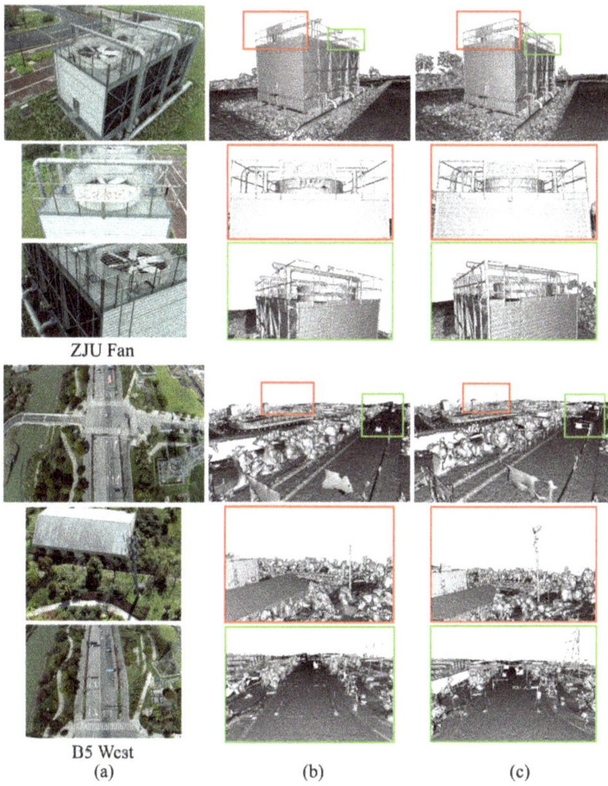

Figure 8. Comparison of our method with the third party software RealityCapture, on the cases "ZJU Fan" and "B5 West". (**a**) Some representative source images of each case. (**b**) The reconstructed 3D models using RealityCapture. (**c**) The reconstructed 3D models by our DP-MVS. Some detailed structures are highlighted in the rectangles to show the effectiveness of our proposed method.

4.2. Quantitative Evaluation

Table 1 provides the quantitative comparison of our DP-MVS system with other SOTA methods on the case "ZJU CCE" which captures an academic building and a clock tower occupying an area of almost 3000 m^2. The ground truth (GT) 3D model of "ZJU CCE" was captured by laser scanning for accuracy evaluation on both Root Mean Squared Error (RMSE) and Mean Absolute Error (MAE). For model accuracy evaluation, we use CloudCompare (http://cloudcompare.org accessed on 3 September 2021) to compare the reconstructed meshes with GT: we align the mesh with GT using manual rough registration followed by ICP fine registration, then evaluate the mesh-point-to-GT-plane distances. This routine is achieved with CloudCompare's built-in functions. We can see from the model accuracy evaluation in Table 1 that compared to OpenMVS and COLMAP, our DP-MVS system reconstructs the surface model of the scene with a centimeter-level accuracy, which turns out to be the best in both RMSE and MAE. Also, from the comparison of the finally reconstructed 3D models with other methods on "ZJU CCE" in Figure 9, we can see that our approach preserves better geometric details than other methods, especially for thin structures as highlighted in the rectangles.

Table 1. The Root Mean Squared Error (RMSE) and Mean Absolute Error (MAE) of the reconstruction results by our system, OpenMVS [56] and COLMAP [26] on the case "ZJU CCE", whose GT model is used as reference for error computation. We use bold format to highlight the smallest errors among all the methods.

Case	Measures [cm]	OpenMVS	COLMAP	DP-MVS
ZJU CCE	RMSE/MAE	4.337/2.084	4.476/2.275	**3.698/1.859**

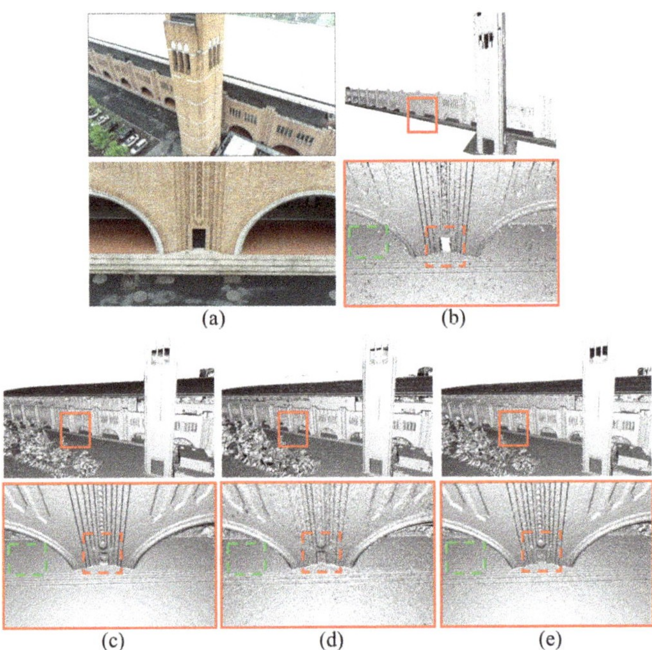

Figure 9. Comparison of our method with other SOTA methods on the case "ZJU CCE", whose true 3D model is captured by a 3D laser scanner as GT. (**a**) Some representative source images. (**b**) The GT model captured by the 3D laser scanner. (**c**) The reconstructed surface model of Open-MVS [56]. (**d**) The reconstruction result of COLMAP [26]. (**e**) The reconstructed 3D model by our DP-MVS. Some detailed structures are highlighted in the rectangles to show the effectiveness of our reconstruction method.

We further evaluate our fused point clouds on the high resolution multi-view datasets of ETH3D benchmark [61]. Table 2 lists the F1-score, accuracy and completeness of the point clouds estimated by ACMM [50], OpenMVS, COLMAP and DP-MVS. ACMM obtains higher accuracy than our DP-MVS, because it performs depth map estimation with geometric consistency guidance twice and final median filter with multiple scales to suppress depth noises, but its detailed structures are also lost. OpenMVS generates point clouds with more noise and higher redundancy, resulting in lower F1-score and accuracy. COLMAP achieves higher accuracy at the cost of lower completeness by filtering the points with low confidence and large reprojection error. In comparison, our proposed system outperforms other methods in terms of F1-score and completeness because of our detail preserving depth estimation with even distribution of four-neighboring hypotheses.

Table 2. Evaluation on high resolution multi-view datasets of ETH3D benchmark. It shows F1-score, accuracy and completeness at different error levels (2 cm and 10 cm), with bold format highlighting the best evaluation among all the methods including ACMM [50], OpenMVS [56], COLMAP [26] and DP-MVS.

Dataset	Error	Measure	ACMM	OpenMVS	COLMAP	DP-MVS
Training dataset	2 cm	F1-score	78.86	76.15	67.66	**80.11**
		accuracy	90.67	78.44	**91.85**	83.56
		completeness	70.42	74.92	55.13	**77.56**
	10 cm	F1-score	91.70	92.51	87.61	**94.77**
		accuracy	98.12	95.75	**98.75**	95.95
		completeness	86.40	89.84	79.47	**93.72**
Test dataset	2 cm	F1-score	80.78	79.77	73.01	**83.11**
		accuracy	90.65	81.98	**91.97**	84.05
		completeness	74.34	78.54	62.98	**82.73**
	10 cm	F1-score	92.96	92.86	90.40	**95.68**
		accuracy	98.05	95.48	**98.25**	95.55
		completeness	88.77	90.75	84.54	**95.85**

4.3. Time Statistics

Table 3 gives the time consumption of the depth map estimation, fusion and meshing of our DP-MVS and other SOTA methods. The experiments are conducted on a server platform with a 14-Core Intel Xeon E5-2680 CPU, 8 GeForce 1080Ti GPUs, and 500 GB memory. It can be seen that our pipeline is the most efficient on high-resolution images, which is more than twice faster than OpenMVS and COLMAP. Note that the time consumption of our depth map fusion step is extremely more efficient because of our memory-efficient fusion strategy as mentioned in Section 3.3, which also verifies the practical usefulness of our cluster-based depth map fusion strategy for large-scale scenes with multiple high-resolution images as input, for which other SOTA works might have both time and memory limitations.

Table 3. We report detailed time consumptions of our DP-MVS system and other SOTA methods including OpenMVS [56] and COLMAP [26] in all the steps of cases "B5 Tower", "B5 West", "Qiaoxi Street" and "B5 Wire". All the time consumptions are calculated by minutes, with bold format highlighting the fastest time of all the methods.

Case	#Images	Stages	OpenMVS	COLMAP	DP-MVS
B5 Tower	1163	Depth Estimation	757.57	135.795	**54.207**
		Fusion	99.55	1810.244	**48.139**
		Meshing	132.65	**125.49**	158.803
		Total	989.77	2071.529	**261.149**
B5 West	513	Depth Estimation	312.56	148.484	**26.833**
		Fusion	47.53	214.713	**8.413**
		Meshing	163.62	64.92	**63.191**
		Total	523.71	428.117	**98.437**
Qiaoxi Street	305	Depth Estimation	168.87	69.313	**11.318**
		Fusion	18.33	42.538	**3.862**
		Meshing	156	**54.1**	56.227
		Total	343.2	165.951	**71.407**
B5 Wire	1251	Depth Estimation	728.58	224.183	**40.826**
		Fusion	99.57	1168.048	**56.312**
		Meshing	220.62	165.4	**130.764**
		Total	1048.77	1557.631	**227.902**

5. Discussion

Our method reconstructs 3D models with too dense faces even in the planar regions, which results in oversampled mesh topology that gives pressure to both storage and rendering. How to further optimize and simplify the reconstructed 3D models with more optimal and more compact topology is a problem worth studying in future. Besides, our DP-MVS method focuses on how to preserve detailed structures, but does not consider too much about how to preserve good surface structures for textureless regions or non-lambertian surfaces, which are as usual as detailed structures in the natural scenes. How to jointly consider and handle these problems to develop a more powerful multi-view reconstruction strategy remains to be our future work.

6. Conclusions

In this work, we propose a detail preserving large-scale scene reconstruction pipeline called DP-MVS. We first present a detail preserving multi-view stereo method to generate rich detailed structures such as thin objects in the estimated depth maps. Then, a cluster-based depth map fusion method is proposed to handle large-scale high-resolution images with limited memory. Moreover, we alter the conventional Delaunay triangulation method by imposing new visibility constraint and density constraint to extract complete detailed geometry. The effectiveness of the proposed DP-MVS method for large-scale scene reconstruction is validated in our experiments.

Author Contributions: Conceptualization, L.Z. and Z.Z.; methodology, L.Z., Z.Z. and H.J.; software, L.Z., Z.Z. and H.S.; validation, H.S.; formal analysis, G.Z.; investigation, L.Z.; resources, G.Z. and H.B.; data curation, L.Z.; writing—original draft preparation, L.Z., H.J. and G.Z.; writing—review and editing, H.J. and G.Z.; visualization, Z.Z.; supervision, G.Z. and H.B.; project administration, H.J.; funding acquisition, G.Z. and H.B. All authors have read and agreed to the published version of the manuscript.

Funding: This research was partially supported by the National Key Research and Development Program of China under Grant 2020YFF0304300, and the National Natural Science Foundation of China (No. 61822310).

Institutional Review Board Statement: Not applicable.

Informed Consent Statement: Not applicable.

Data Availability Statement: Not applicable.

Acknowledgments: The authors wish to thank Jinle Ke, Chongshan Sheng, Li Zhou and Yi Su for their kind helps in the experiments and the development of the 3D model visualization system.

Conflicts of Interest: The authors declare no conflict of interest.

References

1. Seitz, S.M.; Curless, B.; Diebel, J.; Scharstein, D.; Szeliski, R. A Comparison and Evaluation of Multi-View Stereo Reconstruction Algorithms. In Proceedings of the IEEE Conference on Computer Vision and Pattern Recognition, New York, NY, USA, 17–22 June 2006; Volume 1, pp. 519–528.
2. Seitz, S.M.; Dyer, C.R. Photorealistic scene reconstruction by voxel coloring. *Int. J. Comput. Vis.* **1999**, *35*, 151–173. [CrossRef]
3. Vogiatzis, G.; Esteban, C.H.; Torr, P.H.; Cipolla, R. Multiview stereo via volumetric graph-cuts and occlusion robust photo-consistency. *IEEE Trans. Pattern Anal. Mach. Intell.* **2007**, *29*, 2241–2246. [CrossRef]
4. Sinha, S.N.; Mordohai, P.; Pollefeys, M. Multi-view stereo via graph cuts on the dual of an adaptive tetrahedral mesh. In Proceedings of the IEEE International Conference on Computer Vision, Rio de Janeiro, Brazil, 14–21 October 2007; pp. 1–8.
5. Faugeras, O.; Keriven, R. Variational Principles, Surface Evolution, PDE's, Level Set Methods and the Stereo Problem. *IEEE Trans. Image Process.* **1998**, *7*, 336–344. [CrossRef] [PubMed]
6. Esteban, C.H.; Schmitt, F. Silhouette and stereo fusion for 3D object modeling. *Comput. Vis. Image Underst.* **2004**, *96*, 367–392. [CrossRef]

7. Hiep, V.H.; Keriven, R.; Labatut, P.; Pons, J.P. Towards high-resolution large-scale multi-view stereo. In Proceedings of the IEEE Conference on Computer Vision and Pattern Recognition, Miami, FL, USA, 20–25 June 2009; pp. 1430–1437.
8. Li, S.; Siu, S.Y.; Fang, T.; Quan, L. Efficient multi-view surface refinement with adaptive resolution control. In Proceedings of the European Conference on Computer Vision, Amsterdam, The Netherlands, 11–14 October 2016; Springer: Cham, Switzerland, 2016; pp. 349–364.
9. Romanoni, A.; Matteucci, M. Mesh-based camera pairs selection and occlusion-aware masking for mesh refinement. *Pattern Recognit. Lett.* **2019**, *125*, 364–372. [CrossRef]
10. Cremers, D.; Kolev, K. Multiview stereo and silhouette consistency via convex functionals over convex domains. *IEEE Trans. Pattern Anal. Mach. Intell.* **2010**, *33*, 1161–1174. [CrossRef]
11. Lhuillier, M.; Quan, L. A quasi-dense approach to surface reconstruction from uncalibrated images. *IEEE Trans. Pattern Anal. Mach. Intell.* **2005**, *27*, 418–433. [CrossRef] [PubMed]
12. Goesele, M.; Snavely, N.; Curless, B.; Hoppe, H.; Seitz, S.M. Multi-view Stereo for Community Photo Collections. In Proceedings of the IEEE International Conference on Computer Vision, Rio de Janeiro, Brazil, 14–21 October 2007; pp. 1–8.
13. Lowe, D.G. Distinctive image features from scale-invariant keypoints. *Int. J. Comput. Vis.* **2004**, *60*, 91–110. [CrossRef]
14. Furukawa, Y.; Ponce, J. Accurate, dense, and robust multiview stereopsis. *IEEE Trans. Pattern Anal. Mach. Intell.* **2009**, *32*, 1362–1376. [CrossRef] [PubMed]
15. Wu, T.P.; Yeung, S.K.; Jia, J.; Tang, C.K. Quasi-dense 3D reconstruction using tensor-based multiview stereo. In Proceedings of the 2010 IEEE Computer Society Conference on Computer Vision and Pattern Recognition, San Francisco, CA, USA, 13–18 June 2010; pp. 1482–1489.
16. Furukawa, Y.; Curless, B.; Seitz, S.M.; Szeliski, R. Towards internet-scale multi-view stereo. In Proceedings of the 2010 IEEE Computer Society Conference on Computer Vision and Pattern Recognition, San Francisco, CA, USA, 13–18 June 2010; pp. 1434–1441.
17. Strecha, C.; Fransens, R.; Van Gool, L. Combined depth and outlier estimation in multi-view stereo. In Proceedings of the IEEE Conference on Computer Vision and Pattern Recognition, New York, NY, USA, 17–22 June 2006; Volume 2, pp. 2394–2401.
18. Goesele, M.; Curless, B.; Seitz, S.M. Multi-view stereo revisited. In Proceedings of the IEEE Conference on Computer Vision and Pattern Recognition, New York, NY, USA, 17–22 June 2006; Volume 2, pp. 2402–2409.
19. Merrell, P.; Akbarzadeh, A.; Liang, W.; Mordohai, P.; Frahm, J.M.; Yang, R.; Nister, D.; Pollefeys, M. Real-time Visibility-Based Fusion of Depth Maps. In Proceedings of the IEEE International Conference on Computer Vision, Rio de Janeiro, Brazil, 14–21 October 2007; pp. 1–8.
20. Zach, C.; Pock, T.; Bischof, H. A globally optimal algorithm for robust TV-L1 range image integration. In Proceedings of the IEEE International Conference on Computer Vision, Rio de Janeiro, Brazil, 14–21 October 2007; pp. 1–8.
21. Kuhn, A.; Mayer, H.; Hirschmüller, H.; Scharstein, D. A TV prior for high-quality local multi-view stereo reconstruction. In Proceedings of the International Conference on 3D Vision, Tokyo, Japan, 8–11 December 2014; Volume 1, pp. 65–72.
22. Liu, Y.; Cao, X.; Dai, Q.; Xu, W. Continuous depth estimation for multi-view stereo. In Proceedings of the IEEE Conference on Computer Vision and Pattern Recognition, Miami, FL, USA, 20–25 June 2009; pp. 2121–2128.
23. Li, Z.; Zuo, W.; Wang, Z.; Zhang, L. Confidence-Based Large-Scale Dense Multi-View Stereo. *IEEE Trans. Image Process.* **2020**, *29*, 7176–7191. [CrossRef]
24. Bradley, D.; Boubekeur, T.; Heidrich, W. Accurate multi-view reconstruction using robust binocular stereo and surface meshing. In Proceedings of the IEEE Conference on Computer Vision and Pattern Recognition, Anchorage, AK, USA, 23–28 June 2008; pp. 1–8.
25. Campbell, N.D.; Vogiatzis, G.; Hernández, C.; Cipolla, R. Using multiple hypotheses to improve depth-maps for multi-view stereo. In Proceedings of the European Conference on Computer Vision, Marseille, France, 12–18 October 2008; Springer: Cham, Switzerland, 2008; pp. 766–779.
26. Schönberger, J.L.; Zheng, E.; Pollefeys, M.; Frahm, J.M. Pixelwise View Selection for Unstructured Multi-View Stereo. In Proceedings of the European Conference on Computer Vision, Amsterdam, The Netherlands, 11–14 October 2016; Springer: Cham, Switzerland, 2016; pp. 501–518.
27. Shen, S. Accurate multiple view 3D reconstruction using patch-based stereo for large-scale scenes. *IEEE Trans. Image Process.* **2013**, *22*, 1901–1914. [CrossRef]
28. Tola, E.; Strecha, C.; Fua, P. Efficient large-scale multi-view stereo for ultra high-resolution image sets. *Mach. Vis. Appl.* **2012**, *23*, 903–920. [CrossRef]
29. Tola, E.; Lepetit, V.; Fua, P. DAISY: An efficient dense descriptor applied to wide-baseline stereo. *IEEE Trans. Pattern Anal. Mach. Intell.* **2009**, *32*, 815–830. [CrossRef] [PubMed]
30. Li, J.; Li, E.; Chen, Y.; Xu, L.; Zhang, Y. Bundled depth-map merging for multi-view stereo. In Proceedings of the 2010 IEEE Computer Society Conference on Computer Vision and Pattern Recognition, San Francisco, CA, USA, 13–18 June 2010; pp. 2769–2776.

31. Xue, J.; Chen, X.; Hui, Y. Efficient Multi-View 3D Dense Matching for Large-Scale Aerial Images Using a Divide-and-Conquer Scheme. In Proceedings of the Chinese Automation Congress, Xi'an, China, 30 November–2 December 2018; pp. 2610–2615.
32. Mostegel, C.; Fraundorfer, F.; Bischof, H. Prioritized multi-view stereo depth map generation using confidence prediction. *ISPRS J. Photogramm. Remote Sens.* **2018**, *143*, 167–180. [CrossRef]
33. Wei, M.; Yan, Q.; Luo, F.; Song, C.; Xiao, C. Joint bilateral propagation upsampling for unstructured multi-view stereo. *Vis. Comput.* **2019**, *35*, 797–809. [CrossRef]
34. Wang, Y.; Guan, T.; Chen, Z.; Luo, Y.; Luo, K.; Ju, L. Mesh-Guided Multi-View Stereo With Pyramid Architecture. In Proceedings of the IEEE Conference on Computer Vision and Pattern Recognition, Seattle, WA, USA, 13–19 June 2020; pp. 2039–2048.
35. Yao, Y.; Luo, Z.; Li, S.; Fang, T.; Quan, L. MVSNet: Depth Inference for Unstructured Multi-view Stereo. In Proceedings of the European Conference on Computer Vision, Munich, Germany, 8–14 September 2018; pp. 767–783.
36. Yao, Y.; Luo, Z.; Li, S.; Shen, T.; Fang, T.; Quan, L. Recurrent mvsnet for high-resolution multi-view stereo depth inference. In Proceedings of the IEEE Conference on Computer Vision and Pattern Recognition, Long Beach, CA, USA, 15–20 June 2019; pp. 5525–5534.
37. Chen, R.; Han, S.; Xu, J.; Su, H. Point-based multi-view stereo network. In Proceedings of the IEEE International Conference on Computer Vision, Seoul, Korea, 27 October–2 November 2019; pp. 1538–1547.
38. Luo, K.; Guan, T.; Ju, L.; Huang, H.; Luo, Y. P-MVSNet: Learning patch-wise matching confidence aggregation for multi-view stereo. In Proceedings of the IEEE International Conference on Computer Vision, Seoul, Korea, 27 October–2 November 2019; pp. 10452–10461.
39. Gu, X.; Fan, Z.; Zhu, S.; Dai, Z.; Tan, F.; Tan, P. Cascade cost volume for high-resolution multi-view stereo and stereo matching. In Proceedings of the IEEE Conference on Computer Vision and Pattern Recognition, Seattle, WA, USA, 13–19 June 2020; pp. 2495–2504.
40. Kuhn, A.; Sormann, C.; Rossi, M.; Erdler, O.; Fraundorfer, F. DeepC-MVS: Deep confidence prediction for multi-view stereo reconstruction. In Proceedings of the International Conference on 3D Vision, Fukuoka, Japan, 25–28 November 2020; pp. 404–413.
41. Yang, X.; Zhou, L.; Jiang, H.; Tang, Z.; Wang, Y.; Bao, H.; Zhang, G. Mobile3DRecon: Real-time Monocular 3D Reconstruction on a Mobile Phone. *IEEE Trans. Vis. Comput. Graph.* **2020**, *26*, 3446–3456. [CrossRef] [PubMed]
42. Yang, X.; Jiang, G. A Practical 3D Reconstruction Method for Weak Texture Scenes. *Remote Sens.* **2021**, *13*, 3103. [CrossRef]
43. Stathopoulou, E.K.; Battisti, R.; Cernea, D.; Remondino, F.; Georgopoulos, A. Semantically Derived Geometric Constraints for MVS Reconstruction of Textureless Areas. *Remote Sens.* **2021**, *13*, 1053. [CrossRef]
44. Yan, F.; Xia, E.; Li, Z.; Zhou, Z. Sampling-Based Path Planning for High-Quality Aerial 3D Reconstruction of Urban Scenes. *Remote Sens.* **2021**, *13*, 989. [CrossRef]
45. Liu, Y.; Cui, R.; Xie, K.; Gong, M.; Huang, H. Aerial Path Planning for Online Real-Time Exploration and Offline High-Quality Reconstruction of Large-Scale Urban Scenes. *ACM Trans. Graph.* **2021**, *40*, 226:1–226:16.
46. Pepe, M.; Fregonese, L.; Crocetto, N. Use of SfM-MVS approach to nadir and oblique images generated throught aerial cameras to build 2.5 D map and 3D models in urban areas. *Geocarto Int.* **2019**. [CrossRef]
47. Barnes, C.; Shechtman, E.; Finkelstein, A.; Goldman, D.B. PatchMatch: A randomized correspondence algorithm for structural image editing. *ACM Trans. Graph.* **2009**, *28*, 24. [CrossRef]
48. Zheng, E.; Dunn, E.; Jojic, V.; Frahm, J.M. Patchmatch based joint view selection and depthmap estimation. In Proceedings of the IEEE Conference on Computer Vision and Pattern Recognition, Columbus, OH, USA, 23–28 June 2014; pp. 1510–1517.
49. Galliani, S.; Lasinger, K.; Schindler, K. Massively parallel multiview stereopsis by surface normal diffusion. In Proceedings of the IEEE International Conference on Computer Vision, Santiago, Chile, 7–13 December 2015; pp. 873–881.
50. Xu, Q.; Tao, W. Multi-scale geometric consistency guided multi-view stereo. In Proceedings of the IEEE Conference on Computer Vision and Pattern Recognition, Long Beach, CA, USA, 15–20 June 2019; pp. 5483–5492.
51. Xu, Q.; Tao, W. Multi-view stereo with asymmetric checkerboard propagation and multi-hypothesis joint view selection. *arXiv* **2018**, arXiv:1805.07920.
52. Xu, Q.; Tao, W. Planar prior assisted patchmatch multi-view stereo. *Proc. AAAI Conf. Artif. Intell.* **2020**, *34*, 12516–12523. [CrossRef]
53. Romanoni, A.; Matteucci, M. TAPA-MVS: Textureless-aware patchmatch multi-view stereo. In Proceedings of the IEEE International Conference on Computer Vision, Seoul, Korea, 27 October–2 November 2019; pp. 10413–10422.
54. Xu, Z.; Liu, Y.; Shi, X.; Wang, Y.; Zheng, Y. MARMVS: Matching Ambiguity Reduced Multiple View Stereo for Efficient Large Scale Scene Reconstruction. In Proceedings of the IEEE Conference on Computer Vision and Pattern Recognition, Seattle, WA, USA, 13–19 June 2020; pp. 5981–5990.
55. Schönberger, J.L.; Frahm, J.M. Structure-from-Motion Revisited. In Proceedings of the IEEE Conference on Computer Vision and Pattern Recognition, Las Vegas, NV, USA, 27–30 June 2016; pp. 4104–4113.
56. Cernea, D. OpenMVS: Multi-View Stereo Reconstruction Library. Available online: https://cdcseacave.github.io/openMVS (accessed on 3 September 2021).
57. Li, S.; Yao, Y.; Fang, T.; Quan, L. Reconstructing thin structures of manifold surfaces by integrating spatial curves. In Proceedings of the IEEE Conference on Computer Vision and Pattern Recognition, Salt Lake City, UT, USA, 18–23 June 2018; pp. 2887–2896.
58. Labatut, P.; Pons, J.P.; Keriven, R. Robust and efficient surface reconstruction from range data. *Comput. Graph. Forum* **2009**, *28*, 2275–2290. [CrossRef]

59. Vu, H.H.; Labatut, P.; Pons, J.P.; Keriven, R. High accuracy and visibility-consistent dense multiview stereo. *IEEE Trans. Pattern Anal. Mach. Intell.* **2011**, *34*, 889–901. [CrossRef] [PubMed]
60. Waechter, M.; Moehrle, N.; Goesele, M. Let there be color! Large-scale texturing of 3D reconstructions. In Proceedings of the European Conference on Computer Vision, Zurich, Switzerland, 6–12 September 2014; pp. 836–850.
61. Schops, T.; Schonberger, J.L.; Galliani, S.; Sattler, T.; Schindler, K.; Pollefeys, M.; Geiger, A. A multi-view stereo benchmark with high-resolution images and multi-camera videos. In Proceedings of the IEEE Conference on Computer Vision and Pattern Recognition, Honolulu, HI, USA, 21–26 July 2017; pp. 3260–3269.

Article

Automatic Reconstruction of Building Façade Model from Photogrammetric Mesh Model

Yunsheng Zhang [1,2], Chi Zhang [2], Siyang Chen [3] and Xueye Chen [1,*]

[1] Key Laboratory of Urban Land Resources Monitoring and Simulation, Ministry of Land and Resources, Shenzhen 518034, China; Zhangys@csu.edu.cn

[2] School of Geoscience and Info-Physics, Central South University, Changsha 410083, China; zhangchi_csu@csu.edu.cn

[3] Department of Earth Observation Science, Faculty of Geo-Information and Earth Observation (ITC), University of Twente, 7514 AE Enschede, The Netherlands; s.chen-3@utwente.nl

* Correspondence: xueye31@163.com

Abstract: Three-dimensional (3D) building façade model reconstruction is of great significance in urban applications and real-world visualization. This paper presents a newly developed method for automatically generating a 3D regular building façade model from the photogrammetric mesh model. To this end, the contour is tracked on irregular triangulation, and then the local contour tree method based on the topological relationship is employed to represent the topological structure of the photogrammetric mesh model. Subsequently, the segmented contour groups are found by analyzing the topological relationship of the contours, and the original mesh model is divided into various components from bottom to top through the iteration process. After that, each component is iteratively and robustly abstracted into cuboids. Finally, the parameters of each cuboid are adjusted to be close to the original mesh model, and a lightweight polygonal mesh model is taken from the adjusted cuboid. Typical buildings and a whole scene of photogrammetric mesh models are exploited to assess the proposed method quantitatively and qualitatively. The obtained results reveal that the proposed method can derive a regular façade model from a photogrammetric mesh model with a certain accuracy.

Keywords: photogrammetric mesh model; building façade; 3D reconstruction; least square fitting

Citation: Zhang, Y.; Zhang, C.; Chen, S.; Chen, X. Automatic Reconstruction of Building Façade Model from Photogrammetric Mesh Model. *Remote Sens.* **2021**, *13*, 3801. https://doi.org/10.3390/rs13193801

Academic Editors: Wanshou Jiang, San Jiang and Xiongwu Xiao

Received: 17 June 2021
Accepted: 18 September 2021
Published: 22 September 2021

Publisher's Note: MDPI stays neutral with regard to jurisdictional claims in published maps and institutional affiliations.

1. Introduction

The three-dimensional (3D) façade model of urban buildings plays a crucial role in many fields, including urban planning, solar radiation calculations, noise emission simulations, virtual reality, sustainable development research, and disaster simulation [1–3]. The automatic reconstruction of building façades has always been a significant research topic in the fields of photogrammetry and remote sensing, as well as computer vision and computer graphics; nevertheless, due to the intricacy of urban scenes, the automatic reconstruction of urban building façades is still a challenging task.

In past decades, a number of researchers have tried on the (semi-)automatic reconstruction of façade models for generating LoD3 building models [4]. Images and LiDAR (Light Detection and Ranging) point clouds are two common data used for façade model reconstruction. Several methodologies aiming at the automatic reconstruction of 3D façade models have been established in the past years. Xiao et al. [5] proposed a semi-automatic method to generate façade models along a street from multi-view street images. For this purpose, an ortho-rectified image was initially decomposed and structured into a directed acyclic graph of rectilinear elementary patches by considering architectural bilateral symmetry and repetitive patterns. Then each patch was enhanced by the depth from point clouds, which was derived from the results of structure-from-motion. Müller et al. [6] suggested an image-based façade reconstruction approach method by utilizing an image analysis algorithm to divide the façade into meaningful segments and combine the procedural modeling pipeline of shape grammars to ensure the regularity of the final reconstructed model. Sadeghi et al. [7] presented an approach for façade reconstruction from hand-held laser scanner data based on grammar. The method starts from using RANSAC method to extract façade points, and then protrusion, indentation, and wall points are detected by utilizing a density histogram. After that, façade elements are modeled by employing some rules. Edum-Fotwe et al. [8] proposed a façade reconstruction method from LiDAR data; the algorithm employed a top-down strategy to split the point cloud into surface-element rails in signed-distance-field, then completed the façade model reconstruction. Pu and Vosselman [9] contributed to an approach on integrating terrestrial laser points and images for façade reconstruction. The building façade's general structure from the plane in LiDAR point cloud data was discovered and established, and then the line feature in the images was employed to refine the model and to generate texture. These methods obtained promising results, but they have to utilize the images or point clouds from the terrestrial ground. However, the lower part of the façade is commonly enclosed by various types of vegetation, street signs, cars, and pedestrians, and the obtained point clouds usually suffer from a large number of missing data [10]. This issue may hinder the reconstruction of building façades. It is worth mentioning that TLS often acquires data only on the side of urban streets. The other façade data cannot be readily achieved, making hard to establish a comprehensive building façade model.

Along with the development of Unmanned Aerial Vehicle (UAV) and aerial oblique photogrammetry, it is possible to obtain a high-resolution façade image from UAV by an aerial oblique camera system, and then, a multi-view dense matching methodology is implemented to reconstruct and update the 3D model of urban buildings expressed by the photogrammetric mesh model. The automatic reconstructed model often contains millions of triangles, which brings an onerous burden for storage, web transferring, and visualization. Moreover, due to the problem of occlusion, repetitive texture, and transparent object, there are also some defects in the automatically generated mesh model, which could reduce the visual effects. Hence, some mesh editors such as DP-Modeler and OSketch [11,12] are developed to improve the mesh model by manual work. By this view, the main objective of this research work is to develop a method to reconstruct the regular façade model from the photogrammetry mesh model such that the structure of a single building is preserved. The reconstructed model is potentially employed for visual navigation, online visualization, solar energy estimation, etc.

The current façade modeling methods can be generally categorized into two major types. One is a data-driven method [13–16], while another is a model-driven method [17–19]. There are several data-driven approaches proposed to reconstruct façades model from Airborne-Laser Scanning (ALS) data [20]. The reconstruction is completed by vertically extruding the roof outline to the ground. Thereby, the key problem is the roof outline generating, which can be realized by edge-based methods [21], data clustering [22], region growing [23], model fitting [24], etc. Edge-based methodologies are susceptible to outliers and incomplete edges. The method of data clustering relies on the number of classes defined and the clustering center. The approach based on the region growing is usually influenced by the seed point selection. The RANSAC method is implemented in model fitting, which often results in unwanted false planes. Additionally, the accuracy of the reconstructed façade model based on the roof model boundary is susceptible to eaves. Wu et al. [25] proposed a graph-based method to reconstruct urban building models from ALS data. This method was basically constructed on the hierarchical analysis of the contours to gain the structure of the building, then a bipartite graph matching method was employed to obtain the correspondence between consecutive contours for subsequent surface modeling. The final model heavily relies on the contour's quality. If there exist some noise or artifacts in the point cloud as in the photogrammetric mesh model, the matching and surface modeling process in Ref. [25] would drop the quality of the final model. Thus, it cannot adapt to the under-study photogrammetric mesh model. For data-driven methods based on the ground data, regularity of symmetry is often detected in the source data, and then exploited to regularize the final model.

Façades usually exhibit strong structural regularities, such as piecewise planar segments, parallelism, and orthogonality lines. Generally, model-driven methods employ this prior information about the face structure to constrain the façade modeling. Nan et al. [26–29] generated building details by automatically fitting 3D templates on coarse building models with texture. To this end, the 3D templates were produced by employing a semi-automatic procedure, emerging a template construction tool. The boxes were directly fitted to the imperfect point cloud based on the Manhattan-World hypothesis, and then the best subset is selected to achieve reconstruction. Larfage et al. [17] proposed urban buildings reconstruction method by detecting and optimizing 3D blocks on a Digital Surface Model (DSM).

Since the mesh models based on the aerial oblique images often contain noise, herein, a model-driven approach is proposed. The façade of the under-study building is assumed to be composed of several cuboids. The photogrammetric mesh models are iteratively divided into various components from bottom to top by the segmented contour group. Subsequently, each component is fitted by a set level of cuboids, and then we will arrive at the final façade model.

The organization of the paper is as follows: In Section 2, the proposed method for façade modeling is explained described in some detail. In Section 3, the performance of our proposed method is evaluated through a scene of photogrammetric mesh model. In Section 4, some discussions are provided. Finally, the main conclusions are given (i.e., Section 5).

2. Methods

2.1. Overview of the Approach

Generally, a given scene of the photogrammetric mesh model can be classified into façade mesh models of individually single buildings and others. The main goal of the proposed method in the present study is to automatically produce a 3D regular building façade from the photogrammetric façade mesh model (hereafter noted as photogrammetric façade mesh). The workflow of the proposed approach is displayed in Figure 1. It mainly includes three parts in the following:

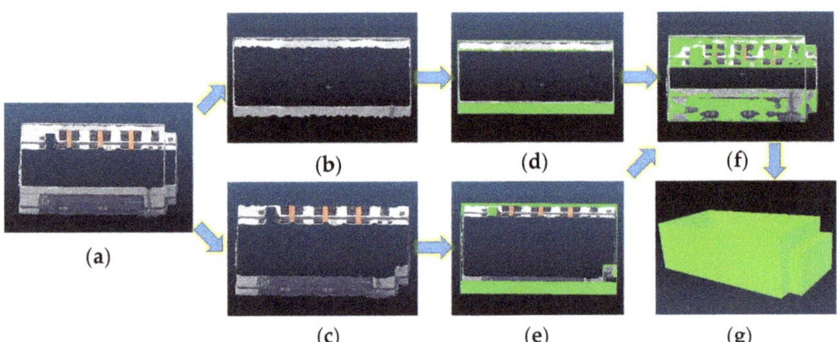

Figure 1. The workflow of the proposed approach: (**a**) input of the photogrammetric mesh model; (**b,c**) segmented components; (**d,e**) minimum circumscribed cuboids; (**f**) adjusting model; (**g**) final 3D building façade model.

(1) Firstly, the photogrammetric mesh model is decomposed into components based on the contour line. The closed contours on irregular triangular networks are tracked, and local contour trees are exploited to find the segmented contour groups by analyzing the topological relationship between the contours of the photogrammetric mesh model.

Subsequently, such a model is segmented from bottom to top into diverse components through an iterative process.

(2) The photogrammetric mesh model components are approximated by minimum circumscribed cuboids iteratively.

(3) The parameters of the cuboid model are adjusted by means of a least square algorithm to ensure the accuracy of the façade model.

2.2. Component Decomposition Based on Contours Analysis

Assume that the building façade is composed of several cuboids; hence, the first step is to recognize the façade component abstracted by a cuboid. To this end, the photogrammetric mesh model is divided into various parts by analyzing the topological relationship of contours, and then each component is distinctly reconstructed. Generally, the photogrammetric mesh model is segmented from bottom to top by a segmented contour pair.

2.2.1. Contour Segment Pair Generation

If point clouds are used, as in Ref. [25], a linear Triangulated Irregular Network (TIN) interpolation method has to be performed firstly to obtain TIN. In contrast, the photogrammetric mesh model in the presented study is represented by a continuous TIN, the contour line tracking is directly performed on the TIN exploitation of the original data to avoid the loss of accuracy of data interpolation [30]. For the contour lines tracking, the initial elevation Z is set as the lowest elevation of the photogrammetric mesh model for each building, while the contour interval D is set according to the vertical accuracy of the photogrammetric mesh model. Subsequently, each contour line is carefully tracked. In general, there would be two types of contours: open and closed contours. Only closed ones are retained for subsequent processing.

After producing the contour lines, a building can be represented by contour clusters abstracted by cuboids. To split the contour lines into separate parts, the contours are transferred to a graph-based localized contour tree [25,31]. The tree consists of a root node, several internal nodes (branches), and several terminal nodes (leaf). The closed contour is represented as a node in the structure, while the relationship between contours is denoted by an edge between the nodes in the tree-based structure.

The local contour trees are constructed from bottom to top based on the contour elevations. For instance, let us take into account a complex building as demonstrated in Figure 2a. The local contour tree (Figure 2b) is initialized by contour A1 with the lowest elevation as the root node. Then, the adjacent contour A2 is identified and added as the child node of contour A1. These steps are iterated until the highest contour B6 is included. During the adding process, when meeting n ($n > 1$) contours for a given height value, n branches will be constructed. Figure 2a shows that there are two contours (contour B1 and C1) for the fourth height value. Thus, two subtrees are generated from A3. In these trees, only the contours whose topological relations have not changed exist in the same structure. These contours are represented by a subtree in the contour tree. Finally, the contour tree illustrated in Figure 2b is obtained, where the same color part indicates the same structure of the photogrammetric mesh model. Node A3 has two sub-nodes B1 and C1, and node C3 has a sub-node D1, indicating a separation relationship in the sense of topological representation. After producing the contour tree, the segmented contour pair is attained between subtrees. Therefore, the segmented contour pairs of the photogrammetric mesh model in Figure 2a are A3–B1, A3–C1, and C3–D1.

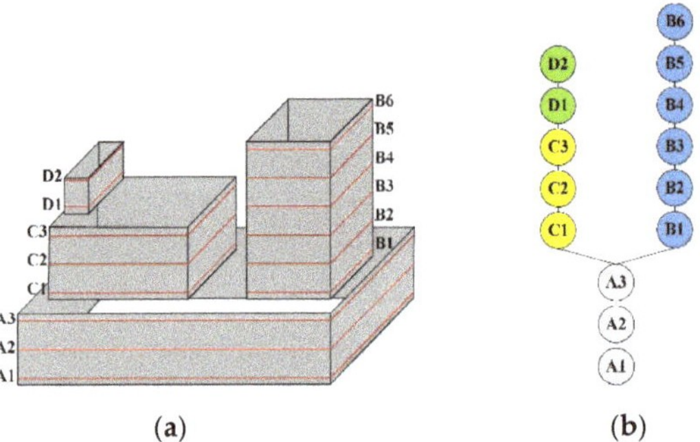

(a) (b)

Figure 2. An illustration of the local contour tree of the photogrammetric mesh model: (a) contours tracking results; (b) local contour tree generation result.

2.2.2. Decomposition of Components

After generating the contours trees, the photogrammetric mesh model is subdivided to mesh clusters based on the obtained segmented contour pair. For the local contour tree as shown in Figure 2b, firstly, the lowest elevation contour pair A3–B1 (Figure 3a) is exploited to remove the triangles placed between the contours A3 and B1. Then, the remaining triangles are clustered into three components of the photogrammetric mesh model. As demonstrated in Figure 3a, the gray part of the model, which is lower than the A3–B1 elevation of the segmented contour pair, is successfully segmented. Thereafter, the components of the photogrammetric mesh model are subdivided, which are higher than the elevation of the segmented contour pair A3–B1. Due to the lower elevation contour of the next group of the segmented contour pair with the lowest elevation A3–C1 is the same as those of A3, this segmented contour pair (A3–C1) is then skipped. Subsequently, the remaining cluster by the next segmented contour pair C3–D1 is subdivided, then the yellow component of the photogrammetric mesh model (see Figure 3a) is successfully segmented. This process is repeatedly carried out until there is no segmented contour group, and then, the original photogrammetric mesh model is subdivided into basic components. The final obtained results are illustrated in Figure 3a.

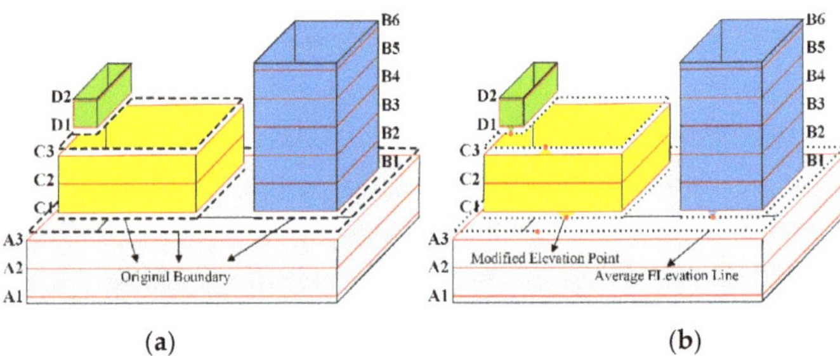

(a) (b)

Figure 3. Illustration of decomposition of the photogrammetric mesh model: (a) decomposed photogrammetric mesh model components; (b) photogrammetric mesh model components after modification.

During the component decomposing, the triangles between different trees are removed, resulting in a gap between the subsequent generating models (i.e., the gap between A3 and C1 in Figure 3a). To resolve the aforementioned issue, the elevation of the closest point to the segmented contour pair in the photogrammetric mesh model component should be appropriately reformed to the average elevation of the segmented contour pair. The photogrammetric mesh model components after the points' modification are presented in Figure 3b.

2.3. Cuboid Abstraction

After decomposing the photogrammetric mesh model into separated components, a set of cuboids is exploited to fit each component. At first, a region growing method is applied to the current component mesh model to produce super-facets. Then the least square algorithm is utilized to fit the normal vector of the largest super-facet. After that, the coordinate axis is transformed to the calculated normal vector, and the coordinates of mesh vertexes are centralized to lessen the subsequent iteratively minimum circumscribed cuboid fitting process.

To reconstruct the complex building model, the mesh model components are abstracted to several levels of minimum circumscribed cuboids. The cuboid abstraction performs iteratively, as the corresponding workflow is shown as Figure 4.

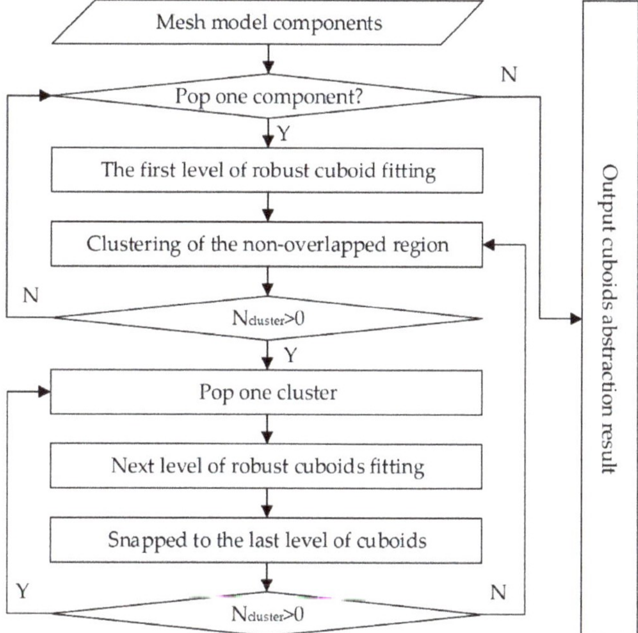

Figure 4. The workflow of the cuboid abstraction processing.

The abstraction starts from popping one component from the separated components. If no components are left, the abstraction result is exported to the following processing; otherwise, an iteratively robust cuboid fitting process is performed. For the current component, the first level circumscribed cuboid is fitted to most outside of the component. There could be some noise in the original photogrammetric mesh model. For instance, if all points are exploited to fit the façade model, bias may exist in the façade parameters. A robust fitting strategy is proposed to eliminate possible noise points. Firstly, the distance between each point to the closest plane of the fitted cuboid is calculated. When the distance is larger than a given threshold value of T_d (T_d is experimentally set equal to 0.2 m in the present study), the points are removed, and the remaining ones are utilized to fit a new plane again for the corresponding side of the cuboid. By taking Figure 5 as an example, it can be seen in Figure 5a (the top view of model component) that there is one protuberance on the north side. The original cuboid is a rectangle with green color, which does not well fit to the point cloud. After removing the possible noise part, a new cuboid is fitted and marked as yellow color in Figure 5b, and the fitted model snapped the point cloud well.

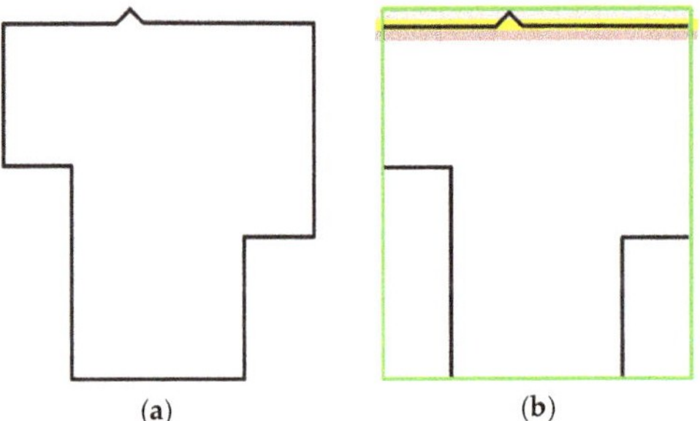

(a)   (b)

Figure 5. Illustration of the robust cuboid fitting segmentation: (**a**) top view of the model component; (**b**) fitting result.

After generating the first level of cuboid, the average distance between the vertexes of a triangle to the nearest plane of the circumscribed cuboid is evaluated. The points whose distances are larger than a given threshold value T_2 (T_2 is experimentally set equal to 0.2 m) are grown to gather the non-overlapping regions to region groups. The regions with values less than a predefined value on the vertex number and areas are overlooked, and the predefined value of the vertex number and areas are determined according to the target detail of the model. For the remaining groups, robust cuboid fitting processing is performed to derive the next level of cuboids. After generating the next level of cuboid using the remained non-overlapping region, there will be a slight bias from the previous level of the cuboid. In Figure 6a, the top view illustrates the whole process since the façade is vertical to the ground. As observed, the corner of the current cuboid is not on the first level of the cuboid. To avoid this problem, as shown in Figure 6b, the coordinates of the current level of the cuboid are extended to intersect with the nearby cuboid sides, and the new intersect point will be used to replace the original cuboid corner to guarantee the close of the model.

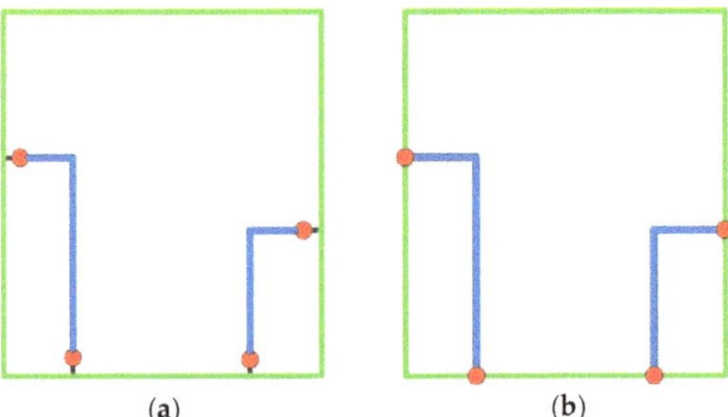

(a)   (b)

Figure 6. Top view of extending the endpoint of the non-overlapping region to the nearest plane: (**a**) the two endpoints (red circle) before extension; (**b**) the two endpoints after extension.

The same procedure is repeated until there would be no non-overlapping region in the current component. Further, the component fitting process is repeatedly carried out until no component is left.

By taking into account Figure 7a as the input component, Figure 7a–e illustrates the process of the iteratively cuboid fitting process step-by-step. If a minimum circumscribed cuboid is directly fitted to the original model component, the first level of the circumscribed cuboid is produced (i.e., the green cuboid in Figure 7b). The cuboid does not sit adjacent to the original mesh very well. By removing possible noise points or small objects on the façades, the remaining points would fit the model as displayed in Figure 7c. It appears that the cuboid is closer to the original model after these modifications. For the non-overlapping

125

from Figure 7c, two second levels of circumscribed cuboids are derived, as demonstrated in Figure 7d. All the cuboids are joined together when there are no overlapping areas, as presented in Figure 7e.

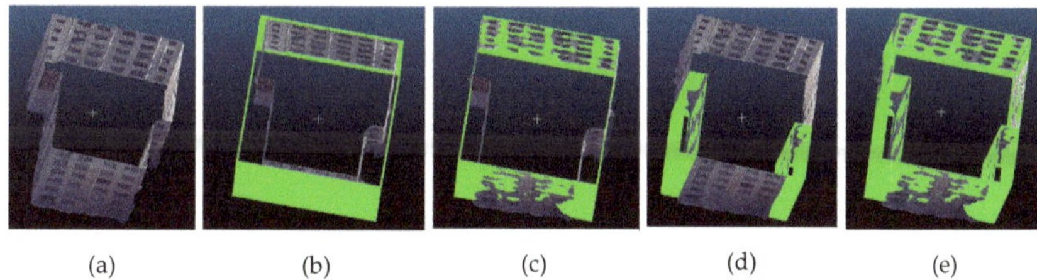

(a)　　　　　　(b)　　　　　　(c)　　　　　　(d)　　　　　　(e)

Figure 7. Illustration of the cuboid abstraction processing: (**a**) input of the original photogrammetric mesh model; (**b**) the first fitted cuboid; (**c**) the modification of the first fitted cuboid; (**d**) the second fitted cuboid; (**e**) combined cuboid abstraction result.

2.4. Parameter Adjustment of Cuboid Model Based on the Least Square Method

As the initial cuboid is attained by a range of transformed coordinates, the resulted cuboid may not fit the initial photogrammetric mesh model very well because of existing noise and noise in the coordinate transform parameters. Thus, the least square method is employed to adjust the cuboid fitting the model to the initial photogrammetric mesh model. Each cuboid can be specified by six parameters (X_0, Y_0, Z_0, W, L, H), as the façade only considers the plane coordinates, Z_0 and H are kept fixed throughout the adjustment process. The adjustment of the model parameters is commonly accomplished by minimizing the distance between the initial model (i.e., the results of the cuboid abstraction process) and the photogrammetric mesh model by a least square algorithm. The adjust mode is defined as Equation (1).

$$\begin{cases} v_1 = (X_0 + \delta X) - X_I \\ v_2 = (Y_0 + \delta Y) + (W_0 + \delta W) - Y_I \\ v_3 = (X_0 + \delta X) + (L_0 + \delta L) - X_I \\ v_4 = (Y_0 + \delta Y) - Y_I \end{cases} \tag{1}$$

where $(\delta X, \delta Y, \delta W, \delta L)$ denote the adjusted cuboid parameters, (X_0, Y_0, W_0, L_0) represent the initial cuboid parameters, (X_1, Y_1) are the coordinates of the vertexes of the involved triangles.

After obtaining the error equations, it can be solved by implementing the traditional least square approach. The error equations associated with Equation (1) is formatted in the matrix form as follows:

$$V = Ax - L \tag{2}$$

in which:

$$V = \begin{bmatrix} v1 \\ v2 \\ v3 \\ v4 \end{bmatrix}, \ x = \begin{bmatrix} \delta X \\ \delta Y \\ \delta W \\ \delta L \end{bmatrix}, \ A = \begin{bmatrix} 1 & 0 & 0 & 0 \\ 0 & 1 & 1 & 0 \\ 1 & 0 & 0 & 1 \\ 0 & 1 & 0 & 0 \end{bmatrix}, \ L = \begin{bmatrix} X_I - X_0 \\ Y_I - Y_0 - W_0 \\ X_I - X_0 - L_0 \\ Y_I - Y_0 \end{bmatrix}.$$

After obtaining the error equations displayed in Equation (2), the solution for the unknowns is completed in the following form:

$$x = (A^T A)^{-1} (A^T L) \tag{3}$$

For the first level of the cuboid, four model parameters (X_0, Y_0, W, L) are adjusted. For other levels of cuboids, only the error equations pertinent to sides over-lapped with the initial photometric mesh model are adjusted, while other parameters are kept fixed.

After adjusting the cuboid parameters, there would be some gaps between the subsequent level of the cuboid and its former level, the low level of the cuboid is shifted to the nearest high level of the cuboid.

After performing the adjustment process, the existing planes are chosen from the cuboid and employed to producing the final façade mode.

3. Experiment Results and Analysis

To validate the performance of the proposed method, a set of photogrammetric mesh models is employed to perform the experiments. The mesh model is generated from oblique aerial images taken by SWDC-5 by using ContextCapture [31]. The ground sample distance of the original image is around 0.1 m. The used photogrammetric mesh model is cut from a large part of the scene. Initially, four complex buildings are selected to evaluate the method quantitatively and qualitatively. To further evaluate the actual performance of the proposed method, the whole scene is then reconstructed by the proposed approach.

3.1. Date Description

To check the effect of the proposed façade modeling method, four typical complex buildings, as shown in the subfigures of the first column in Figure 8, are selected. The numbers of vertices and triangles are listed in Table 1. All selected buildings are composed of thousands of triangles.

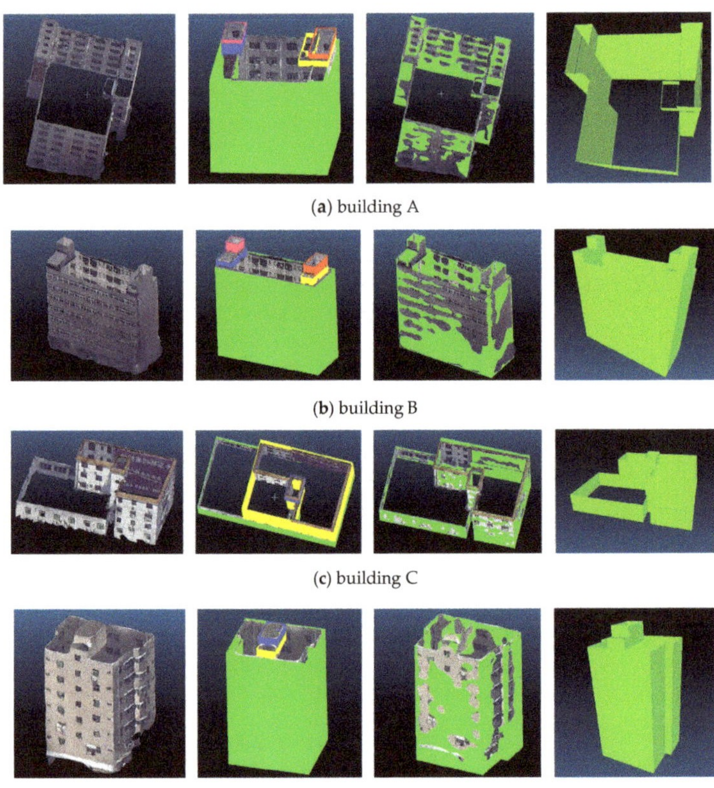

(**a**) building A

(**b**) building B

(**c**) building C

(**d**) building D

Figure 8. The reconstruction results. Each row (from left to right) presents the original photogrammetric mesh model, initial cuboid abstraction set, façade model overlaid on the original date, and 3D building façade model.

Table 1. The basic information of the selected buildings.

Test Data	Number of Vertices	Number of Triangle Facets
Building A	4324	8081
Building B	2647	4899
Building C	3424	6322
Building D	1834	3358

3.2. Resontruction Results and Analysis

The reconstructed results are presented in the subfigures of the fourth column of Figure 8. The depicted results reveal that the proposed method can generate a faithful polygon model. Compared with the original data, the various profile and details of the building are well preserved in the corresponding reconstruction model.

Figure 8 illustrates the detailed reconstruction processes of four complex buildings. To clarify the reconstruction process better, we present the cuboids before the global fitting of model parameters at each level. In the subfigures of second column of Figure 8, the initial cuboid abstraction results are given. The subfigures in the third column of Figure 8 present the final façade model overlaid on the original date. While those of the fourth column of Figure 8 show the final 3D façade model. Concerning building A, it is divided into five components. The first component generates one first-level and two second-level cuboids (presented by green color in the subfigures of the second column of Figure 8a). The second component produces a first-level and a second-level cuboid (highlighted by yellow color in the subfigures of the second column of Figure 8a). Components 3, 4, and 5 produce a cuboid (displayed by blue, red, and pink colors in the subfigures of the second column of Figure 8a). Then, the final 3D building façade model is generated via plane selection and global fitting of the model parameters. Building B is divided into five components, each one generates a first-level cuboid. Similar to building A, building B is successfully reconstructed. As shown by blue and pink cuboids in building A and yellow and red cuboids in building B, this method can successfully reconstruct buildings with minor structural. Building C is divided into three components. The first component provides a first-level and a second-level cuboid. The second component generates a first-, a second-, and a third-level cuboid, while component three produces one cuboid. The model is established by the global fitting of the model parameters and the plane selection. The first-level yellow cuboid of the second component has a small part that does not fit the original data. It is because of this fact that the distance of the unfitted piece from the model is smaller than the set threshold, so the whole model fits the original data very well. Building D is divided into three components. The first component generates first-level and two second-level cuboids, while components three and four produce a cuboid. The model generated by the global fitting of model parameters and plane selection is also realistically reconstructed.

The reconstruction results show the lower part of the original data has a large number of missing parts. Further, there are many abnormal data phenomena, but the method can still generate fairly accurate reconstruction results.

To further assess the proposed method, the experimentally obtained results are quantitatively evaluated. The quality of the reconstructed model is evaluated by checking the average distance between the original data and the nearest surface of the reconstructed model. Figure 9 displays the reconstruction error diagrams for the under-study buildings. After removing the apparent outliers of the original data, the diagrams of the average distance from the original data for four buildings under consideration are presented in Figure 10. The minimum, maximum, and average errors of the four reconstructed 3D façade models in order are 0.066, 0.154, and 0.09 m.

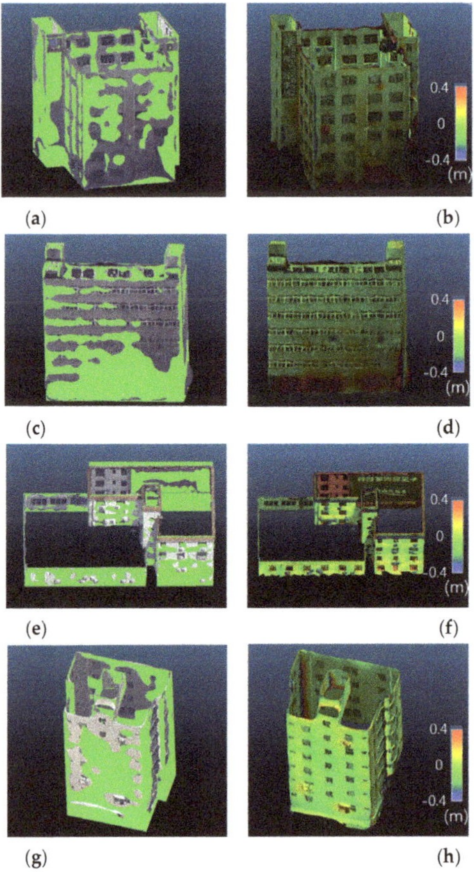

Figure 9. The graphics of reconstruction errors for four buildings: (**a**,**c**,**e**,**g**) façade model overlaid on the original date; (**b**,**d**,**f**,**h**) reconstruction error.

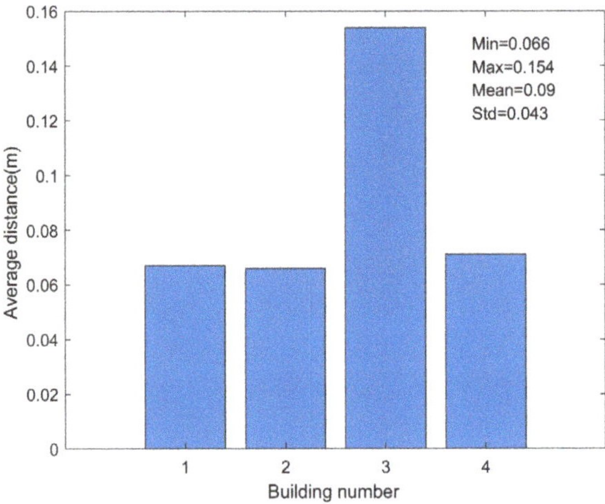

Figure 10. The statistical results of the reconstruction errors for four buildings.

3.3. Reconstruc Result on a Whole Scene

To assess the practicability and robustness of the proposed method more systematically, we apply the proposed method to the whole scene of the photogrammetric mesh model as shown in Figure 11. The scene is composed of 13 buildings. Figure 12 presents the reconstructed 3D façade model. The obtained results further indicate that the proposed method can generate trustable polygon models from the photogrammetric mesh model in the complex scene, and the reconstructed results are in reasonably agreement with the original data. The final 3D façade model is compared to the original photogrammetric mesh model to assess the reconstruction result; Figure 13 illustrates the average distance from the original data to the derived façade model. The minimum, maximum, and average errors of the thirteen reconstructed 3D façade models, respectively, are 0.066, 0.2, and 0.124 m. The accuracy of the whole scene is somehow lower than that of the test data. This issue is mainly because some buildings in the experimental scene have balconies, resulting in more deformation in the scene.

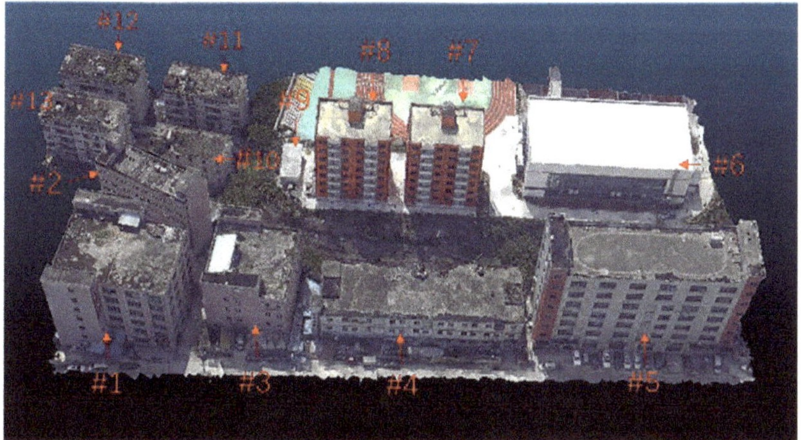

Figure 11. A scene of photogrammetric mesh model.

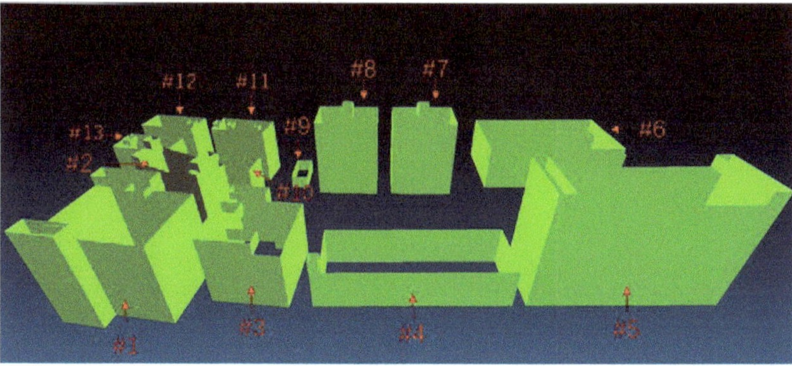

Figure 12. The façade reconstruction result.

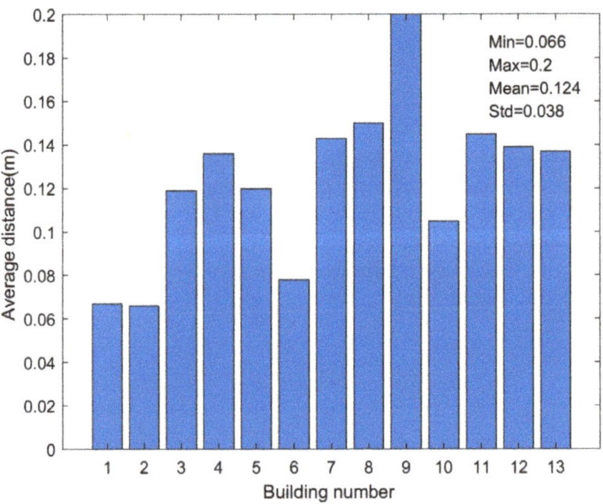

Figure 13. The statistical results of the reconstruction errors for a scene.

4. Discussion

4.1. Comparison

The main objective of the present paper is to construct a regular façade model from a photogrammetric mesh model derived from oblique aerial images by Structure from Motion (SfM) and a Multi-View Stereo (MVS) pipeline. Since the relevant methods in the literature often generate models from LiDAR data, having different sources from the proposed method, a direct and thorough comparison to these methods would be difficult. By this view, some presented mean reconstructed errors in typical methods are provided in Table 2 for the sake of comparison. For the selected typical four buildings, the average distance between the reconstructed 3D façade models and the vertex of the input photogrammetric mesh model is 0.09 m, while for the whole scene, the mean distance is 0.124 m. The presented results reveal that the proposed method can successfully reconstruct the building façade model from the photogrammetric mesh model compared to the previously obtained results from the LiDAR data.

Table 2. A comparison study on previously obtained results by relevant studies.

Methods	Data Source	Error (Meters)
Li et al. [27]	3D point cloud generated from UAV images by SFM and MVS	0.15
Song et al. [32]	Airborne LiDAR data	0.17–0.34
Wu et al. [33]	Airborne LiDAR point clouds	0.32
Lafarge and Mallet, [34]	Digital surface model	0.1–0.24

Contours extracted from the photogrammetric mesh model are used to segment the different part of a building. However, the surface link between the correspondence contour points between consecutive contours, as used in Ref. [25], cannot adapt to the photogrammetric mesh model well. This is due to fact that there may be some noise in the contours derived from the under-study photogrammetric mesh model. The proposed method in this paper exploited cuboids to fit the mesh model and the least square adjustment to ensure the quality; as a result, the effect of noise in the source mesh model is reduced.

4.2. Limitation of the Proposed Method

The proposed method assumes that the building façades are composed of several cuboids, which cannot adapt to some complex buildings. For future works, some other elements to fit the model can be considered.

Although the proposed method takes into account outliers in the mesh model, when facing a facade model with large defects caused by trees, etc., in front of the façade, the proposed method cannot obtain proper results in such a situation. The future work will focus on considering some clues from ground view data sources to ensure the completeness of the derived façade models.

Currently, the proposed method only considers the geometrics of the model; future work would consider the texture.

5. Conclusions

This paper presents a method for automatically generating a 3D regular building façade model from a photogrammetric mesh model. The proposed method mined the advantage of using a TIN structure for expressing the photogrammetric mesh model. Each component is abstracted by a cuboid, which can reduce the effects of small errors in the contours; in addition, a robust cuboid fitting method is proposed to alleviate the noise problem or small parts possible existing in the source photogrammetric mesh model. A least square method is used to adjust the parameters to finally ensure the quality of the reconstructed model. The average error of the reconstructed 3D building façade models is obtained as 0.124 m. The experimental results reveal that the newly developed method can effectively reconstruct the building façade model from the photogrammetric mesh model. Additionally, the proposed method is not affected by numerous data missing from the lower parts of the original data, outliers, and other issues.

The proposed process starts from the final mesh model, which is derived from software consuming more processing power than the dense point cloud from multiple-view images. For future work, the interested scholars can apply the proposed method to a dense point cloud by incorporating a proper contour-tracking algorithm into the current workflow.

Author Contributions: Conceptualization, Y.Z. and X.C.; methodology, Y.Z. and C.Z.; software, C.Z.; validation, C.Z. and S.C.; resources, X.C.; writing—original draft preparation, Y.Z. and C.Z.; writing—review and editing, Y.Z., C.Z. and X.C. All authors have read and agreed to the published version of the manuscript.

Funding: This research was funded by Open Fund of Key Laboratory of Urban Land Resource Monitoring and Simulation, Ministry of Land and Resource (No. KF-2018-03-047), the University Innovative Platform Open Fund of Hunan (No. 19K099).

Acknowledgments: The authors are grateful to the editors for their helpful comments and Lelin Li for his helpful discussion.

Conflicts of Interest: The authors declare no conflict of interest.

References

1. Catita, C.; Redweik, P.; Pereira, J.; Brito, M.C. Extending solar potential analysis in buildings to vertical facades. *Comput. Geosci.* **2014**, *66*, 1–12. [CrossRef]
2. Bagheri, H.; Schmitt, M.; Zhu, X. Fusion of multi-sensor-derived heights and OSM-derived building footprints for urban 3D reconstruction. *ISPRS Int. J. Geo-Inf.* **2019**, *8*, 193. [CrossRef]
3. Zheng, Y.; Weng, Q. Model-driven reconstruction of 3-D buildings using LiDAR data. *IEEE Geosci. Remote Sens. Lett.* **2015**, *12*, 1541–1545. [CrossRef]
4. Xie, L.; Hu, H.; Zhu, Q.; Li, X.; Tang, S.; Li, Y.; Wang, W. Combined rule-based and hypothesis-based method for building model reconstruction from photogrammetric point clouds. *Remote Sens.* **2021**, *13*, 1107. [CrossRef]
5. Xiao, J.; Fang, T.; Tan, P.; Zhao, P.; Ofek, E.; Quan, L. Image-based façade modeling. *ACM Trans. Graph.* **2008**, *27*, 1–10. [CrossRef]
6. Müller, P.; Zeng, G.; Wonka, P.; Van Gool, L. Image-based procedural modeling of façades. *ACM Trans. Graph.* **2007**, *26*, 85. [CrossRef]
7. Sadeghi, F.; Arefi, H.; Fallah, A.; Hahn, M. 3D building Façade reconstruction using handheld laser scanning data. *Int. Arch. Photogramm. Remote Sens. Spat. Inf. Sci.* **2015**, *XL-1/W5*, 625–630. [CrossRef]

8. Edum-Fotwe, K.; Shepherd, P.; Brown, M.; Harper, D.; Dinnis, R. Fast, accurate and sparse, automatic façade reconstruction from unstructured ground laser-scans. In Proceedings of the ACM SIGGRAPH 2016, Anaheim, CA, USA, 24–28 July 2016; Association for Computing Machinery: New York, NY, USA, 2016; pp. 45:1–45:2. [CrossRef]
9. Pu, S.; Vosselman, G. Building façade reconstruction by fusing terrestrial laser points and images. *Sensors* **2009**, *9*, 4525–4542. [CrossRef] [PubMed]
10. Riemenschneider, H.; Krispel, U.; Thaller, W.; Donoser, M.; Havemann, S.; Fellner, D.; Bischof, H. Irregular lattices for complex shape grammar façade parsing. In Proceedings of the IEEE Computer Society Conference on Computer Vision and Pattern Recognition, Providence, RI, USA, 16–21 June 2012; IEEE Computer Society: Washington, DC, USA; pp. 1640–1647.
11. OSketch. Available online: https://www.zhdgps.com/ (accessed on 4 April 2021).
12. DP-Modeler. Available online: http://www.whulabs.com/ (accessed on 4 April 2021).
13. Zhang, H.; Xu, K.; Jiang, W.; Lin, J.; Cohen-Or, D.; Chen, B. Layered analysis of irregular façades via symmetry maximization. *ACM Trans. Graph.* **2013**, *32*, 1–13. [CrossRef]
14. Li, Z.; Zhang, L.; Mathiopoulos, P.T.; Liu, F.; Zhang, L.; Li, S.; Liu, H. A hierarchical methodology for urban façade parsing from TLS point clouds. *ISPRS J. Photogramm. Remote Sens.* **2017**, *123*, 75–93. [CrossRef]
15. Yang, B.; Dong, Z.; Wei, Z.; Fang, L. Extracting complex building façades from mobile laser scanning data. *Acta Geod. Cartogr. Sin.* **2013**, *42*, 411–417.
16. Yan, L.; Hu, Q.W.; Wu, M.; Liu, J.M.; Wu, X. Extraction and simplification of building façade pieces from mobile laser scanner point clouds for 3D street view services. *ISPRS Int. J. Geo-Inf.* **2016**, *5*, 231. [CrossRef]
17. Lafarge, F.; Descombes, X.; Zerubia, J.; Pierrot-Deseilligny, M. Structural approach for building reconstruction from a single DSM. *IEEE Trans. Pattern Anal. Mach. Intell.* **2010**, *32*, 135–147. [CrossRef]
18. Kwak, E.; Habib, A. Automatic representation and reconstruction of DBM from LiDAR data using recursive minimum bounding rectangle. *ISPRS J. Photogramm. Remote Sens.* **2014**, *93*, 171–191. [CrossRef]
19. Henn, A.; Gröger, G.; Stroh, V.; Plümer, L. Model driven reconstruction of roofs from sparse LIDAR point clouds. *ISPRS J. Photogramm. Remote Sens.* **2013**, *76*, 17–29. [CrossRef]
20. Cheng, L.; Gong, J. Building boundary extraction using very high resolution images and LiDAR. *Acta Geod. Cartogr. Sin.* **2008**, *37*, 391–393.
21. Jiang, X.; Bunke, H. Fast segmentation of range images into planar regions by scan line grouping. *Mach. Vis. Appl.* **1994**, *7*, 115–122. [CrossRef]
22. Sampath, A.; Shan, J. Segmentation and reconstruction of polyhedral building roofs from aerial LiDAR point clouds. *IEEE Trans. Geosci. Remote Sens.* **2010**, *48*, 1554–1567. [CrossRef]
23. Chen, D.; Zhang, L.; Li, J.; Liu, R. Urban building roof segmentation from airborne LiDAR point clouds. *Int. J. Remote Sens.* **2012**, *33*, 6497–6515. [CrossRef]
24. Chen, D.; Zhang, L.; Mathiopoulos, P.; Huang, X. A methodology for automated segmentation and reconstruction of urban 3-D buildings from ALS point clouds. *IEEE J. Sel. Top. Appl. Earth Obs. Remote Sens.* **2014**, *7*, 4199–4217. [CrossRef]
25. Wu, B.; Yu, B.; Wu, Q.; Yao, S. A Graph-based approach for 3D building model reconstruction from airborne LiDAR point clouds. *Remote Sens.* **2017**, *9*, 92. [CrossRef]
26. Nan, L.; Jiang, C.; Ghanem, B.; Wonka, P. Template assembly for detailed urban reconstruction. *Comput. Graph. Forum.* **2015**, *34*, 217–228. [CrossRef]
27. Li, M.; Nan, L.; Smith, N.; Wonka, P. Reconstructing building mass models from UAV images. *Comput. Graph.* **2016**, *54*, 84–93. [CrossRef]
28. Li, M.; Nan, L.; Liu, S. Fitting boxes to Manhattan scenes using linear integer programming. *Int. J. Digit. Earth* **2016**, *9*, 806–817. [CrossRef]
29. Li, M.; Wonka, P.; Nan, L. Manhattan-World Urban Reconstruction from Point Clouds. In Proceedings of the 14th European Conference on Computer Vision, Amsterdam, The Netherlands, 11–14 October 2016; Springer: Cham, Switzerland, 2016; pp. 54–69.
30. Zhang, Z.; Zhang, J. *Digital Photogrammetry*, 1st ed.; Wuhan University Press: Wuhan, China, 1997; pp. 83–84.
31. ContextCapture. Available online: https://www.bentley.com/en/products/brands/contextcapture (accessed on 4 April 2021).
32. Song, J.; Wu, J.; Jiang, Y. Extraction and reconstruction of curved surface buildings by contour clustering using airborne LiDAR data. *Optik* **2015**, *126*, 513–521. [CrossRef]
33. Wu, Q.; Liu, H.; Wang, S.; Yu, B.; Beck, R.; Hinkel, K. A localized contour tree method for deriving geometric and topological properties of complex surface depressions based on high-resolution topographical data. *Int. J. Geogr. Inf. Sci.* **2015**, *29*, 2041–2060. [CrossRef]
34. Lafarge, F.; Mallet, C. Creating large-scale city models from 3d-point clouds: A robust approach with hybrid representation. *Int. J. Comput. Vis.* **2012**, *99*, 69–85. [CrossRef]

remote sensing

MDPI

Article

Automatic, Multiview, Coplanar Extraction for CityGML Building Model Texture Mapping

Haiqing He [1,2], Jing Yu [1,2], Penggen Cheng [1,2,*], Yuqian Wang [1,2], Yufeng Zhu [1], Taiqing Lin [3] and Guoqiang Dai [3]

[1] School of Geomatics, East China University of Technology, Nanchang 330013, China; hehaiqing@ecut.edu.cn (H.H.); 201910816004@ecut.edu.cn (J.Y.); neo@ecut.edu.cn (Y.W.); yfzhu@ecut.edu.cn (Y.Z.)

[2] Key Laboratory of Mine Environmental Monitoring and Improving around Poyang Lake, Ministry of Natural Resources, Nanchang 330013, China

[3] Jiangxi Academy of Water Science and Engineering, Nanchang 330029, China; lintaiqing@whu.edu.cn (T.L.); 201910705033@ecut.edu.cn (G.D.)

* Correspondence: pgcheng@ecut.edu.cn

Abstract: Most 3D CityGML building models in street-view maps (e.g., Google, Baidu) lack texture information, which is generally used to reconstruct real-scene 3D models by photogrammetric techniques, such as unmanned aerial vehicle (UAV) mapping. However, due to its simplified building model and inaccurate location information, the commonly used photogrammetric method using a single data source cannot satisfy the requirement of texture mapping for the CityGML building model. Furthermore, a single data source usually suffers from several problems, such as object occlusion. We proposed a novel approach to achieve CityGML building model texture mapping by multiview coplanar extraction from UAV remotely sensed or terrestrial images to alleviate these problems. We utilized a deep convolutional neural network to filter out object occlusion (e.g., pedestrians, vehicles, and trees) and obtain building-texture distribution. Point-line-based features are extracted to characterize multiview coplanar textures in 2D space under the constraint of a homography matrix, and geometric topology is subsequently conducted to optimize the boundary of textures by using a strategy combining Hough-transform and iterative least-squares methods. Experimental results show that the proposed approach enables texture mapping for building façades to use 2D terrestrial images without the requirement of exterior orientation information; that is, different from the photogrammetric method, a collinear equation is not an essential part to capture texture information. In addition, the proposed approach can significantly eliminate blurred and distorted textures of building models, so it is suitable for automatic and rapid texture updates.

Keywords: texture mapping; coplanar extraction; deep convolutional neural network; geometric topology; homography matrix

Citation: He, H.; Yu, J.; Cheng, P.; Wang, Y.; Zhu, Y.; Lin, T.; Dai, G. Automatic, Multiview, Coplanar Extraction for CityGML Building Model Texture Mapping. *Remote Sens.* **2022**, *14*, 50. https://doi.org/10.3390/rs14010050

Academic Editors: Wanshou Jiang, San Jiang and Xiongwu Xiao

Received: 14 November 2021
Accepted: 21 December 2021
Published: 23 December 2021

Publisher's Note: MDPI stays neutral with regard to jurisdictional claims in published maps and institutional affiliations.

1. Introduction

1.1. Background

The development of smart city highly depends on the quality of geospatial data infrastructure, and 3D visualization is a core technology of a digital city [1]. A representative city geography markup language (CityGML) is developed by Open Geospatial Consortium for defining and describing 3D building attributes, such as geometric, topological, semantic, and appearance characteristics, which are very valuable for many applications, such as simulation modeling, urban planning, and map navigation [2]. Texture mapping of building models has always been a hot and significant research topic in the fields of computer vision, photogrammetry, and remote sensing. Nevertheless, due to problems such as ground-object occlusion, texture mapping of CityGML building models is still challenging.

Generally, CityGML can be divided into five levels of detail (LOD), including LOD0 (e.g., regional, landscape), LOD1 (e.g., city, region), LOD2 (e.g., city neighborhoods, projects), LOD3 (e.g., architectural models (exterior), landmarks), and LOD4 (e.g., architectural models (interior)) [3–6]. However, popular map providers, such as Google and Baidu, are currently limited to LOD0-2 as a result of extremely large and complex data processing, as well as high costs and time consumption of data acquisition. In addition, researchers are attempting to build 3D city models using multisource geospatial data (e.g., airborne LiDAR point cloud and photogrammetric mapping) to generate LOD2- and LOD3-level city models [7]. Although these techniques can obtain desirable 3D building models, they still require high potential costs for frequent updates of the texture of building models. However, a single data source usually suffers from several problems, such as object occlusion.

1.2. Related Work

In the previous decades, building large-scale urban models has been broadly studied, including manual, automatic, and human–computer-interaction methods. Evidently, the manual method is not desirable, given its long production cycle and high cost. From the perspective of data sources used for building-texture mapping, many studies mainly focus on photogrammetric data, LiDAR point cloud data, and crowd-sourced data.

1.2.1. Texture Mapping Based on Photogrammetric Data and LiDAR Point Cloud Data

In recent years, the popularization of unmanned aerial vehicles (UAV), oblique cameras, and LiDAR, coupled with the increasing maturity of high-resolution stereo imaging, not only realizes rapid production of large-scale urban models but also gradually shortens the modeling cycle and continuously reduces the cost. Consequently, photogrammetric data and LiDAR point cloud data are also widely used in 3D modeling and texture mapping. Li et al. [8] proposed an optimized combination of graph-based 3D visualization and image-based 3D visualization to realize geographic information system (GIS) 3D visualization. Yalcin et al. [9] suggested creating a 3D city model from aerial images based on oblique photogrammetry. Abayowa et al. [10] presented 3D city modeling based on the fusion of LiDAR data and aerial image data. Through the efforts of the researchers, these data sources can provide users with a multidimensional, multiperspective, and omnidirectional environment to browse, measure, and analyze ground objects, which are suitable for spatial decision-making applications. However, these methods have many problems, such as the large amount of data acquisition, complex processing algorithms, fuzzy texture information, high production cost, and long update cycle. Strong theoretical and technical support is also provided for urban modeling and its application due to the vigorous development of remote sensing, photogrammetry, computer graphics, stereo vision, and machine learning, while research on the continuous expansion of the breadth and depth of the urban model is promoted. Among them, Heo et al. [11] proposed a semi-automatic method for high-complexity 3D city modeling using point clouds collected by ground LiDAR. Wang et al. [12] suggested a method for urban modeling based on oblique photogrammetry and 3DMax plug-in development technology. These methods aim to combine the advantages of multisource data to provide practical, efficient, and semi-automatic urban modeling methods. Zhang et al. [13] presented a rapid-reconstruction method of 3D city model texture based on the principle of oblique photogrammetry, which can automatically extract the texture and uniform color of building façades and perform texture mapping, considering multiple building occlusions. These methods have achieved certain results in response to these problems. However, other problems, such as object occlusion and incomplete texture, are still challenging for fine texture mapping of building models.

1.2.2. Texture Mapping Based on Crowd-Sourced Data

In recent years, crowd-sourced data (such as public images) have been broadly used as alternative or supplementary data sources for many GIS modeling applications. These public images can provide structured map description by tags and attributes, and existing 2D images can be converted into 3D models in batches in terms of related attributes. Therefore, this type of crowd-sourced data has also become an important data source for 3D city refined texture mapping. Many 3D city modeling methods based on crowd-sourced data have also been proposed. For example, Lari et al. [14] introduced a new method for 3D reconstruction of flat surfaces; it aims to improve the interpretability of planes extracted from laser-scanning data by using the spectral information of the overlapping images collected by low-cost aerial surveying and mapping systems. Khairnar et al. [15] used the structural and geographic information retrieved from OpenStreetMap (OSM) to reconstruct the shape of the building. They also used the images obtained from the street view of Google Maps to extract information about the appearance of the building to map textures to building boundaries. Girindran et al. [16] proposed a method to generate low-cost 3D city models from public 2D building data by combining satellite-elevation datasets, confirming a potential solution for the lack of free, high-resolution 3D city models. In addition, the use of other data, combined with public images, has made great breakthroughs in texture mapping. Gong et al. [17] used vehicle-mounted mobile measurement data to supplement and refine building façades by using an enhanced method. Li et al. [18] proposed a seamless reconstruction method for texture optimization based on low-altitude, multi-lens, oblique photography in the production of 3D urban models. Hensel et al. [7] improved the quality of textures on the façades of LOD2 CityGML building models based on deep learning and mixed-integer linear programming. Although these studies have achieved good performance in texture optimization, update cycle, modeling cost, quality and scalability of building models decline in the process of urban modeling because the textures of building models are still not updated promptly.

Generally, most traditional 3D models have been built in the form of pure graphics or geometry, ignoring the semantic and topological relationship between graphics and geometry. These models are limited to 3D visualization and cannot satisfy the requirement of in-depth applications, such as thematic query, spatial analysis, and spatial data mining. CityGML defines the classification of most geographic objects in the city and the relationship between them. It fully considers the geometry, topology, semantics, appearance, and other attributes of the regional model, thereby making up for the traditional 3D models in terms of data sharing and interoperability. In addition, the city's 3D model has become reusable, greatly reducing the cost of the city's 3D modeling [2]. Many studies have been conducted using CityGML building modeling. Deng et al. [19] proposed a relatively complete and high-precision mapping framework between IFC and CityGML in different LOD CityGML models, including the transformation of geometric shapes, coordinate systems, and semantic frameworks. Fan et al. [20] introduced a method to derive LOD2 buildings from the LOD3 model, which separated the different semantic components of the building, with the goal of preserving the features of the floor plan, roof, and wall structure as much as possible. Hensel et al. [7] described the workflow of generating an LOD3 CityGML model (i.e., a semantic building model with a structured appearance) by adding window and door objects to texture LOD2 CityGML building models. Kang et al. [21] developed an automatic multiprocessing LOD geometric mapping method based on screen-buffer scanning, including semantic mapping rules, to improve the efficiency of the mapping task. However, these studies using the CityGML model rarely involved improvement of visualization and interpretability through texture mapping. In addition, most existing texture-mapping methods for remotely sensed imagery and terrestrial images heavily depend on exterior orientation information and normally require a collinear equation to associate the 3D models and image texture. Nevertheless, due to the simplification of the building model and the inaccurate location information of LOD CityGML building models, the commonly used photogrammetric methods cannot satisfy the requirement of texture mapping for the CityGML building model. In addition, textured buildings derived from aerial photogrammetry are often occluded by ground objects, e.g., pedestrians, vehicles, and trees, as shown in Figure 1, resulting in object occlusion and texture distortion.

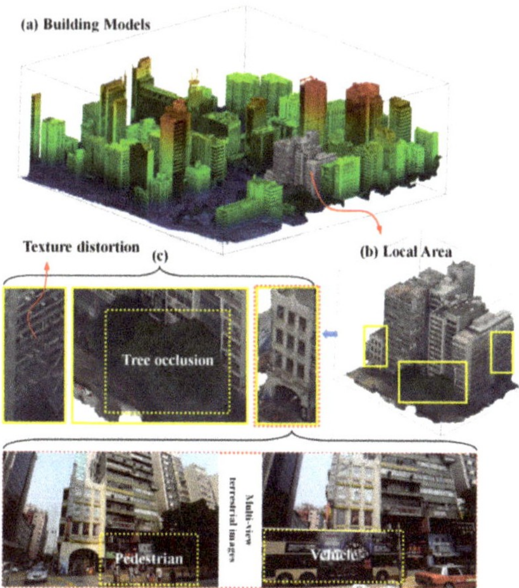

Figure 1. Objection occlusion and texture distortion in building models. (**a**) Example of CityGML building models; (**b**) building models of a local area; (**c**) object occlusion, e.g., tree, pedestrian, and vehicle. Yellow boxes with solid line indicate texture distortion of local areas caused by object occlusion in aerial photogrammetry. Yellow boxes with dotted line indicate object occlusion, e.g., tree, pedestrian, and vehicle.

1.3. Research Objectives

The above-introduced photogrammetric methods seek to perform texture mapping from UAV, LiDAR, and crowd-sourced data through rigorous geometric transformation (e.g., aerial triangulation), which is not suitable for texture mapping of simplified CityGML building models. Additionally, a single data source usually suffers from several problems, such as object occlusion. In this study, we propose a novel approach of texture mapping for 3D building models from multisource data, such UAV remotely sensed imagery and terrestrial images, to alleviate these problems of texture mapping for CityGML building models. This approach does not perform aerial triangulation; instead, only multiview coplanar extraction was explored for texture mapping, without the requirement of exterior orientation information. Inspired by the superiority of deep learning, an object-occlusion detection method combining deep convolutional neural networks and vegetation removal is exploited to filter out pedestrians, vehicles, and trees under complex image background, such as uneven illumination and geometric deformation. Point-feature-based matching under the constraint of building boundaries is conducted to compute the homography matrix of the overlapped image, in which multiview 2D planes are extracted as the candidate textures. Then, geometric topology is derived to accurately delineate the façade boundaries of building models using Hough-transform and iterative least-squares methods. Subsequently, based on the registration of the map and the street view, the untextured or textured building models of CityGML can be mapped or updated using texture information of terrestrial images. Therefore, texture mapping of CityGML building models can be achieved by air-ground integrated data acquisition (e.g., aerial oblique images, ground street-scene images) and processing technologies. Furthermore, the texture of CityGML building models can be automatically and rapidly updated to significantly eliminate blurred and distorted textures caused by object occlusion in aerial photogrammetry.

The main contribution of this work is to propose an approach for texture mapping that is suitable for CityGML building models using 2D remotely sensed and terrestrial images. In this study, deep convolutional neural networks enable high-quality texture to be extracted from complex image backgrounds. Multiview coplanar extraction is defined to extract building façades by perspective transformation without the requirement of exterior orientation information. In addition, geometric topology is used to optimize the façade boundaries of building models for denoising.

The remainder of the paper is organized as follows. Section 2 describes the details of the proposed approach for building façade texture mapping. Sections 3 and 4 present the comparative experimental results in combination with a detailed analysis and discussion. Section 5 concludes this paper and discusses possible future work.

2. Methods

2.1. Overview of the Proposed Approach

Generally, texture mapping on top of the building model using UAV remote imagery is simpler than that on the building façade with terrestrial images because no object occlusion exists. The workflow of the proposed approach focuses on the façade of the building for texture mapping by terrestrial images, as shown in Figure 2; it consists of three stages. In the preprocessing stage, the relevant terrestrial images of the building are gathered from public images through some basic attributes, e.g., GPS position and annotation, and the texture is preferred by excluding object occlusion, e.g., pedestrians, vehicles, and trees, using deep convolutional networks (e.g., NanoDet [22]). In the multiview planar extraction stage, point-based image matching is utilized to compute the homography matrix, which is used to extract texture information in 2D multiview planes under the constraint of building boundaries. In the texture-plane optimization stage, the quadrilateral shape of building façades is defined based on the geometric topology of point and line features and optimized using Hough-transform and iterative least-squares methods. Finally, in the texture-mapping operation, the building façade is mapped from the extracted texture by perspective transformation, including projection, mapping, and resampling.

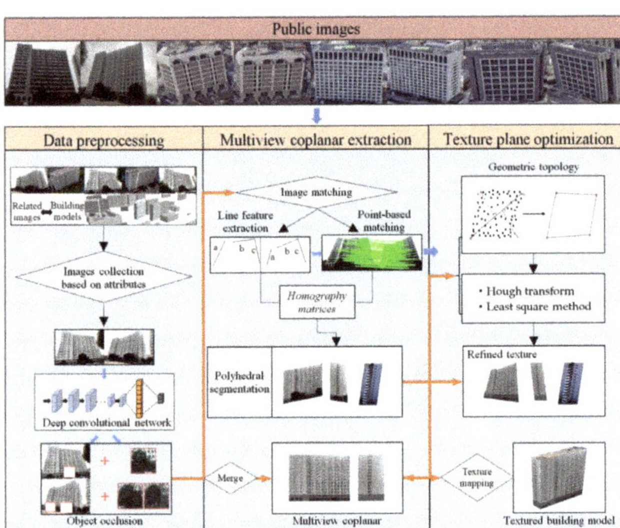

Figure 2. Workflow of the proposed approach.

2.2. Relevant Terrestrial Image Collection Based on Attributes

Although a large number of public images offer an opportunity for texture mapping of building façades easily with low cost, determining which images correspond to which buildings is difficult and time-consuming. Fortunately, many public images are captured by mobile phones with some attributes, e.g., global navigation satellite system (GNSS) position and image annotation, which can be used to filter images that are unrelated to a building. In particular, the image annotation is usually provided with a building name, which can correspond to the building through the online map, and the orientation of an image to the center of the building can be derived in combination with the GPS position. Therefore, we developed an approach of relevant terrestrial image collection based on attributes.

As shown in Figure 3, the related regions, *R* and *R'*, of Buildings 1 and 2 are defined by a given radius, *r*. Then, the subregions, *Region*1–4 and *Region*1'–4', corresponding to each façade of Buildings 1 and 2, respectively, are split by the lines (*l*1,*l*2) and (*l*1',*l*2'), which are defined based on the diagonal of the buildings. Evidently, the candidate terrestrial images corresponding to a building can be obtained on the basis of GPS position and annotation. However, a terrestrial image is usually annotated by only one place name, which may be related to multiple buildings, i.e., the public image *P* may be related to Buildings 1 and 2. Then, we explore the dot-product [23] operation of two vectors, $\vec{PC1}$ and $\vec{PC2}$, to determine whether a public image can provide the potential texture for multiple buildings. In addition, two vectors, $\vec{PC1}$ and $\vec{PC2}$, can be defined based on the GPS position of the public image, *P*, and the centers, *C*1 and *C*2, of Buildings 1 and 2. The dot-product of $\vec{PC1}$ and $\vec{PC2}$ is computed as follows:

$$val\left(\vec{PC1},\vec{PC2}\right) = \vec{PC1}\cdot\vec{PC2} = \left|\vec{PC1}\right|\left|\vec{PC2}\right|cos\theta, \tag{1}$$

where θ is the angle between two vectors, $\vec{PC1}$ and $\vec{PC2}$. Through many experiments and statistical analysis, we conclude that when $\theta > 90$ degrees, a satisfactory texture is difficult to capture due to severe deformation. Hence, when $val\left(\vec{PC1},\vec{PC2}\right) < 0$, if either building is not annotated, then the public image, *P*, cannot be considered a candidate texture for the annotated building.

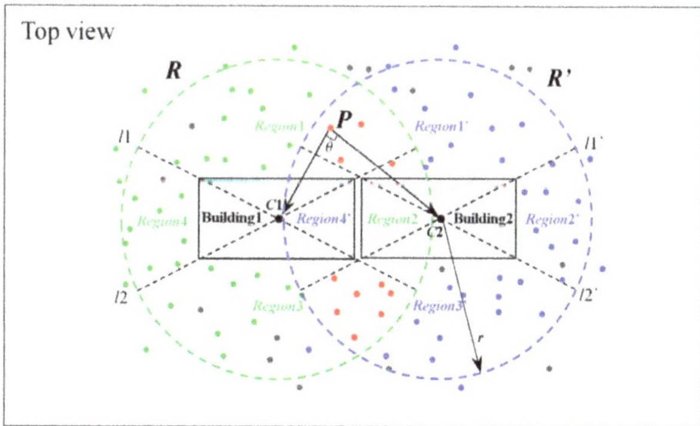

Figure 3. Relevant public-image collection. *C*1 and *C*2 are the centers of Buildings 1 and 2, respectively. *R* and *R'* are the regions corresponding to Buildings 1 and 2, respectively. (*l*1,*l*2) and (*l*1',*l*2') are the splitting lines of *R* and *R'*, respectively. *Region*1-4 and *Region*1'-4' are the subregions split by (*l*1,*l*2) and (*l*1',*l*2'), respectively. Green and blue dots denote the terrestrial images that may be related to Buildings 1 and 2 within the radius, *r*, respectively. Red dots denote the public images that may be related to buildings (i.e., Building 1 and 2), and gray dots denote the terrestrial images that are unrelated to either Building 1 or 2. θ is the angle between two vectors, $\vec{PC1}$ and $\vec{PC2}$.

2.3. Object-Occlusion Detection Based on Deep Learning

Although a large number of public images enables the CityGML building model to perform texture mapping from terrestrial images without extra data acquisition, some problems cannot be ignored, such as texture redundancy and object occlusion. Unfortunately, because public images are acquired from different viewpoints, times, conditions, and cameras, complex nonlinear transformation, such as uneven illumination, deformation, and object mixture often exist. Then, automatically detecting object occlusion and selecting high-quality texture using the conventional feature-based methods is difficult [24–26].

Compared with these methods based on manually-designed features, deep learning can perform better in image classification, pattern recognition, image processing, and other fields [27–29]. Inspired by the progresses and outstanding nonlinear feature extraction achieved in deep learning in recent years, convolutional neural networks can not only extract multiscale and nonlinear features from images but are also insensitive to image translation, scale, viewpoint, and deformation [30,31]. Therefore, we utilized a deep convolutional neural network to detect object occlusion and gather high-quality terrestrial images for texture mapping.

In recent years, many convolutional neural networks, e.g., VGG [32] and GoogleNet [33], have been proposed and perform well for some applications, such as object recognition and classification [34,35]. However, these networks are very deep and large-scale, with tens of millions of parameters. Thus, deep neural networks, such as VGG and GoogleNet, cannot satisfy the requirement of fast object recognition involved in CityGML building-texture mapping. Recently, a project named NanoDet [22] appeared on GitHub; it is an open-sourced and real-time anchor-free detection model, which can provide good performance—as much as that of the YOLO network [36–38]; it is also easy for training and porting. NanoDet is a detection model considering accuracy, efficiency, and model scale; it is achieved by combining some tricks that refer to deep learning literature to obtain a detection model considering accuracy, efficiency, and model scale. Generalized focal loss and box regression are used in NanoDet to reduce a large number of convolutional operations and significantly improve efficiency. Although NanoDet is a lightweight model, its performance is similar to that of the state-of-the-art networks [22]. Therefore, to avoid complex training from scratch, we explore a transfer-learning strategy based on NanoDet to evaluate object occlusion and determine high-quality public images for texture mapping, considering the performance and efficiency of the deep neural network. In addition, typical objects, such as pedestrians and vehicles, can be easily detected. Other types of object occlusion (e.g., trees) are not easily inferred by most convolutional neural networks, such as DanoDet, because of uncertainty and irregular distribution. Furthermore, as illustrated in Figure 4, we introduce a gamma-transform green leaf index, named *GGLI* [39], to detect tree occlusion. Then, an approach combining DanoDet and *GGLI* is proposed to evaluate object occlusion, and the area of object occlusion can be calculated. Subsequently, low-quality public images with high occlusion ratios can be excluded without being performed for texture mapping.

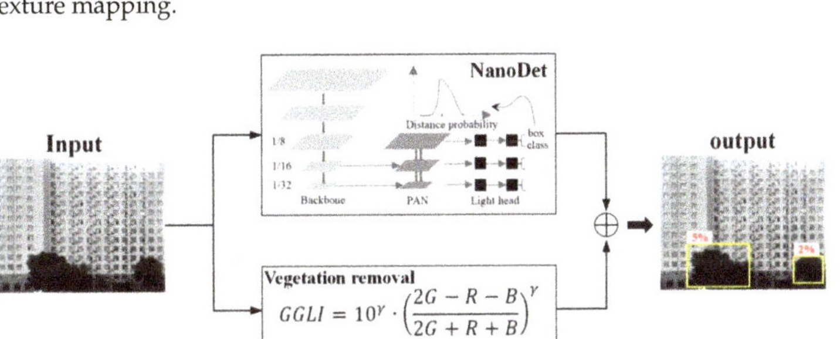

Figure 4. Object-occlusion detection combined with NanoDet and vegetation removal. γ denotes a gamma value, and R, G, and B are the three components of RGB color.

2.4. Multiview Coplanar Extraction

The actual textured 3D building model derived from photogrammetric technologies (including oblique photogrammetry and laser scanning) can finely characterize the geometric building structure. However, limited by the error of building-model reconstruction, especially for simplified CityGML building models, such as the LOD2 CityGML models, it cannot perform automatic texture mapping using photogrammetric technologies from the terrestrial images without the support of exterior orientation information. Unfortunately, in most cases, the camera parameters and pose of public images are unknown; thus, texture mapping from public images becomes more difficult. In general, the façades of most buildings are composed of several approximate planes. In particular, homography transformation is usually used to describe the relationship between two images of some points on the common plane and broadly used for photogrammetry and computer vision, such as image correction, image mosaic, camera-pose estimation, and visual simultaneous localization and mapping (SLAM) [40,41]. Therefore, based on these characteristics of CityGML building façades and homography transformation, we developed a multiview coplanar extraction approach from the candidate terrestrial images by homography matrix.

As opposed to the commonly used photogrammetric methods, some parameters, such as interior parameters and exterior orientation elements, are not the prerequisites for deriving spatial correspondence between public images and the CityGML building model in this study; that is, the collinear equation is not an essential condition for texture mapping. In other words, compared with the commonly used photogrammetric methods, the proposed multiview coplanar extraction based on homography matrix is more available and is an alternative method for texture mapping of CityGML building models. A single homography matrix, i.e., global homography matrix, cannot be simply applied to define the transformation of two views for extraction of textures of multiple building façades because a public image may cover multiple planes of a building. Therefore, we exploit multiple local homography matrices, named L_H, to model the multiple façades of a building, as shown in Figure 5. The mathematical formula can be expressed as follows:

$$L\_H(I_1, I_2, \ldots, I_n) = \{\{H_p | p \in (I_1, I_2, \ldots, I_n)\}\}, \tag{2}$$

$$H_p = \begin{bmatrix} h_{00} & h_{01} & h_{02} \\ h_{10} & h_{11} & h_{12} \\ h_{20} & h_{21} & 1 \end{bmatrix}, \tag{3}$$

where $I_1, I_2, \ldots, I_n$ are the candidate public images, $1 \sim n$, for texture mapping; H_p denotes a homography matrix of a local plane in $(I_1, I_2, \ldots, I_n)$; and $h_{00} \sim h_{21}$ are the matrix elements of H_p. Therefore, L_H may involve more than one homography matrix.

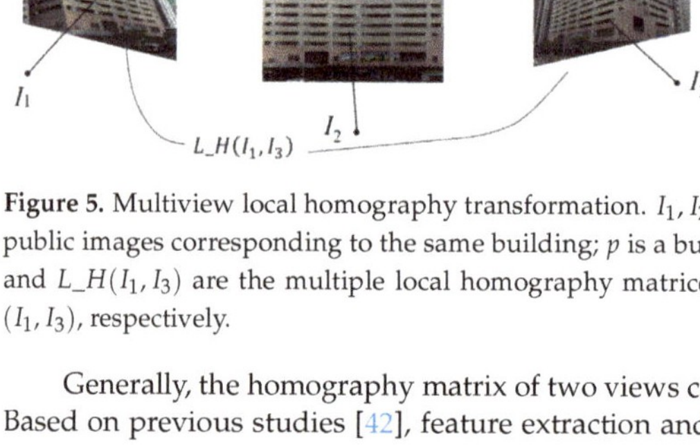

Figure 5. Multiview local homography transformation. I_1, I_2, I_3 denote three multiview candidate public images corresponding to the same building; p is a building façade. $L\_H(I_1, I_2)$, $L\_H(I_2, I_3)$, and $L\_H(I_1, I_3)$ are the multiple local homography matrices of image pairs (I_1, I_2), (I_2, I_3), and (I_1, I_3), respectively.

Generally, the homography matrix of two views can be obtained by image matching. Based on previous studies [42], feature extraction and matching are performed using a sub-Harris operator coupled with the scale-invariant feature-transform algorithm, which can find evenly distributed corresponding points to compute the homography matrix. In homogeneous coordinates, the homography transformation between a point, $X(x_i, y_i, 1)$, of a public image, I, and the corresponding point, $X'(x'_i, y'_i, 1)$, of the matched image, I', can be described by a mathematical formula, $X' = H_p X$, which is also the perspective transformation. The homography matrix, H_p, has 8 degrees of freedom; thus, at least four matching pairs are required to solve this matrix. Then, the homography transformation, in terms of matches, can be expressed as follows:

$$\begin{bmatrix} x'_i \\ y'_i \\ 1 \end{bmatrix} \cong L\_H_p(I, I') \begin{bmatrix} x_i \\ y_i \\ 1 \end{bmatrix}, \{(x_i, y_i) \in F_I, (x'_i, y'_i) \in F_{I'}\}, \tag{4}$$

where F_I and $F_{I'}$ denote the set of features in images I and I', respectively.

For the case of n matches in a plane, p, of images I and I', using the least-squares method, Equation (4) can be expressed using an alternative formula, $Af = 0$. Then, the coefficients, h_{00}–h_{21}, are calculated by a nonlinear optimization of $\min\|Af\|^2$. Here, A and f are expressed as follows:

$$A = \begin{bmatrix} x_1, y_1, 1, 0, 0, 0, -x_1 x'_1, -y_1 x'_1, -x'_1 \\ 0, 0, 0, x_1, y_1, 1, -x_1 y'_1, -y_1 y'_1, -y'_1 \\ \vdots \\ x_n, y_n, 1, 0, 0, 0, -x_n x'_n, -y_n x'_n, -x'_n \\ 0, 0, 0, x_n, y_n, 1, -x_n y'_n, -y_n y'_n, -y'_n \end{bmatrix}, \tag{5}$$

$$f = [h_{00}, h_{01}, h_{02}, h_{10}, h_{11}, h_{12}, h_{20}, h_{21}, 1]^{\mathrm{T}}. \tag{6}$$

Note that only one global homography matrix can be obtained based on the previous studies; therefore, we propose a strategy to define multiple planes that may exist in two paired views by extracting coplanar features. Although the mathematical transformation of each plane in a terrestrial image can be derived by L_H, the sub-Harris corners are insufficient to form the geometric shape of the façades of a building. Extracting the textures on each plane is still a problem because so far, no accurate boundary of the polygon on building façades is delineated. Generally, a large number of line features is distributed on the building façades. In addition, due to the advantages of easy extraction and strong anti-noise ability, line features are extracted to obtain abundant geometric description of the façades, and the corresponding points on the lines between two paired views are determined by the calculated L_H. To further determine the coplanar features on the same building, based on the similarity of geometry and texture on the same façade, we use the feature descriptors, namely RGB-SIFT descriptors [43], to exclude features not on the same façade or outliers by clustering.

2.5. Texture-Plane Quadrilateral Definition Based on Geometric Topology

The geometric boundary of a façade is assumed to be consistent with the quadrilateral, which is warped due to perspective transformation. On the basis of the spatial distribution of the coplanar features, we subsequently perform texture-plane extraction based on geometric topology. Specifically, as shown in Figure 6, the two farthest points, X_a, X_b, are initially determined as the two initial diagonal corners of the quadrilateral façade from the coplanar point set $S(X)$. Then, the line equation, l_{ab}, between points X_a and X_b can be expressed as follows:

$$\alpha x + \beta y + \delta = 0, \tag{7}$$

where the coefficients, α, β, δ, can be calculated by the coordinates of points X_a, X_b; (x, y) is the coordinate of a coplanar point. The two other corners, X_c, X_d, of the façade can be defined based on the condition, i.e., Equation (8), that the farthest vertical distance from the point set, $S(X)$, on both sides of line l_{ab}.

$$\begin{cases} d_{X_c \to l_{ab}} = \max(d|(x,y) \in S(X) \cap \{\alpha x + \beta y + \delta > 0\}) \\ d_{X_d \to l_{ab}} = \max(d|(x,y) \in S(X) \cap \{\alpha x + \beta y + \delta < 0\}) \end{cases}, \tag{8}$$

where d is the distance between a coplanar point (x,y) and line l_{ab}, and the calculation formula is as follows.

$$d = \left| (\alpha x + \beta y + \delta)/\sqrt{\alpha^2 + \beta^2} \right|. \tag{9}$$

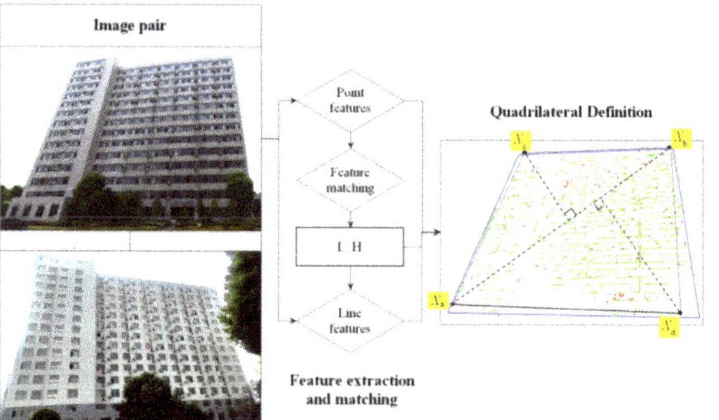

Figure 6. Initial quadrilateral definition of a plane. The red and green dots denote the corresponding points obtained from point features and line features, respectively. Blue lines represent the ground truth of the boundary of a façade, and the black line is the initial quadrilateral boundary.

Although the initial boundary of a façade can be defined based on the four anchor points, i.e., X_a, X_b, X_c, X_d, a clear error is found on the boundary because the contribution of coplanar points on the edge of the façade is not considered. Then, we use Hough-transform and iterative least-squares methods together to optimize the initial boundary obtained by the four anchor points. This optimization consists of the following steps:

(1) Along the straight lines, l_{ab}, l_{ad}, l_{bc}, and l_{bd}, point sets $S^X_{l_{ab}}$, $S^X_{l_{ad}}$, $S^X_{l_{bc}}$, and $S^X_{l_{bd}}$ with the closest vertical distance to the corresponding straight lines are found.

(2) Hough-transform algorithm is conducted to derive the mathematical formulas (i.e., $y = kx + \epsilon$, where k, ϵ denote slope and intercept of lines l_{ab}, l_{ad}, l_{bc}, and l_{bd} from $S^X_{l_{ab}}$, $S^X_{l_{ad}}$, $S^X_{l_{bc}}$, and $S^X_{l_{bd}}$, respectively.

(3) Iterative weighted least-squares method [44] is explored to optimize each mathematical formula of lines l_{ab}, l_{ad}, l_{bc}, and l_{bd}, and the error correction, $\hat{x}$, is expressed as follows:

$$\hat{x} = \left(A^{\mathsf{T}} P A \right)^{-1} A^{\mathsf{T}} P L, \tag{10}$$

$$P = diag(P_1, P_2, \ldots, P_n), \tag{11}$$

where $\hat{x} = \begin{bmatrix} \delta k \\ \delta \epsilon \end{bmatrix}$; $A = \begin{bmatrix} x_1 & 1 \\ \vdots & \vdots \\ x_n & 1 \end{bmatrix}$; $B = \begin{bmatrix} y_1 \\ \vdots \\ y_n \end{bmatrix}$; n is the number of points in $S^X_{l_{ab}}$, $S^X_{l_{ad}}$, $S^X_{l_{bc}}$, and $S^X_{l_{bd}}$; P is the diagonal weight matrix; and $P_i \propto 1/d$, which is updated after each line-formula optimization.

2.6. Sub-Image Mosaic for Object-Occlusion Filling

This study aims at texture mapping for low-quality textured and untextured CityGML building models and attempts to solve the problem of texture occlusion through multiview images. Although the multiview texture of a façade can be captured based on this method, object occlusion caused by pedestrians, trees, and vehicles may lead to missing partial texture in a single image. Fortunately, multiview public images obtained from different perspectives may capture textures at different angles. Thus, the missing local texture of a façade may be filled by the unobstructed area; that is, the unobstructed sub-images obtained from multiview textures can be mosaicked to solve missing texture in the masked area introduced in Section 2.3.

The details of the Algorithm 1 for texture extraction are shown in the following section.

Algorithm 1: Texture extraction based on sub-image mosaic

Input: $S(I)$ is the set of candidate terrestrial images for one façade, *num* is the size of $S(I)$, (x, y) is the coplanar point, and $S(T)$ is the texture set.
Parameters: Gamma-transform green leaf index, $GGLI$, multiple local homography matrices, L_H. k, ϵ are the coefficients of a line.
Output: Texture, T, of the façade
1: **for** $i = 1$ to *num* **do**
2:     Perform object detection using NanoDet
3:     Compute $GGLI$
4:     Remove area $R_o(i)$ of object occlusion
5: **end for**
6: **for** $i = 1$ to *num* $- 1$ **do**
7:     **for** $j = i + 1$ to *num* **do**
8:         Perform feature extraction and matching based on sub-Harris operator
9:         Compute L_H
10:         Define initial quadrilateral $\mathbb{R}^q \leftarrow (X_a, X_b, X_c, X_d)$
11:         Compute k, ϵ based on Hough transform
12:         Compute $\hat{x} \leftarrow \begin{bmatrix} \delta k \\ \delta \epsilon \end{bmatrix}$ based on least square method
13:         Update $k, \epsilon \leftarrow \hat{x}$
14:         Refine $(l_{ab}, l_{ad}, l_{bc}, l_{bd})$ and $\mathbb{R}^q$
15:         Repeat steps from 12 to 14 until error convergence or the maximum iteration number is reached
16:         Add $\mathbb{R}^q$ into $S(T)$
17:     **end for**
18: **end for**
19: Merge $S(T)$ into T

---

3. Experiment Results and Analysis

A set of CityGML building models is used to perform the experiments. The datasets mainly include two categories, as follows: (1) the untextured building models downloaded from the commercial map providers, such as Baidu, and (2) the textured building models derived from the photogrammetric method. Initially, three building models are selected to evaluate the method quantitatively and qualitatively. To further evaluate the performance of texture update, five textured building models are selected to evaluate the proposed approach by replacing low-quality texture. In addition, the public images are collected to capture texture from street-view images managed by the commercial map providers. Only relatively regular and simplified building models, such as LOD2 CityGML building models, are selected to evaluate the proposed approach because this study mainly focuses on the texture mapping of nondetailed building models.

3.1. Data Description

To validate the effect of the proposed texture-mapping method, the three typical untextured building models and the corresponding multiview texture images, including UAV remotely sensed and terrestrial images, as shown in the subfigures of the first and second columns in Figure 7, are selected. These building models are characterized by simplified geometric structure, different styles, different heights, and different façades. The candidate texture images with multiple perspectives include object occlusion, such as pedestrians, trees, vehicles, and other nonbuilding objects.

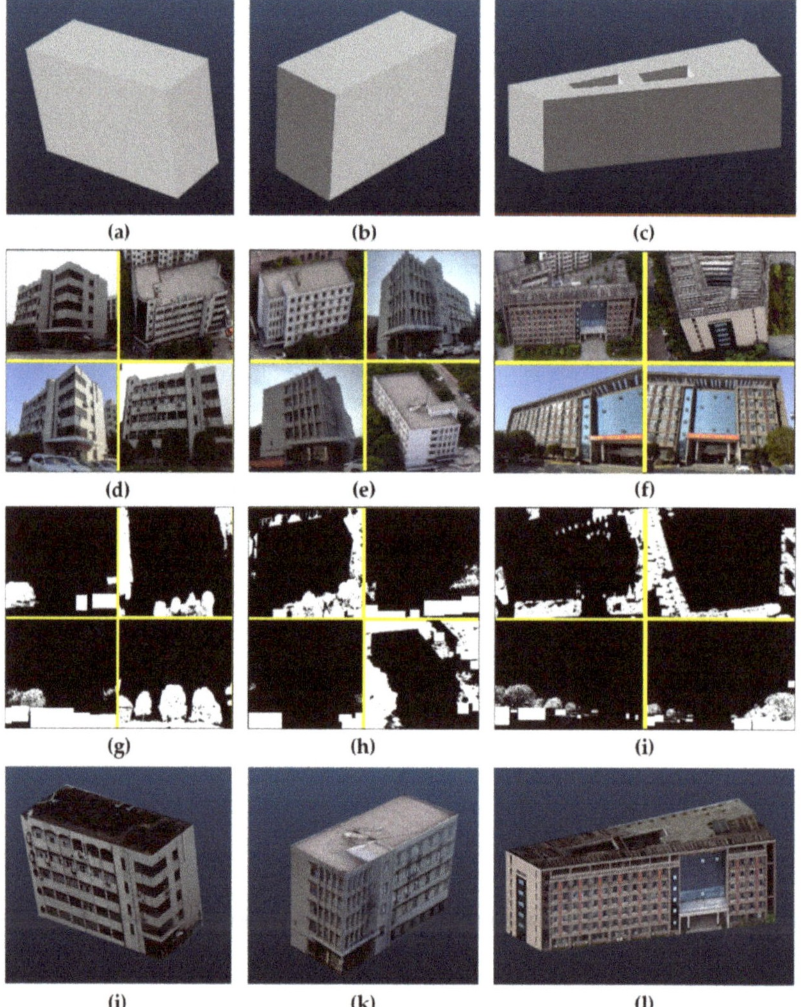

Figure 7. Typical untextured building models and the multiview texture images. (**a**–**c**) are three untextured building models; (**d**–**f**) are examples of texture images with different viewpoints corresponding to (**a**–**c**), respectively. (**g**–**i**) are the object occlusions, marked by white regions, detected by combining NanoDet and GGLI corresponding to (**a**–**c**), respectively. (**j**–**l**) are three textured building models by texture mapping using the proposed approach corresponding to (**a**–**c**), respectively.

In addition, the three textured building models derived by photogrammetric mapping and the corresponding multiview texture images, as shown in the subfigures of the first and second columns in Figure 8, are selected. Different from the untextured building models in Figure 7, the textured building models are characterized by detailed geometric structure. In the experiments, texture mapping of these textured building models has low-quality, which is probably caused by photogrammetric error, noneliminated object occlusion, or imaging quality. These models are specially selected to validate the performance of the proposed approach for improving texture quality.

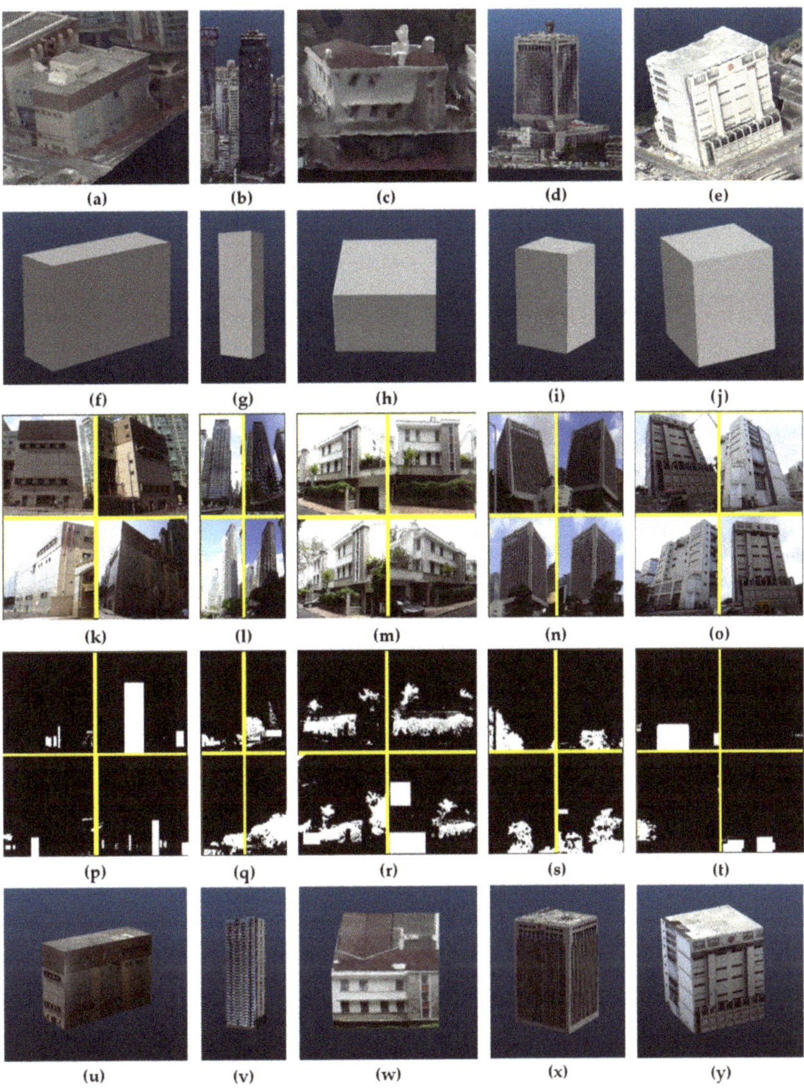

Figure 8. Typical textured building models and multiview texture images. (**a–e**) are five textured building models, and (**f–o**) are examples of texture images with different viewpoints and LOD CityGML building models corresponding to (**a–e**). (**p–t**) are object occlusions, marked by white regions, detected by combining NanoDet and GGLI corresponding to (**a–e**). (**u–y**) are five textured building models by texture mapping using the proposed approach corresponding to (**a–e**).

3.2. Qualitative Performance Evaluation

In the experiments, as shown in Figures 7 and 8, object occlusions, such as pedestrians, vehicles, trees, and other objects, can be effectively removed by jointly using NanoDet and GGLI. The façades of the simplified building models, such as LOD CityGML building models, e.g., Figure 7a–c, can be textured by the multiview coplanar extraction from multiple public images (e.g., Figure 7d,e,f) obtained from different viewpoints. The texture on the top façade is captured from UAV orthophoto based on the previous studies [39]. The façades surrounding these buildings can be textured from public images in which the texture hidden by object occlusion (e.g., Figure 7g–i) can be uncovered, given that the textures may appear in multiview images because of the different depths of objects and buildings. Therefore, object occlusion can be effectively removed by merging multiple texture planes, which are then defined based on quadrilateral geometric topology to delineate the boundary of textures for mapping, as illustrated in Figure 7j–l.

Although actual textured 3D building models are derived by UAV mapping, such as oblique photogrammetry in some places, as shown in Figure 8a–e, low-quality textures characterized by low-resolution and warped surfaces are inevitably mapped to the building façades because of object occlusion, low precise building geometric models, and missing texture information. Generally, low-quality building models should be improved and updated by manual processing and mapping operations, which is a tedious and time-consuming task. Fortunately, the LOD CityGML building models corresponding to these actual textured 3D building models are provided by the commercial map providers, such as Baidu. Then, a simplified LOD CityGML building model can be considered an alternative for the low-quality textured 3D building, as shown in Figure 8f–g. Similar to Figure 7, object-occlusion removal shown in Figure 8p–t and multiview coplanar extraction from public images (e.g., Figure 8k–o) are conducted for texture mapping. On the contrary, although the alternative building models, such as the simplified LOD CityGML building models, cannot provide the detailed geometric structure of buildings, they significantly improve the geometric shape and texture-mapping quality of building façades. For examples, in Figure 8a,c, the warped façades are replaced by regular planes, and the texture quality of the building is also optimized by terrestrial images with higher resolution and more abundant detail.

Compared with the visualization of the textured building models, the proposed approach is suitable for performing texture mapping on regular building models, such as LOD CityGML building models. In addition, in some cases, the low-quality geometric structure of building façades can be optimized by regular planes. Not all façades of building models perform texture mapping when relevant terrestrial images, such as B3–B6, are limited.

3.3. Performance Evaluation of Object-Occlusion Detection

One advantage of the proposed approach is that it has outstanding performance in object-occlusion detection. To evaluate the performance of combining NanoDet and GGLI to detect object occlusion, state-of-the-art deep neural networks, including R-CNN [45], Faster-R-CNN [46], YOLO [37], and NanoDet [35], are selected for comparison and analysis. A metric, namely overall accuracy (OA), is used to quantitatively assess performance, and OA is computed as follows:

$$OA = (TP + TN)/(TP + FN + TN + FP), \tag{12}$$

where TP, FN, TN, and FP are defined as accurately detected object-occluded regions, inaccurately detected non-object-occluded regions, accurately detected non-object-occluded regions, and inaccurately detected object-occluded regions, respectively. The texture images, namely Datasets 1–8, corresponding to the building models in Figures 7 and 8 are collected to compare the performance of object occlusion with the state-of-the-art networks. Table 1 presents the comparative results of OA values using R-CNN, Faster-R-CNN, YOLO,

NanoDet, and our method (i.e., the combination of NanoDet and GGLI). The combination of NanoDet and GGLI achieves a better performance than the five other deep learning networks in terms of OA values. Compared with the four other methods, our method significantly improves the accuracy of object-occlusion detection by vegetation removal based on previous studies, such as GGLI. However, the other deep learning networks do not have the ability to detect vegetation, such as trees. As shown in Figure 8k,o corresponding to Datasets 4 and 8, the tree occlusion is less than in other datasets. Thus, the accuracy of object-occlusion detection, represented by OA values, is close to our proposed method.

Table 1. Comparisons of OA values using Faster-R-CNN, YOLO, NanoDet, and our method.

Dataset	R-CNN	Faster-R-CNN	YOLO	NanoDet	NanoDet + GGLI
Dataset1	66.0	66.9	71.6	71.2	92.9
Dataset2	53.3	56.9	70.4	67.2	88.3
Dataset3	61.6	68.3	74.1	64.7	85.9
Dataset4	86.1	95.9	96.0	98.9	99.6
Dataset5	50.4	57.9	68.4	59.6	87.6
Dataset6	57.7	59.1	63.6	74.9	92.6
Dataset7	43.3	48.9	50.1	78.7	87.8
Dataset8	73.7	81.1	83.2	87.9	92.8

3.4. Performance Evaluation of Multiview Coplanar Extraction

The results of multiview coplanar extraction using the proposed approach are evaluated by the quantitative metrics, i.e., recall, precision, and intersection over union (IoU), which can be computed as [47]

$$\text{Recall} = (\mathbb{R}^{GT} \cap \mathbb{R}^{q})/\mathbb{R}^{GT}, \tag{13}$$

$$\text{Precision} = (\mathbb{R}^{GT} \cap \mathbb{R}^{q})/\mathbb{R}^{q}, \tag{14}$$

$$\text{IoU} = (\mathbb{R}^{GT} \cap \mathbb{R}^{q})/(\mathbb{R}^{GT} \cup \mathbb{R}^{q}), \tag{15}$$

where $\mathbb{R}^{GT}$ and $\mathbb{R}^{q}$ are the ground truth delineated by manual operation and the quadrilateral region extracted by multiview coplanar extraction, respectively.

Point-feature-based matching is a popular method used to compute the geometric transformation between images. However, building facades often have weak textures. Thus, point features may be insufficient to reconstruct the boundary of the texture quadrilateral region. Line features are extracted to obtain coplanar features to evaluate the performance of combining point and line features to detect the boundary of the texture quadrilateral region. We compare the results obtained by point-based and point-line-based methods. Figure 9 depicts the comparative results of Recall, Precision, and IoU values calculated from the public images, corresponding to the eight building models, including three untextured and five textured models. The point-line-based method for quadrilateral-region detection achieves a better performance than the point-based method in terms of the Recall, Precision, and IoU values; that is, the texture boundaries obtained from the point-line-based method are closer to the ground-truth texture regions. The point-line-based method is suitable for achieving this goal due to the following reason: linear objects, such as building edges and window edges, are abundant and easy to extract from building façades and can be used to support boundary detection.

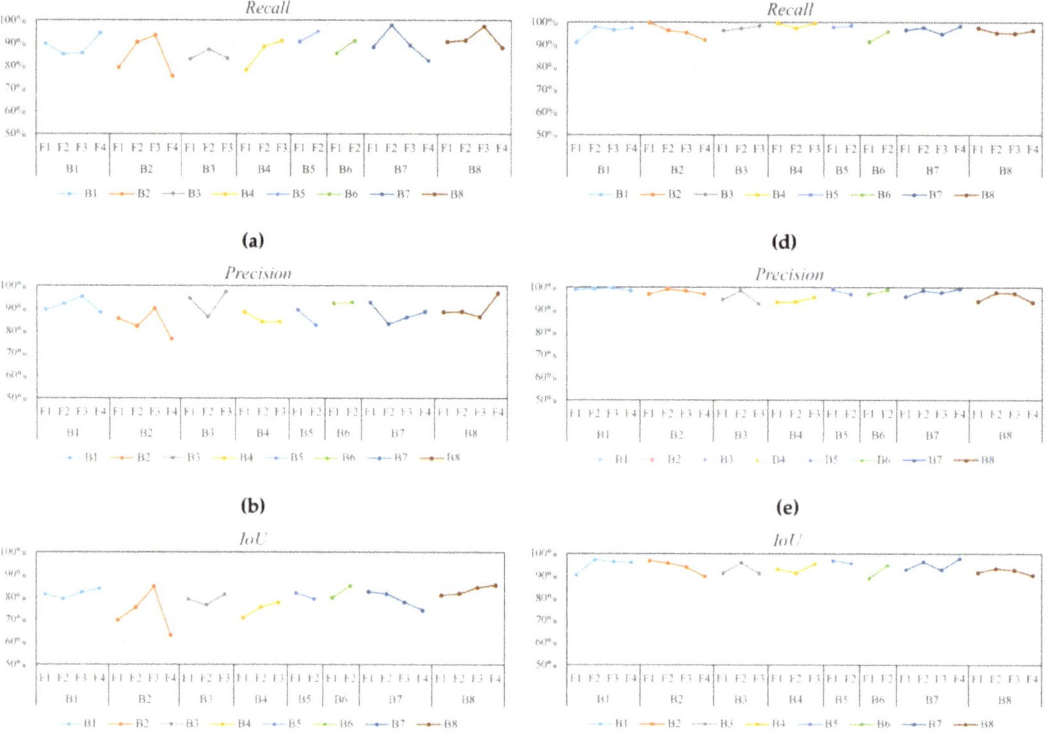

Figure 9. Comparisons of Recall, Precision, and IoU values calculated from the public images corresponding to the three untextured and five textured models shown in Figures 7 and 8. (**a–c**) are the Recall, Precision, and IoU values obtained by point-based quadrilateral-region detection. (**d–f**) are the Recall, Precision, and IoU values obtained by point-line-based quadrilateral-region detection. B1–B8 indicate the number of building models, and F1–F4 indicate the façades of building models.

3.5. Quality Evaluation of Updated Texture

Visual comparison of Figure 8a–e,u–y shows that some geometric details and textures on the façades of building models are seriously blurred and distorted in Figure 8a–e. They are optimized and substituted using patches obtained from the high-resolution terrestrial images by the proposed approach in Figure 8u–y. In addition, to further evaluate the superiority of the proposed approach, a metric, namely a Tenengrad function based on gradient without reference image [48], is used to quantitatively compare the texture quality before and after optimization. The Tenengrad value, *Ten*, of an image, *I*, is computed as follows:

$$Ten = \sum_y \sum_x |G(x,y)|(G(x,y) > T), \tag{16}$$

$$G(x,y) = \sqrt{G_x^2(x,y) + G_y^2(x,y)}, \tag{17}$$

in which $G(x,y)$ is the gradient of a pixel $I(x,y)$, and $G_x(x,y)$ and $G_y(x,y)$ are gradients in the horizontal and vertical directions, respectively. T is a given threshold. The comparative results of five building models in Figure 8a–e are shown in Figure 10, which indicates that the texture after optimization has higher quality and clearer details than that prior to optimization in terms of *Ten* values, and it can also significantly eliminate blurred and distorted textures.

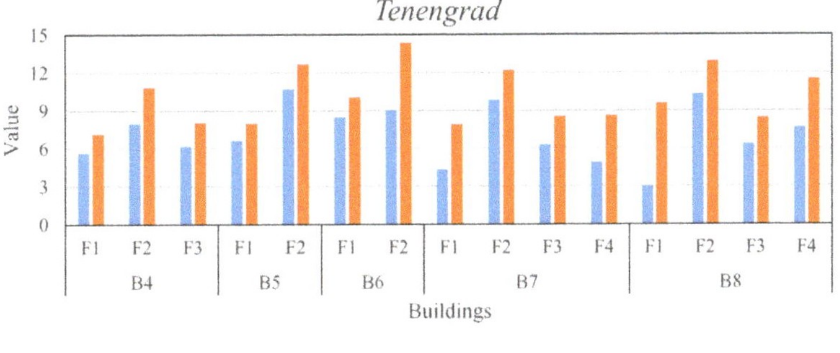

Figure 10. Comparisons of *Ten* values calculated from the textures on the façades of five building models in Figure 8a–e. Dark blue and orange columns denote the textures before and after optimization, respectively. B4–B8 indicate the number of building models, and F1–F4 indicate the façades of building models.

4. Discussion

On the basis of experimental results of texture mapping, the proposed approach can be considered an alternative for performing texture mapping for regular building models, such as simplified LOD CityGML building models. In particular, as opposed to the commonly used photogrammetric method for texture mapping, reconstructing high-quality textures for the façades of building models using exterior orientation information is not imperative. The effectiveness of the proposed texture-mapping approach can be explained by a number of reasons. First, high-resolution terrestrial images gathered based on spatial relevancy derived from spatial location and attributes, such as GPS position and image annotation, can provide multidata for texture mapping. As shown in Figures 7d–f and 8k–o, the higher-resolution terrestrial images compared with aerial photography can be used to reconstruct the textures for the façades of building models. Second, as illustrated in Figures 7g–i and 8k–t, the abundant terrestrial images offer the opportunity to collect higher-quality textures by effectively filtering out object occlusions, such as pedestrians, vehicles, and trees by deep learning and vegetation removal. Third, multiview coplanar extraction based on multiple local homography matrices enables texture mapping for simplified or regular building models, such as LOD CityGML building models, in multiple 2D spaces without the support of interior parameters and exterior orientation elements. It even allows nonprofessional practitioners to perform texture mapping with high-resolution terrestrial images. Finally, the point-line-based method for quadrilateral-region detection is available to capture the optimal building façade boundaries of patches for texture mapping.

The essence of texture mapping is the two-dimensional parameterization of 3D building models; that is, a one-to-one correspondence between 2D texture space and 3D building façades should be established. In this study, multiview coplanar extraction is definitely proposed to establish this correspondence. However, this study concentrates on texture mapping for simplified or regular building models, such as LOD2 CityGML building models, which are popular in street view maps (e.g., Google, Baidu). Therefore, it may not be suitable to perform texture mapping for some complex buildings with abundant building details or complex geometric structure.

5. Conclusions

We present a framework to effectively perform texture mapping for LOD CityGML building models by extracting high-quality textures from terrestrial images. First, terrestrial images corresponding to the target building are collected from public images based on spatial relevancy. Second, integration of deep learning and GGLI is used to filter out object occlusions (e.g., pedestrians, vehicles, and trees) and obtain non-occluded building candidate texture distributions. Third, point-line-based coplanar features are extracted to characterize multiple planes in 2D space under the constraint of multiple local homography matrices, and the initial boundaries of the building models are obtained from four anchor points. Fourth, geometric topology is conducted to optimize the initial boundaries of texture patches based on a strategy combining Hough-transform and iterative least-squares methods. Finally, abundant candidate texture patches are mosaicked to obtain high-quality object-occlusion filling. The statistical and visualization results indicate that the proposed methods can effectively perform texture mapping of CityGML building models. The framework also shows higher-quality textures for all experimental building models, including untextured and textured models, according to quantitative and qualitative comparisons and analyses. The results prove the high capability of the proposed approach in texture mapping for CityGML building models from 2D terrestrial images.

The proposed texture-mapping approach relies greatly on the regular geometric shape of building models, in which the façades are composed of multiple rectangles. At present, the proposed approach focuses on texture mapping of simplified or regular building models, such as LOD2 CityGML building models. It does not optimize the geometric structure of the façades of building models. However, it cannot satisfy the requirement of texture mapping for some building models with high levels of detail, such as LOD3 CityGML building models.

In future studies, we will attempt to improve the proposed approach by optimizing the geometric structure on the façades of building models using multiscale and multiview coplanar extraction and improve the performance of texture mapping for complex building models, such as LOD3 CityGML building models.

Author Contributions: Conceptualization, H.H. and J.Y.; methodology, H.H. and P.C.; software, Y.W. and Y.Z.; validation, J.Y., T.L. and G.D.; formal analysis, P.C.; investigation, H.H. and Y.Z.; resources, Y.W.; data curation, T.L. and G.D.; writing—original draft preparation, H.H.; writing—review and editing, J.Y.; visualization, H.H.; supervision, P.C.; project administration, Y.W.; funding acquisition, H.H. All authors have read and agreed to the published version of the manuscript.

Funding: This research was funded by the "National Natural Science Foundation of China, grant number 41861062 and 41401526", "Fuzhou Youth Science and Technology Leading Talent Program, grant number 2020ED65", "Jiangxi University Teaching Reform Research Project, grant number JXJG-18-6-11", "Science and Technology Project of Jiangxi Provincial Department of Water Resources, grant number 202123TGKT12", and "Jiangxi 03 Special Project and 5G Project, grant number 20204ABC03A04".

Data Availability Statement: Data available on request.

Acknowledgments: The authors thank Mengyun Ling for providing datasets.

Conflicts of Interest: The authors declare no conflict of interest.

References

1. Shan, J.; Li, Z.X.; Zhang, W.Y. Recent progress in large-scale 3D city modeling. *Acta Geod. Cartogr. Sin.* **2019**, *48*, 1523–1541.
2. Gröger, G.; Kolbe, T.H.; Nagel, C.; Häfele, K.H. *OGC City Geography Markup Language (CityGML) Encoding Standard*; Open Geospatial Consortium: Rockville, MD, USA, 2012.
3. Kolbe, T.H. Representing and Exchanging 3D City Models with CityGML. In *3D Geo-Information Sciences*; Springer: New York, NY, USA, 2009.
4. Kutzner, T.; Chaturvedi, K.; Kolbe, T.H. CityGML 3.0: New Functions Open Up New Applications. *PFG—J. Photogramm. Remote Sens. Geoinf. Sci.* **2020**, *88*, 43–61. [CrossRef]
5. Eriksson, H.; Harrie, L. Versioning of 3D City Models for Municipality Applications: Needs, Obstacles and Recommendations. *ISPRS Int. J. Geo-Inf.* **2021**, *10*, 55. [CrossRef]
6. Pepe, M.; Costantino, D.; Alfio, V.S.; Vozza, G.; Cartellino, E. A Novel Method Based on Deep Learning, GIS and Geomatics Software for Building a 3D City Model from VHR Satellite Stereo Imagery. *ISPRS Int. J. Geo-Inf.* **2021**, *10*, 697. [CrossRef]
7. Hensel, S.; Goebbels, S.; Kada, M. Facade reconstruction for textured Lod2 Citygml models based on deep learning and mixed integer linear programming. *ISPRS Ann. Photogramm. Remote Sens. Spat. Inf. Sci.* **2019**, *IV-2/W5*, 37–44. [CrossRef]
8. Li, D. 3D visualization of geospatial information: Graphics based or imagery based. *Acta Geod. Cartogr. Sin.* **2010**, *39*, 111–114.

9. Yalcin, G.; Selcuk, O. 3D City Modelling with Oblique Photogrammetry Method. *Procedia Technol.* **2015**, *19*, 424–431. [CrossRef]
10. Abayowa, B.O.; Yilmaz, A.; Hardie, R.C. Automatic registration of optical aerial imagery to a LiDAR point cloud for generation of city models. *SPRS J. Photogramm. Remote Sens.* **2015**, *106*, 68–81. [CrossRef]
11. Heo, J.; Jeong, S.; Park, H.K.; Jung, J.; Han, S.; Hong, S.; Sohn, H.-G. Productive high-complexity 3D city modeling with point clouds collected from terrestrial LiDAR. *Comput. Environ. Urban. Syst.* **2013**, *41*, 26–38. [CrossRef]
12. Wang, Q.D.; Ai, H.B.; Zhang, L. Rapid city modeling based on oblique photography and 3ds Max technique. *Sci. Surv. Mapp.* **2014**, *39*, 74–78. [CrossRef]
13. Zhang, C.S.; Zhang, W.L.; Guo, B.X.; Liu, J.C.; Li, M. Rapidly 3D Texture Reconstruction Based on Oblique Photography. *Acta Geod. Cartogr. Sin.* **2015**, *44*, 782–790.
14. Lari, Z.; El-Sheimy, N.; Habib, A. A new approach for realistic 3D reconstruction of planar surfaces from laser scanning data and imagery collected onboard modern low-cost aerial mapping systems. *Remote Sens.* **2017**, *9*, 212. [CrossRef]
15. Khairnar, S. An Approach of Automatic Reconstruction of Building Models for Virtual Cities from Open Resources. Master's Thesis, University of Windsor, Windsor, ON, Canada, 2019.
16. Girindran, R.; Boyd, D.S.; Rosser, J.; Vijayan, D.; Long, G.; Robinson, D. On the Reliable Generation of 3D City Models from Open Data. *Urban Sci.* **2020**, *4*, 47. [CrossRef]
17. Gong, J.Y.; Cui, T.T.; Shan, J.; Ji, S.P.; Huang, Y.C. A Survey on Façade Modeling Using LiDAR Point Clouds and Image Sequences Collected by Mobile Mapping Systems. *Geomat. Inf. Sci. Wuhan Univ.* **2015**, *40*, 1137–1143.
18. Li, M.; Zhang, W.L.; Fan, D.Y. Automatic Texture Optimization for 3D Urban Reconstruction. *Acta Geod. Cartogr. Sin.* **2017**, *46*, 338–345.
19. Deng, Y.; Cheng, J.C.; Anumba, C. Mapping between BIM and 3D GIS in different levels of detail using schema mediation and instance comparison. *Autom. Constr.* **2016**, *67*, 1–21. [CrossRef]
20. Fan, H.; Meng, L. A three-step approach of simplifying 3D buildings modeled by CityGML. *Int. J. Geogr. Inf. Sci.* **2012**, *26*, 1091–1107. [CrossRef]
21. Kang, T.W.; Hong, C.H. IFC-CityGML LOD mapping automation using multiprocessing-based screen-buffer scanning including mapping rule. *KSCE J. Civ. Eng.* **2017**, *22*, 373–383. [CrossRef]
22. NanoDet. Super Fast and Light Weight Anchor-Free Object Detection Model: Real-Time on Mobile Devices. Available online: https://github.com/RangiLyu/nanodet (accessed on 14 November 2021).
23. Bazi, Y.; Bashmal, L.; Al Rahhal, M.M.; Al Dayil, R.; Al Ajlan, N. Vision Transformers for Remote Sensing Image Classification. *Remote Sens.* **2021**, *13*, 516. [CrossRef]
24. Wu, B.; Nevatia, R. Simultaneous Object Detection and Segmentation by Boosting Local shape Feature Based classifier. In Proceedings of the 2007 IEEE Conference on Computer Vision and Pattern Recognition—CVPR'07, Minneapolis, MN, USA, 17–22 June 2007; pp. 1–8.
25. Wu, B.; Nevatia, R. Detection and Segmentation of Multiple, Partially Occluded Objects by Grouping, Merging, Assigning Part Detection Responses. *Int. J. Comput. Vis.* **2008**, *82*, 185–204. [CrossRef]
26. Pena, M.G. A Comparative Study of Three Image Matching Algorithms: SIFT, SURF, and FAST. Master's Thesis, Utah State University, Logan, UT, USA, 2011.
27. Druzhkov, P.N.; Kustikova, V.D. A survey of deep learning methods and software tools for image classification and object detection. *Pattern Recognit. Image Anal.* **2016**, *26*, 9–15. [CrossRef]
28. Pritt, M.; Chern, G. Satellite Image Classification with Deep Learning. In Proceedings of the 2017 IEEE Applied Imagery Pattern Recognition Workshop (AIPR), Washington, DC, USA, 10–12 October 2017; pp. 1–7.
29. Wang, P.; Fan, E.; Wang, P. Comparative analysis of image classification algorithms based on traditional machine learning and deep learning. *Pattern Recognit. Lett.* **2021**, *141*, 61–67. [CrossRef]
30. Kauderer-Abrams, E. Quantifying translation-invariance in convolutional neural networks. *arXiv* **2017**, arXiv:1801.01450. Available online: https://arxiv.fenshishang.com/pdf/1801.01450.pdf (accessed on 14 November 2021).
31. Rodríguez, M.; Facciolo, G.; Von Gioi, R.G.; Musé, P.; Morel, J.-M.; Delon, J. Sift-Aid: Boosting Sift with an Affine Invariant Descriptor Based on Convolutional Neural Networks. In Proceedings of the 2019 IEEE International Conference on Image Processing (ICIP), Taipei, Taiwan, 22–25 September 2019; pp. 4225–4229.
32. Simonyan, K.; Zisserman, A. Very deep convolutional networks for large-scale image recognition. *arXiv* **2014**, arXiv:1409.1556. Available online: https://arxiv.fenshishang.com/pdf/1409.1556.pdf(2014.pdf (accessed on 14 November 2021).
33. Szegedy, C.; Liu, W.; Jia, Y.; Sermanet, P.; Reed, S.; Anguelov, D.; Erhan, D.; Vanhoucke, V.; Rabinovich, A. Going Deeper with Convolutions. In Proceedings of the IEEE Conference on Computer Vision and Pattern Recognition, Boston, MA, USA, 7–12 June 2015; pp. 1–9.
34. Geirhos, R.; Janssen, D.H.J.; Schütt, H.H.; Rauber, J.; Bethge, M.; Wichmann, F.A. Comparing deep neural networks against humans: Object recognition when the signal gets weaker. *arXiv* **2017**, arXiv:1706.06969. Available online: https://arxiv.fenshishang.com/pdf/1706.06969.pdf (accessed on 14 November 2021).
35. Afzal, M.Z.; Kölsch, A.; Ahmed, S.; Liwicki, M. Cutting the Error by Half: Investigation of Very Deep Cnn and Advanced Training Strategies for Document Image Classification. In Proceedings of the 2017 14th IAPR International Conference on Document Analysis and Recognition (ICDAR), Kyoto, Japan, 9–15 November 2017; pp. 883–888.

36. Redmon, J.; Farhadi, A. Yolov3: An incremental improvement. *arXiv* **2018**, arXiv:1804.02767. Available online: https://arxiv.fenshishang.com/pdf/1804.02767.pdf (accessed on 14 November 2021).
37. Bochkovskiy, A.; Wang, C.Y.; Liao, H.Y.M. Yolov4: Optimal speed and accuracy of object detection. *arXiv* **2020**, arXiv:2004.10934. Available online: https://arxiv.fenshishang.com/pdf/2004.10934.pdf (accessed on 14 November 2021).
38. Wang, C.Y.; Bochkovskiy, A.; Liao, H.Y.M. Scaled-yolov4: Scaling Cross Stage Partial Network. In Proceedings of the IEEE/CVF Conference on Computer Vision and Pattern Recognition, Nashville, TN, USA, 19–25 June 2021; pp. 13029–13038.
39. He, H.; Zhou, J.; Chen, M.; Chen, T.; Li, D.; Cheng, P. Building Extraction from UAV Images Jointly Using 6D-SLIC and Multiscale Siamese Convolutional Networks. *Remote Sens.* **2019**, *11*, 1040. [CrossRef]
40. Sun, Y.; Zhao, L.; Huang, S.; Yan, L.; Dissanayake, G. Line matching based on planar homography for stereo aerial images. *ISPRS J. Photogramm. Remote Sens.* **2015**, *104*, 1–17. [CrossRef]
41. Kim, J.-I.; Kim, T. Comparison of Computer Vision and Photogrammetric Approaches for Epipolar Resampling of Image Sequence. *Sensors* **2016**, *16*, 412. [CrossRef]
42. Vincent, E.; Laganiére, R. Detecting Planar Homographies in an Image Pair. In Proceedings of the 2nd International Symposium on Image and Signal Processing and Analysis (ISPA 2001) in Conjunction with 23rd International Conference on Information Technology Interfaces, Pula, Croatia, 19–21 June 2001; pp. 182–187.
43. Ai, D.-N.; Han, X.-H.; Ruan, X.; Chen, Y.-W. Color Independent Components Based SIFT Descriptors for Object/Scene Classification. *IEICE Trans. Inf. Syst.* **2010**, *E93-D*, 2577–2586. [CrossRef]
44. Zhang, L.; Yang, L.; Lin, H.; Liao, M. Automatic relative radiometric normalization using iteratively weighted least square regression. *Int. J. Remote Sens.* **2008**, *29*, 459–470. [CrossRef]
45. Girshick, R.; Donahue, J.; Darrell, T.; Malik, J. Rich Feature Hierarchies for Accurate Object Detection and Semantic Segmentation. In Proceedings of the IEEE Conference on Computer Vision and Pattern Recognition, Columbus, OH, USA, 23–28 June 2014; pp. 580–587.
46. Ren, S.; He, K.; Girshick, R.; Sun, J. Faster r-Cnn: Towards real-time object detection with region proposal networks. In *Advances in Neural Information Processing Systems*; MIT Press: Cambridge, MA, USA, 2015; pp. 91–99.
47. Bah, M.D.; Hafiane, A.; Canals, R. CRowNet: Deep network for crop row detection in UAV images. *IEEE Access* **2019**, *8*, 5189–5200. [CrossRef]
48. Hu, S.; Li, Z.; Wang, S.; Ai, M.; Hu, Q.A. A Texture Selection Approach for Cultural Artifact 3D Reconstruction Considering Both Geometry and Radiation Quality. *Remote Sens.* **2020**, *12*, 2521. [CrossRef]

remote sensing

Article

Developing a Method to Extract Building 3D Information from GF-7 Data

Jingyuan Wang [1,2], Xinli Hu [1,2,3,*], Qingyan Meng [1,2,3], Linlin Zhang [1,2,3], Chengyi Wang [1,2], Xiangchen Liu [1,2] and Maofan Zhao [1,2]

[1] Aerospace Information Research Institute, Chinese Academy of Sciences, Beijing 100094, China; wangjingyuan19@mails.ucas.ac.cn (J.W.); mengqy@radi.ac.cn (Q.M.); zhangll@radi.ac.cn (L.Z.); wangcy@radi.ac.cn (C.W.); liuxc@radi.ac.cn (X.L.); zhaomaofan19@mails.ucas.ac.cn (M.Z.)

[2] University of Chinese Academy of Sciences, Beijing 100049, China

[3] Key Laboratory of Earth Observation of Hainan Province, Hainan Research Institute, Aerospace Information Research Institute, Chinese Academy of Sciences, Sanya 572029, China

* Correspondence: huxl@radi.ac.cn; Tel.: +86-010-6485-2195

Abstract: The three-dimensional (3D) information of buildings can describe the horizontal and vertical development of a city. The GaoFen-7 (GF-7) stereo-mapping satellite can provide multi-view and multi-spectral satellite images, which can clearly describe the fine spatial details within urban areas, while the feasibility of extracting building 3D information from GF-7 image remains under-studied. This article establishes an automated method for extracting building footprints and height information from GF-7 satellite imagery. First, we propose a multi-stage attention U-Net (MSAU-Net) architecture for building footprint extraction from multi-spectral images. Then, we generate the point cloud from the multi-view image and construct normalized digital surface model (nDSM) to represent the height of off-terrain objects. Finally, the building height is extracted from the nDSM and combined with the results of building footprints to obtain building 3D information. We select Beijing as the study area to test the proposed method, and in order to verify the building extraction ability of MSAU-Net, we choose GF-7 self-annotated building dataset and a public dataset (WuHan University (WHU) Building Dataset) for model testing, while the accuracy is evaluated in detail through comparison with other models. The results are summarized as follows: (1) In terms of building footprint extraction, our method can achieve intersection-over-union indicators of 89.31% and 80.27% for the WHU Dataset and GF-7 self-annotated datasets, respectively; these values are higher than the results of other models. (2) The root mean square between the extracted building height and the reference building height is 5.41 m, and the mean absolute error is 3.39 m. In summary, our method could be useful for accurate and automatic 3D building information extraction from GF-7 satellite images, and have good application potential.

Keywords: GF-7 image; building footprint; building height; multi-view; deep learning; point cloud

Citation: Wang, J.; Hu, X.; Meng, Q.; Zhang, L.; Wang, C.; Liu, X.; Zhao, M. Developing a Method to Extract Building 3D Information from GF-7 Data. *Remote Sens.* **2021**, *13*, 4532. https://doi.org/10.3390/rs13224532

Academic Editor: Sander Oude Elberink

Received: 16 September 2021
Accepted: 1 November 2021
Published: 11 November 2021

Publisher's Note: MDPI stays neutral with regard to jurisdictional claims in published maps and institutional affiliations.

1. Introduction

The structure of urban areas in both two and three dimensions has a significant impact on local and global environments [1]. As the basic element of a city, buildings are the main sites of production and housing. The three-dimensional (3D) information of buildings portrays the horizontal and vertical morphological characteristics of a city, both of which play a crucial role in urban construction and management for sustainable development. Research on the 3D information extraction of urban buildings can serve the research fields of urban climate [2–5], urban expansion [6,7], pollutant dispersion [8], urban 3D reconstruction [9–12], urban scene classification [13], energy consumption [14], and population assessment [15–17]. Therefore, large-scale and high-precision 3D information extraction of urban buildings is essential for a comprehensive understanding of urban development.

With the development of remote sensing equipment, remote sensing technology provides an effective tool for surveying and mapping buildings at the urban scale. Due to the limited availability of 3D data, most studies on urban building extraction focus on the two-dimensional level [18–25], and only a small number of studies focus on the 3D structure of buildings [26–32]. Huang et al. [26] used ZY-3 data combined with A-map (a map service provider of China) building height data and proposed a multi-view, multi-spectral, and multi-objective neural network (called M^{3}Net) to extract large-scale building footprints and heights, and verified the applicability of the extraction method in various cities. Wang et al. [27,28] proposed an inversion method of building heights using GLAS data assisted by QuickBird imagery and used satellite-borne LiDAR full waveform data to extract building height within a laser spot footprint. Li et al. [29] realized the extraction of building height with a resolution of 500 m based on Sentinel-1 data, and verified results in most cities of the United States. Qi et al. [30] estimated the height of buildings based on the shadows of buildings from Google Earth images. It is more economical to use shadow information to estimate the height of buildings. However, this method is susceptible to many restrictions, such as building heights, shadow effects, and viewing angles. Liu et al. [31] used a random forest method to extract building footprints from ZY-3 multi-spectral satellite images and combined this approach with the digital surface model (DSM) constructed by ZY-3 multi-view images to estimate building heights. However, the accuracy of building footprint extraction using random forest method is low, and the estimated height of a building is easily affected by the height of the ground's surface.

In summary, although previous studies have made some progress in building 3D information extraction, there are still the following limitations:

1. Building semantic segmentation accuracy is not high, and there are many problems, such as unclear edges of buildings and difficulty in extracting large buildings [22–24,33].
2. Most high-resolution building height information extraction is limited to a small scale, and there is a lack of large-scale high-resolution building height extraction methods [12,26–31].
3. The GaoFen-7 (GF-7) multi-view satellite image can describe the vertical structure of a ground object well. However, there are few studies on the extraction of building information from GF-7 satellite images, and satellite vertical structure extraction capabilities still require evaluation.

To fill this knowledge gap on urban building 3D information estimation over large areas, we developed a building footprint and height extraction method and assessed the quality of the results from GF-7 imagery.

Our research is divided into three parts. First, we use deep learning methods to extract building footprints from GF-7 multi-spectral images. To solve the problem of accuracy in terms of building footprint extraction, we propose a multi-stage attention U-Net (MSAU-Net). Second, this study used the multi-view images of GF-7 to construct the point cloud of the study area and performed point cloud filtering process to obtain the ground point. The DSM, the digital elevation model (DEM), and the normalized digital surface model (nDSM) of the study area are generated from the point cloud. Afterward, the building footprint extraction results of the study area are superimposed with the nDSM data to generate a 3D product of the building. Finally, this study verified the accuracy of the building footprint extraction and compared our network with other deep learning methods; we then collected actual building height values in the study area as the reference buildings to verify the accuracy of estimated building height information.

The remainder of this paper is arranged as follows. Section 2 introduces the GF-7 data and study area. Our methodology is presented in Section 3. The results and discussion are reported in Section 4. Finally, conclusions are drawn in Section 5.

2. Data and Study Area

GF-7 was successfully launched in November 2019. It is China's first civilian sub-meter stereo surveying and mapping satellite equipped with a two-line array scanner. The ground sample distance (GSD) is 0.8 m for the oblique panchromatic cameras viewing in a forward direction (26°), 0.65 m for the oblique panchromatic cameras viewing in a backward direction (−5°), and 2.6 m for the infrared multi-spectral scanner. The GF-7 satellite has many applications, such as natural resource surveys, basic surveying and mapping, and urban 3D building model generation.

Our study area is located in Beijing, China. As the capital of China, Beijing has a complete urban infrastructure and dense, built-up environments. The GF-7 satellite image we selected was captured on 16 October 2020. The scope of the study area we selected is shown in the red box in Figure 1. The study area covers the central area of Beijing, including landmark high-rise buildings, large building groups, middle- and high-rise residential areas, low-rise residential areas, urban green spaces, and other typical features covering an area of 169 square kilometers. For the task of building footprint extraction base with deep learning, we selected the area in the light blue and the yellow box in Figure 1 to make the training and test dataset, which we refer to as the "GF-7 self-annotated building dataset". The area in the light blue box was used for the training and verification datasets, while the area in the yellow box was used for the test dataset. We chose a field surveying the height data of 213 buildings as the reference building height data for the evaluation of building heights. The reference building location is shown in Figure 1 below.

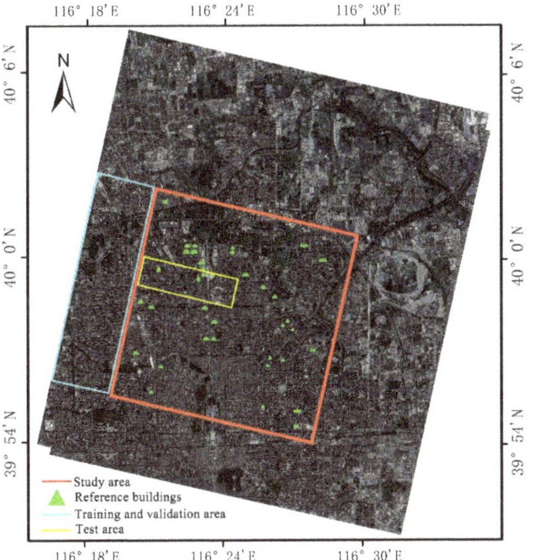

Figure 1. GF-7 multi-spectral and multi-view image of the study area.

3. Methodology

3.1. Overview

The 3D information extraction method of the building in this study is shown in Figure 2. First, we fused the GF-7 backward-view multi-spectral image with the backward-view panchromatic image and proposed MSAU-Net to extract the urban building footprint from the pan sharpening result. We modified the traditional decoder–encoder network structure, used ResNet34 as the backbone feature extraction network, and integrated an attention block in the skip connection part of the network. The attention mechanism was used to improve the building extraction ability of the neural network. Second, the point cloud of the study area was constructed from the multi-view images of GF-7, and then the DSM of the study area was constructed based on the point cloud. Then, we used a cloth simulation algorithm (CSF) [34] to filter the point cloud to obtain the ground point and used it to construct the DEM of the study area. Then, the nDSM was constructed to represent the height of off-terrain objects. Finally, the building footprint extraction results were superimposed with the nDSM to generate building height. In the accuracy assessment part of our study, the test dataset and the reference building height value were used to verify the accuracy of the 3D information of the building.

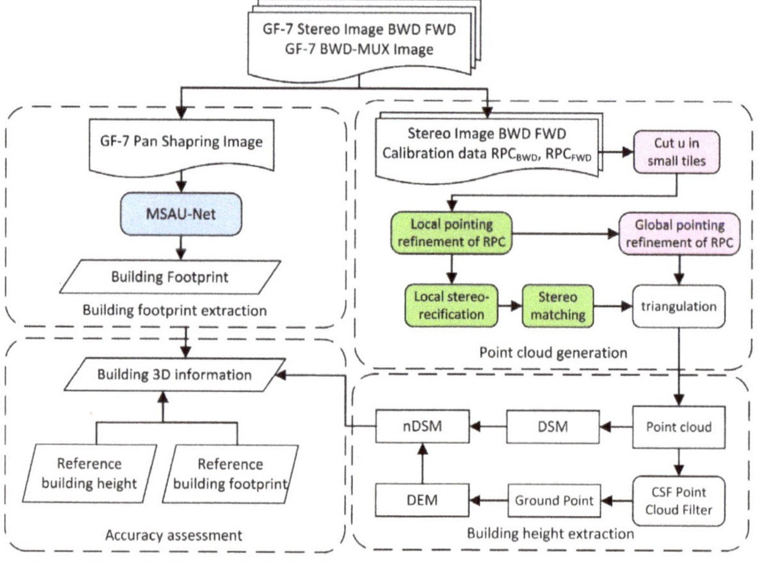

Figure 2. Workflow of the building footprint and building height extraction.

3.2. Building Footprint Extraction

This paper designs the MSAU-Net that can coordinate global and local context information to improve the results of building extraction. This section will describe the proposed network architecture and its components. Our model is based on U-Net [35]. We incorporate spatial attention and channel attention in the skip connection part of the original network. To avoid excessive parameters, our model uses ResNet-34 [36] as the backbone of the feature extraction network. This is because ResNet-34 has suitable feature extraction abilities and its parameter and calculation cost are small. Figure 3 show the structure of the proposed MSAU-Net.

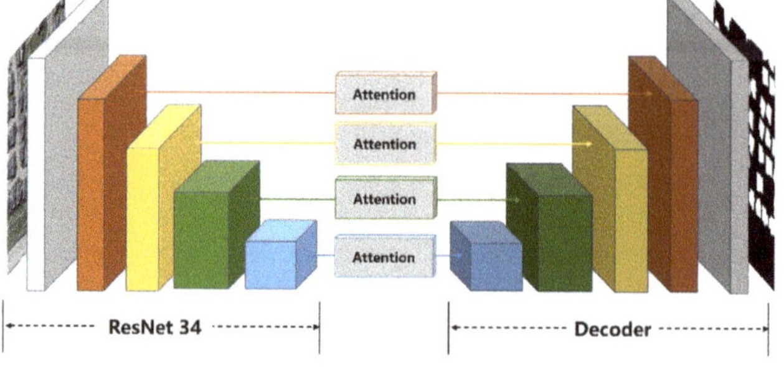

Figure 3. Structure of proposed network.

3.2.1. Attention Block

Some studies [37–39] showed that making full use of long-range dependencies can improve the performance of a network. However, U-Net only uses convolution and pooling operations, which limits the acquisition of long-range dependencies. Choosing a large convolution kernel can increase the receptive field size of a network, but it can also increase GPU memory occupation. An attention mechanism can effectively refine feature maps to improve the performance of neural networks, and it has become a common method in semantic segmentation problems. However, an attention mechanism will generate computational cost and increase GPU memory usage.

Figure 4 shows the structure of the attention block. The attention block includes the channel attention module and the spatial attention module. The following sections will describe the spatial attention and channel attention modules in detail.

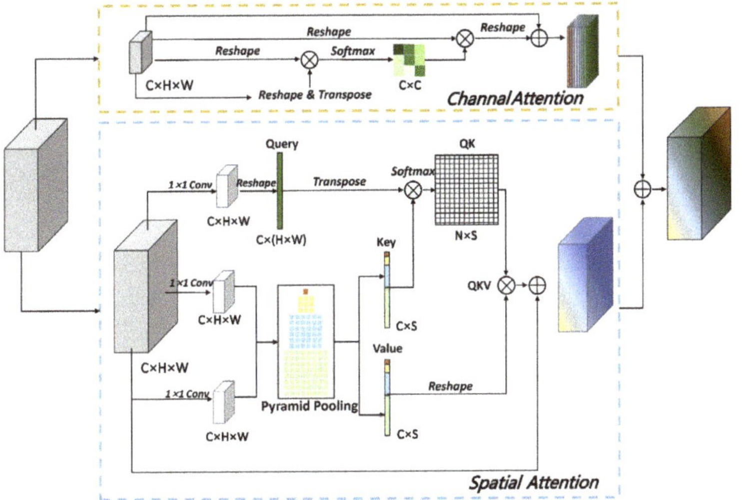

Figure 4. Structure of the attention block.

1. Spatial Attention Block

Due to the small spectral difference between buildings, roads, sports fields, etc., only using convolution operations is insufficient to obtain long-distance dependencies, as this approach easily causes classification errors. This study introduces the non-local module [40] to obtain the long-distance dependence in spatial dimension of remote sensing images, which makes up for the problem of the small receptive field of convolution operations. The non-local module is an especially useful technique for semantic segmentation. However, it has also been criticized for its prohibitive graphics processing unit (GPU) memory consumption and vast computation cost. Inspired by [41–43], to achieve a trade-off between accuracy and extraction efficiency, spatial pyramid pooling was used to reduce the computational complexity and GPU memory consumption of the spatial attention module. Figure 4 shows the structure of the spatial attention module.

A feature map X of the input size (C × H × W, where C represents the number of channels in the feature map, H represents the height of the feature map, and W represents the width) was used in a 1 × 1 convolution operation to obtain the Query, Key, and Value branches. After entering the Q branch, the feature map with a size of C × H × W was flattened into a two-dimensional vector with a size of C × N, where N = H × W. Feature map Q was transposed to obtain a feature vector Q' with a size of N × C. After the feature map entered branch K, the feature map with a size of C × H × W was obtained through spatial pyramid pooling to achieve a reduction in dimensionality. The spatial pyramid pooling operation is shown in Figure 5 below. The spatial pyramid pooling module performed the maximum pooling of the input feature map with a window size of n × n to obtain the feature map with a size of C × n × n. The feature map with a size of C × n × n was used to represent the sampling results of representative anchor points in each area of the origin feature map. Then, all the feature maps after the spatial pyramid pooling were flattened and concatenated to obtain a feature vector with a size of C × S, where S was determined by the size and number of the selected pooling windows. For example, in this article, the pooling widow is 1 × 1, 3 × 3, 6 × 6, and 8 × 8, and S is equal to:

$$S = \sum_{n \in \{1,3,6,8\}} n^2 = 110$$

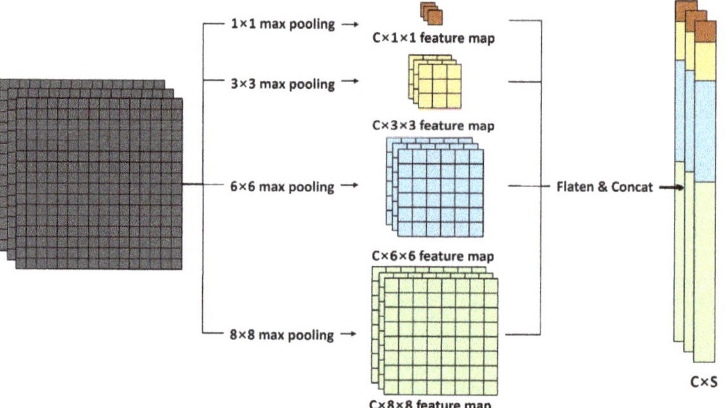

Figure 5. Structure of spatial pyramid pooling.

After the feature map, X entered the Query and Key branches, and the feature vectors Q' with a size of N × C and K' with a size of C × S are matrix multiplied to obtain feature map QK'. Feature map QK' was normalized by SoftMax to obtain the attention map QK. The purpose of this was to calculate the relationship between each pixel in feature vector Q' and each pixel in K'. In this way, we can obtain a feature map of C × S size, which represents the attention relationship between the Query pixel and the feature anchor point in the Key, and represents the long-range dependency in the image.

The Value branch is similar to the Key branch. Feature map X inputs the Value branch can obtain feature vector V' with a size of C × S. After the feature vector was transposed, it was multiplied with attention map QK to generate feature map QKV with a size of C × H × W. Then, feature map QKV and origin feature map X were merged using element-wise summation to obtain the result of the spatial attention module.

2.   Channel Attention Block

In the process of building extraction, each channel of high-level feature maps can be regarded as a response to the specific features of a building, and different channels are related to each other. By extracting the long-range dependence between channel dimension feature maps, we can emphasize the interdependence of the feature maps and improve the feature representation. Therefore, this study used a channel attention module to model the long-range dependence relationship of channel dimensions. The structure of the channel attention module is shown in Figure 4.

The channel attention map was calculated from the original feature map X with a size of C × H × W. Specifically, feature map X was flattened into a feature vector of C × N size (N = H × W). Then, matrix multiplication operations were performed on the feature vector, and the transposition of the feature vector and SoftMax normalization were applied to obtain the channel attention map with a size of $C \times C$.

The channel attention map represents the long-range dependence between the channel dimension of the feature maps. After obtaining the channel attention map, we performed a matrix multiplication operation on input feature map X and the channel attention map to obtain the feature map with a size of $C \times H \times W$. After that, the result was multiplied by learnable scale factor α and merged with origin feature map X using element-wise summation to obtain the result of the channel attention module.

3.2.2. Training Strategy

In order to attain better building footprint extraction results from GF-7 images, we performed pre-training on the Wuhan University (WHU) [44] building dataset to get the initial pre-training weights. Then, we selected the area in the light blue box in Figure 1 to make training and verification samples. In this paper, the training epoch was set at 120 and 80 for WHU building dataset and GF-7 self-annotated building dataset, the batch size parameter (the number of samples during each training iteration at the same time) was set to 8, the initial learning rate was 0.01, and the input image size was 512×512. The learning rate gradually decreases with the increase in training generations to optimize the model. In the training process, sample enhancement processing was performed, including random scale scaling, rotation, flipping, and blur processing.

3.3. Point Cloud Generation

This section uses a stereo pipeline [45–47] to generate point cloud from the backward- and forward-view panchromatic GF-7 images. The generation process is shown in Figure 2, and this section will briefly introduce the process of point cloud generation. Since the imaging method of the satellite is push-broom imaging, it was determined that the epipolar line is hyperbolic [46,47]. Research [47] has proven that, when an image is cut into small tiles, a push-broom geometric imaging model can be approximately regarded as a pinhole model; after that, it uses standard stereo image rectification and stereo-matching tools to process the small tiles. However, due to errors in the RPC parameters of satellite images, local and global corrections need to be performed according to the satellite image RPC parameters and feature point matching results to improve the accuracy of the point cloud.

First, the original image performed block processing according to the RPC parameters given by the satellite image to divide the original image into 512×512 tiles. The push-broom imaging model can be regarded as a pinhole model in a 512×512 size area. Due to the limited accuracy of camera calibration, there is bias in the RPC functions. This bias will cause the global offset of the images; for some purposes, this bias can be ignored [45]. However, the epipolar constraint is derived from the RPC parameters, so it has to be as precise as possible. Thus, the relative errors between the RPC parameters of the multi-view images must be corrected. The local correction method also approximates the push-broom imaging model as a pinhole camera model in small tiles. This study used SIFT [48] to extract and match the feature points in each tile. According to the feature point matching result and combined with the RPC parameter, the translation parameter of the satellite image can be calculated to realize local correction. However, for the whole study area, the local correction will fail, and it must integrate the results of local corrections for global corrections. The global correction method is used to calculate the center of feature points in each tile and combine the local correction results to calculate the affine transformation of the satellite image.

After obtaining the local correction result, stereo image rectification was performed in each tile. The natural method for constructing the epipolar constraint of a stereo image is to use image feature points to perform image correction. However, for satellite imagery, since the distance from the imaging plane to the ground is much larger than the ground fluctuations, it will cause a large error in fundamental matrix F, i.e., the degradation of fundamental matrix F. Additionally, in special cases, the set of feature points are on the same plane, such as the ground. Fundamental matrix F cannot be calculated. Therefore, this paper uses SRTM terrain as prior knowledge and uses local correction results and satellite imagery RPC parameters combined with SRTM information to construct virtual matching points instead of feature matching points. In each tile, virtual points were constructed, estimated the height of the three-space points from the SRTM information, and used the RPC parameter to back-project the point into the multi-view images. In this way, the image of virtual matching points coordinates can be obtained to estimate fundamental matrix F. According to fundamental matrix F, two rectifying affine transformations of the stereo image were extracted to perform image rectification in each tile.

For each rectified tile, a disparity map was calculated by applying a stereo matching algorithm from the stereo rectified image. The SRTM information was used to estimate the initial disparity range. This study chose the classic semi-global stereo matching (SGM) algorithm [49] for stereo matching because of its performance. The disparities are then converted into the point correspondence of the original image coordinates. Combined with the local and global correction results, the ground point coordinates were iteratively calculated to generate point cloud. For more detailed point cloud generation, please refer to the relevant part of the research [45].

3.4. Building Height Extraction

After obtaining the point cloud of the study area, the inverse distance weight interpolation method was used to generate the DSM. However, due to the undulations on the ground, to obtain the height of the building, the elevation value of the lower surface of the building should be extracted from the point cloud.

The point cloud of the study area was filtered to classify ground points and non-ground points. The point cloud generated by satellite imagery is different from the point cloud generated by LiDAR. The point cloud is relatively sparse. Due to viewing angle limitations, there are more hollow areas. This study chose two filtering methods, cloth simulation filtering (CSF) [34] and morphological filtering [50], for filtering processing, and it was found that cloth simulation filtering can achieve better experimental results for the relatively sparse point cloud generated by satellite images.

The main idea of the CSF filtering method is to invert the point cloud and then simulate the process of rigid cloth covering the inverted surface. CSF then analyzed the relationship between the cloth node and the point cloud, determined the position of the cloth node, and separated the ground point by comparing the distance between the original point cloud and the generated cloth. Since this research focuses on buildings, the point cloud of buildings presents a planar distribution far away from the ground points. In the cloth simulation filtering, the cloth with higher hardness is selected for point filtering. In this way, CSF can achieve a better filtering result.

After obtaining the ground point cloud of the study area, the inverse distance weight interpolation method is also used to generate the DEM of the study area. Then, DSM and DEM were performed for difference processing to generate the nDSM. Combined with the results produced in Section 3.1, the building footprint results are superimposed with nDSM. Building heights were assigned as the maximum value of nDSM after removing the outliers of nDSM within each building footprint.

3.5. Evaluation Metrics

In order to test the feasibility of our building 3D information extraction method, this study verified the accuracy of the building footprint and building height results, respectively. Experimental results and accuracy verification are shown in Section 4. This section will introduce the accuracy evaluation method and the indicator calculation method.

To quantitatively evaluate and compare the segmentation performance of footprint extraction, five widely used metrics, i.e., overall accuracy (OA), intersection-over-union (IOU), precision rate, recall, and F1 score, were calculated based on the error matrix:

$$OA = \frac{TP + TN}{TP + TN + FP + FN} \tag{1}$$

$$IoU = \frac{TP}{TP + FP + FN} \tag{2}$$

$$precision = \frac{TP}{TP + FP} \tag{3}$$

$$recall = \frac{TP}{TP + FN} \tag{4}$$

$$F1 = 2 \times \frac{precision \times recall}{precision + recall} \tag{5}$$

where TP is true positive, TN is true negative, FP is false positive, and FN is false negative.

Height accuracy was verified by comparing reference buildings and estimated building heights and selecting the mean absolute error (MAE) and the root mean error (RMSE) as evaluation indicators. The specific formulas are as follows:

$$MAE = \frac{1}{N} \sum_{i=1}^{N} \left| \hat{h}_i - h_i \right| \tag{6}$$

$$RMSE = \sqrt{\frac{1}{N} \sum_{i=1}^{N} \left(\hat{h}_i - h_i \right)^2} \tag{7}$$

where $\hat{h}_i$ denotes the predicted height at building i, h_i denotes the corresponding ground truth height, and N denotes the total number of buildings.

4. Results and Discussion

4.1. Performance of Building Footprint Extraction

In order to verify the performance of building footprint extraction, classic networks such as PSPNet [37], FCN [51], DeepLab v3 + [52], SegNet [53], and U-Net [35] were used for comparison. Experimental results of the WHU building segmentation dataset and the GF-7 self-annotated building dataset are as follows. Experiments are conducted on a computer that has an Intel®Core™ i9-10980XE GPU @3.00 GHz and 64 GB memory. The GPU type used in this computer is RTX 3090 with 24 GB GPU memory.

4.1.1. WHU Building Dataset

The WHU building dataset consists of an aerial image dataset and two satellite image datasets. It has become a benchmark dataset for testing the performance of building footprint extraction bases with deep learning because of the high quality of data annotation. This study uses the WHU aerial dataset to test our model. The WHU aerial dataset contains 8188 non-overlapping images (512 × 512 tiles with spatial resolution 0.3 m), covering 450 square kilometers of Christchurch, New Zealand. Among them, 4736 tiles (containing 130,500 buildings) are separated for training, 1036 tiles (containing 14,500 buildings) are separated for validating, and the rest, 2416 tiles (containing 42,000 buildings), are used for testing. The proposed deep learning of the MSAU-Net is implemented using PyTorch in the Window platform. After 120 epochs (3.8 h of training time), our network achieves a better result on the WHU dataset (Table 1). The changing losses and IOU of the WHU building dataset with the increasing epochs are shown in Figure 6.

Table 1. Experimental results of the WHU building dataset.

Method	OA (%)	IOU (%)	Precision (%)	Recall (%)	F1-Score (%)
PSPNet	98.55	87.67	92.49	94.39	93.45
FCN	97.42	79.48	89.73	87.42	88.56
DeepLab v3+	96.84	73.55	78.79	91.71	84.76
SegNet	98.06	84.01	91.40	91.21	91.31
U-Net	98.56	87.94	93.84	93.33	93.58
MSAU-Net	98.74	89.31	94.18	94.52	94.35

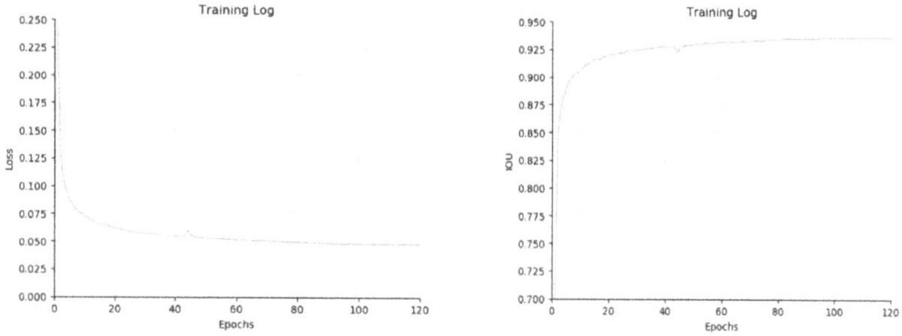

Figure 6. Plots showing the loss and IOU of the proposed model for training the WHU building dataset. The training loss (**a**) and the IOU (**b**) change when the epochs increase.

Four representative experimental results were selected for qualitative assessment of the various building extraction methods. In Figure 7, original image 1 shows a densely distributed group of small buildings. Our model can suitably maintain the appearance of buildings. The red box in original image 2 is a container-like object that is easily confused with buildings. Compared with U-Net, our model can effectively avoid recognizing objects such as buildings due to its increased long-range dependency. Original picture 3 shows large buildings. It can be seen that PSPNet, SegNet, and ours can produce better experimental results. However, while PSPNet and SegNet easily cause blurred boundaries, our model can keep the details of the building's boundary. The red box in original picture 4 shows buildings with an unusual shape. It can be seen that our method can maintain the unusual shape of buildings. In summary, due to its increasing long-range dependence, our model can effectively extract building footprints from fine-resolution remote sensing images. In relation to large buildings and unusually shaped buildings, our method can enhance the integrity and accuracy of a building's shape. This is highly important for the process of building footprint extraction from GF-7 multi-spectral images.

The experimental results of the WHU building dataset are shown in Table 1. From Table 1, it is clear that our method shows a significant improvement in IOU and F1-score. The OA (overall accuracy), precision, and recall are slightly improved. However, OA describes the proportion of correctly classified pixels to total pixels. The IOU indicator describes the proportion of correctly classified building pixels to the total number of pixels in all building categories (including ground truth and predicted buildings). F1-score integrates accuracy and recall. Therefore, F1-score and IOU indicators are more convincing metrics. The WHU building dataset experimental result shows that the building footprint extraction ability of our model is better than other models.

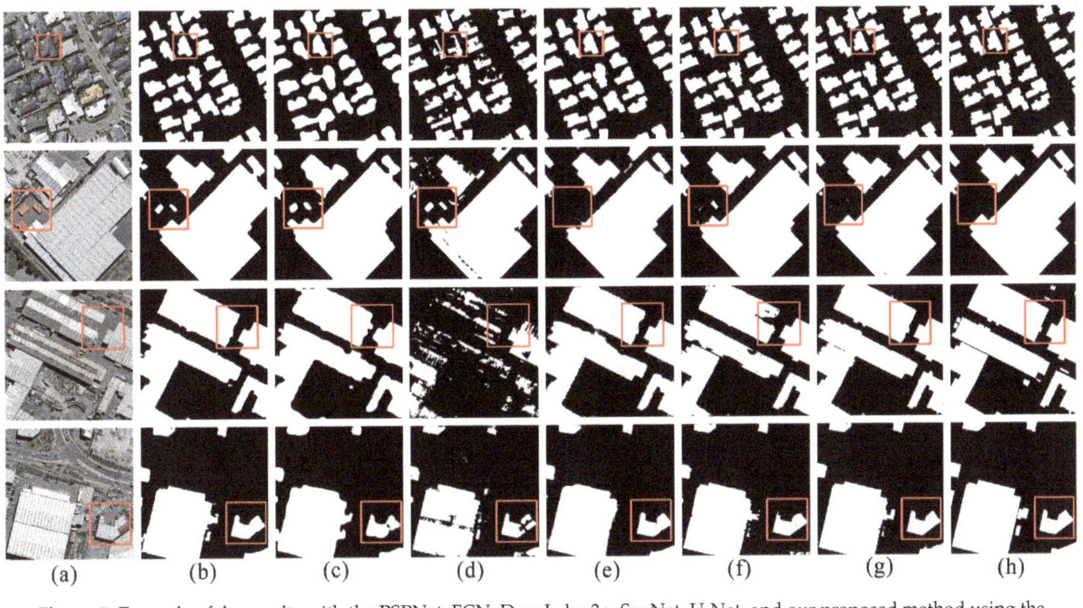

Figure 7. Example of the results with the PSPNet, FCN, DeepLab v3+, SegNet, U-Net, and our proposed method using the WHU building dataset: (**a**) Original image. (**b**) PSPNet. (**c**) FCN. (**d**) DeepLab v3+. (**e**) SegNet. (**f**) U-Net. (**g**) Proposed model. (**h**) Ground truth.

4.1.2. GF-7 Self-Annotated Building Dataset

For the test of building footprint extraction, this study uses the GF-7 self-annotated building dataset to train and test the model. The GF-7 self-annotated building dataset contains 384 non-overlapping images (512 × 512 tiles with spatial resolution 0.65 m), covering 41.2 square kilometers of Beijing. Among them, 300 tiles (containing 4369 buildings) are separated for training, while 38 tiles (containing 579 buildings) are separated for validation. In order to verify the performance of building footprint extraction from GF-7 images, this study selected typical buildings in the study area to establish our test set (contains 46 non-overlapping 512 × 512 images, 886 buildings). During the MSAU-Net training, the training epoch was set at 80 for the GF-7 self-annotated building dataset, and the training time was 1.1 h. The changing losses and IOU of the GF-7 self-annotated building dataset with the increasing epochs are shown in Figure 8.

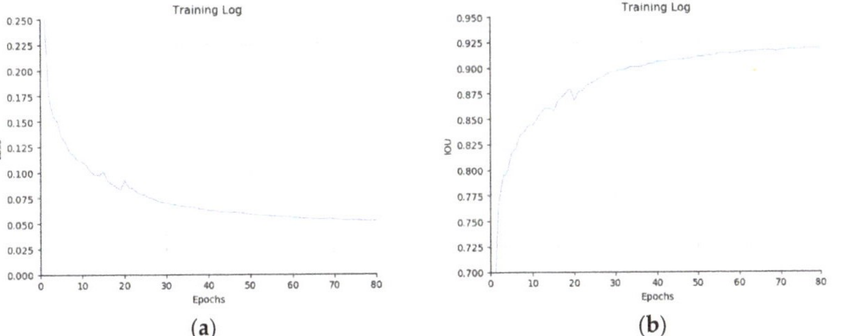

Figure 8. Plots showing the loss and IOU of the proposed model for training the GF-7 self-annotated building dataset. The training loss (**a**) and the IOU (**b**) change when the epochs increase.

Similarly, four representative areas were selected to display the results of the GF-7 self-annotated building dataset for qualitative assessment (Figure 9). Original image 1 is a typical building group in the study area. From the experimental results, our method can maintain the appearance of buildings. Original picture 2 shows that, for large buildings, our method can maintain the integrity of a building footprint due to the increased long-range dependence. The red box of original image 3 is a building with an unusual shape. Our method can obtain a relatively better experimental result than other models. The red box of original image 4 is a landmark building in the study area (the 2008 Olympic venue, Water Cube). From the experimental results, our method can maintain the integrity of the Water Cube.

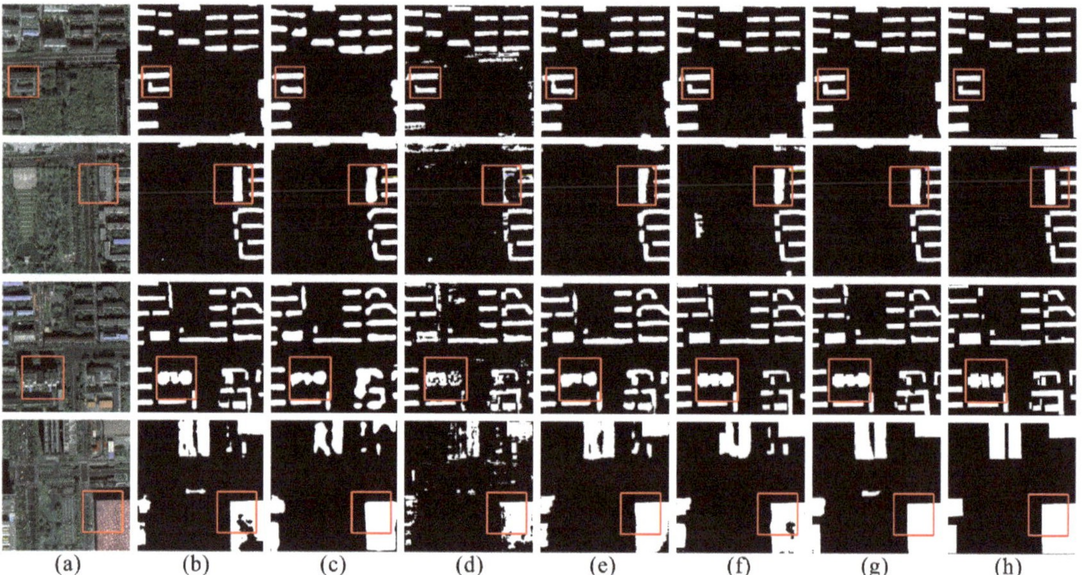

Figure 9. Example of the results with the PSPNet, FCN, DeepLab v3+, SegNet, U-Net, and our proposed method the GF-7 self-annotated building Dataset: (**a**) Original image. (**b**) PSPNet. (**c**) FCN. (**d**) DeepLab v3+. (**e**) SegNet. (**f**) U-Net. (**g**) Proposed model. (**h**) Ground truth.

The experimental results of the GF-7 self-annotated building segmentation dataset are shown in Table 2. As can been from Table 2, our model has significantly improved IOU and F1-score. However, OA and recall are slightly improved. Since the GF-7 multi-spectral image resolution is 2.6 m, compared with the WHU building dataset with a resolution of 0.3 m, building footprint extraction is more complicated, and it is prone to confusion between building areas and non-building areas. Therefore, compared with the results of the WHU building dataset (Table 1), the IOU indicator on the GF-7 (Table 2) is lower. Experimental results show that our model can attain a better performance in relation to building footprints from GF-7 images.

Table 2. Experimental results of the GF-7 self-annotated building segmentation dataset.

Method	OA (%)	IOU (%)	Precision (%)	Recall (%)	F1-Score (%)
PSPNet	94.66	75.27	81.98	90.18	85.89
FCN	93.09	70.21	82.16	82.84	82.50
DeepLab v3+	91.53	62.55	71.40	83.46	76.96
SegNet	94.16	74.04	84.03	86.03	85.08
U-Net	95.17	77.58	84.21	90.70	87.33
MSAU-Net	95.74	80.27	87.46	90.71	89.06

In order to display the accuracy of the results more intuitively, we display the predicted results in color (Figure 10). The green area represents true positive, the grey area represents false negative, the blue area represents false positive, and the red area represents true negative. When the green area (true positive) is in the majority, and the red area (true negative) and the blue area (false positive) are in the minority, the extraction effect is good. From Figure 10, compared with the results of the other five methods, the ratio of the red part and blue part in the extraction result of our method is significantly reduced.

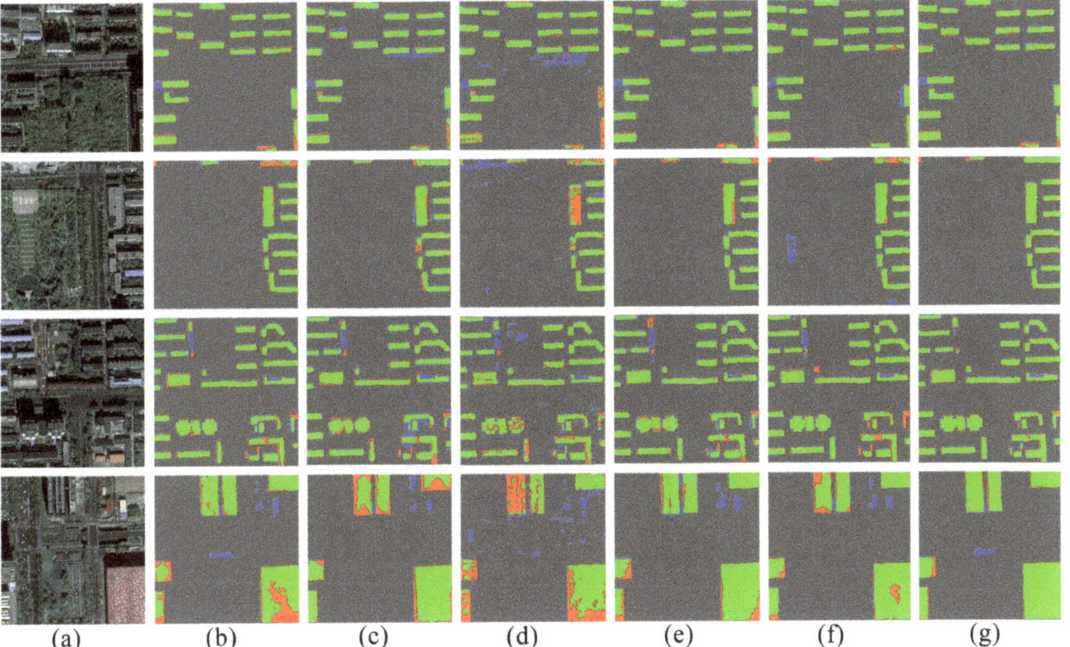

Figure 10. Example of the results with the PSPNet, FCN, DeepLab v3+, SegNet, U-Net, and our proposed method using the GF-7 self-annotated building dataset: (**a**) Original image. (**b**) PSPNet. (**c**) FCN. (**d**) DeepLab v3+. (**e**) SegNet. (**f**) U-Net. (**g**) Proposed model.

4.2. Performance of Building Height Extraction

Figure 11 shows the results of point cloud generation. The results show that the point cloud generation results are relatively sparse but can reflect surface elevation information. In Figure 11c, for single large buildings, the point cloud results are better, as they present a planar distribution far away from the ground points. Additionally, Figure 11a shows that the average seabed in the northeast is lower than the southwest in the study area, which is also in line with the actual geography of Beijing. However, due to the limited viewing angle of satellite images, the point cloud results are poor for dense low-rise buildings, such as the middle and lower parts of the research area. Figure 11d–i show the ground point cloud results and the off-ground point cloud results after CSF. The results show that our method can obtain a relatively complete ground point cloud.

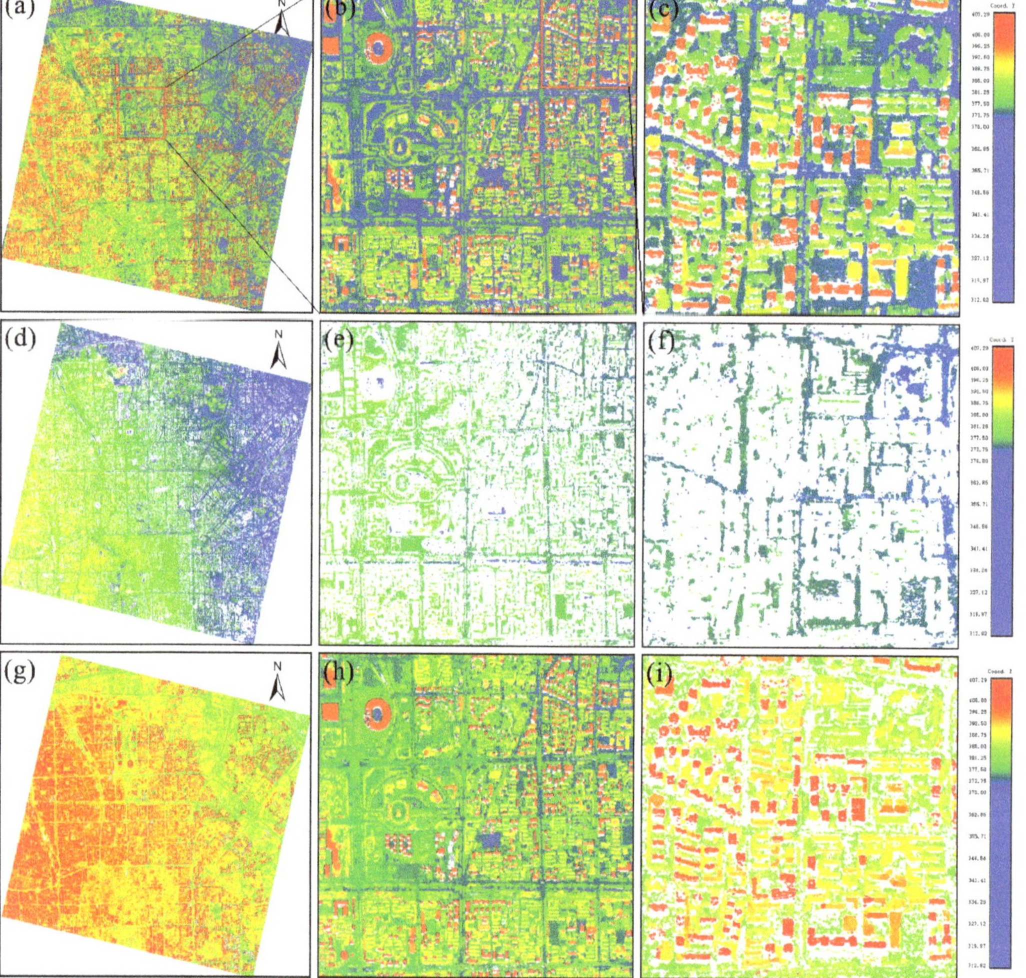

Figure 11. Point cloud generation results in the study area: (**a–c**) point cloud results; (**d–f**) ground point results; (**g–i**) off-ground point cloud results.

The results of the building footprint and height extraction in the study area are shown in Figure 12 to demonstrate the effectiveness of our method. Based on the original image Figure 12a, the corresponding building footprint Figure 12c, point cloud Figure 12e, and building height results Figure 12g are generated; they are enlarged and displayed Figure 12b,d,f,h, respectively. The accuracy of our building footprint extraction results has been quantitatively analyzed in the previous section. It can be seen from Figure 12c,d that our method can obtain relatively complete and accurate building footprint information. Figure 12g is the building height result of the study area; the following section will verify the building height results based on the reference building heights.

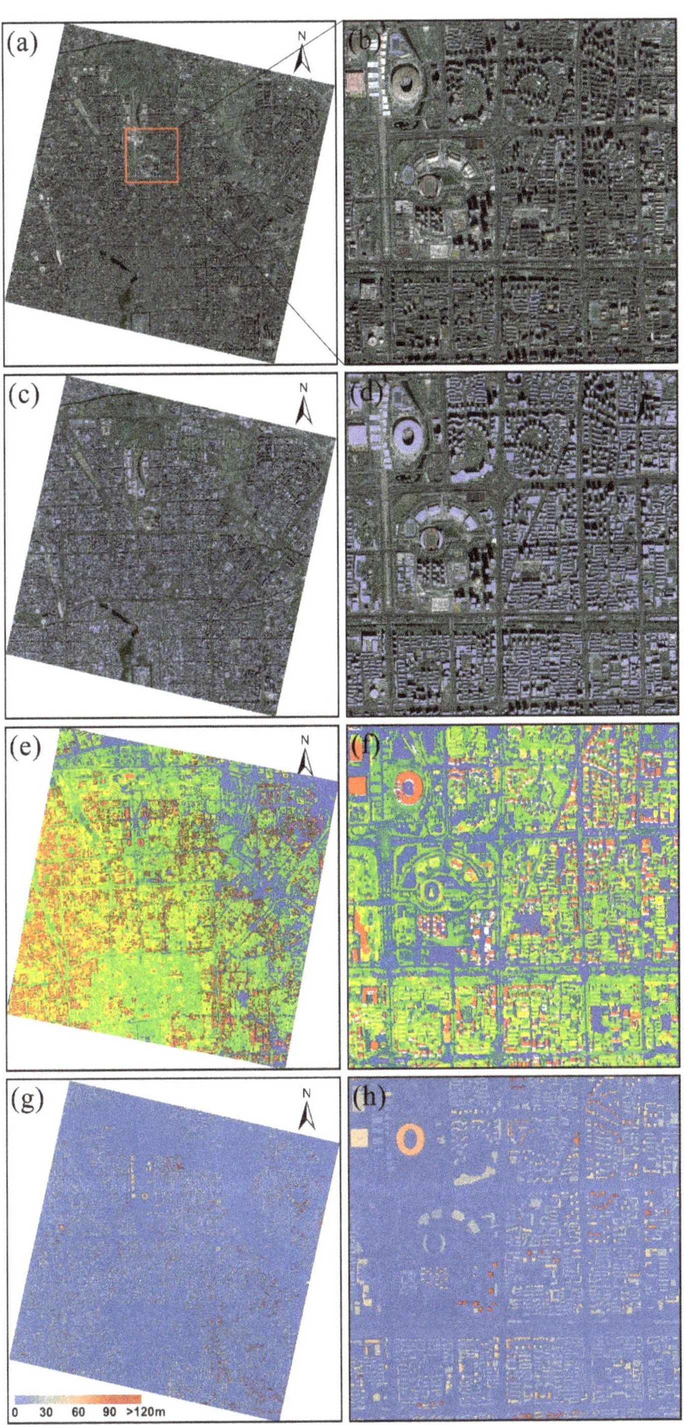

Figure 12. Experimental results in the study area: (**a,b**) original images; (**c,d**) building footprint extraction results; (**e,f**) point cloud results; (**g,h**) building height extraction results.

In order to quantitatively analyze the accuracy of an estimated building height, we measured 213 buildings with different height levels as the reference building height. By comparing the height of the extracted building with the height of the reference building, a quantitative evaluation is carried out according to the RMSE and MAE indicators. The evaluation result is shown in Figure 13. Building height RMSE is 5.41, the average building error is 3.38, and the correlation coefficient is 0.96, all of which point to an excellent experimental result. Linear fitting was performed on the extracted building height, and it can be seen that the line after fitting indicates that the building height value extracted by our method fluctuates up and down in relation to actual building height. Compared with other building height extraction methods based on optical satellite images, our method can achieve better experimental accuracy [26,31].

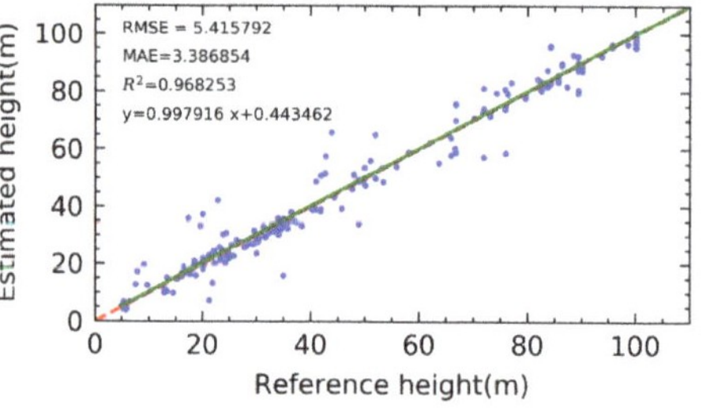

Figure 13. Accuracy of building height predictions.

To analyze whether there was a correlation between the accuracy of the building height value extracted by our method and the actual building height, the building was divided into three groups according to the actual height, i.e., below 30 m, between 30 and 70 m, and higher than 70 m, and the RMSE and MAE were calculated for each group, respectively. The results are shown in Table 3 below. The results show that there is not much difference between the three sets of RMSE, indicating that our height extraction accuracy has little relationship with the height of the building. However, the RMSE of the building height above 30 m is greater than the RMSE of the building height below 30 m, indicating that the estimation of middle- and high-rise buildings is unstable.

Table 3. Experimental results of GF-7 building height.

	Number	RMSE	MAE
Below 30 m	83	4.95	2.83
From 30 m to 70 m	67	5.99	3.91
Above 70 m	63	5.35	3.55
All	213	5.41	3.39

The 3D information results of our method are shown in Figure 14. The experimental results and quantitative verification show that our method could be useful for accurate and automatic 3D building information extraction from GF-7 satellite images. However, for dense, low-rise buildings, such as the center of Beijing (which consists of mostly ancient buildings), due to the diversity of buildings and the viewing angle limitations of satellite imagery, both the result of footprint extraction and the result of point cloud are poor. Therefore, for dense and low-rise buildings, our method cannot get a good 3D information extraction effect.

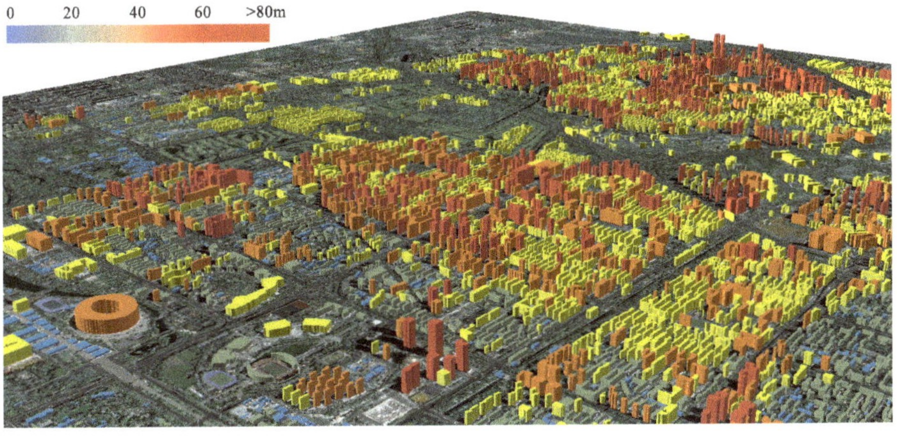

Figure 14. The 3D information results of our method.

5. Conclusions

This study aimed to extract building footprint and height information based on GF-7 satellite images. This was done in view of the existing problems in the current research field: the accuracy of building semantic segmentation is not high; most high-resolution building height information extraction is limited to small scales, and there is a lack of large-scale high-resolution building height extraction methods; GF-7 multi-view satellite images can describe the vertical structure of ground objects, but there is little research on building information extraction from GF-7 satellite images, meaning that satellite building information extraction capabilities are yet to be evaluated fully. Given these issues, we have carried out this research to develop a method for extracting 3D building information from GF-7 satellite images. We proposed a multi-stage U-Net (MSAU-Net) for building footprint extraction from GF-7 multi-spectral images. Then, we generated point cloud data from GF-7 multi-view images and constructed an nDSM to represent the height of off-terrain objects. Building height is generated by combining the results of the building footprint. Finally, we evaluated the accuracy of the extraction results based on reference building information.

We chose the Beijing area as the study area to verify the performance of our proposed method. We tested our model on two datasets: the WHU building dataset and the GF-7 self-annotated building dataset. Our model achieved IOU indicators of 89.31% and 80.27% for the WHU and GF-7 self-annotated datasets, respectively; these values were higher than the IOU indicators of other models. The RMSE between the estimated building height and the reference building height is 5.42 m, and the MAE is 3.39 m, which is higher than other building height extraction methods. The experimental results and quantitative verification show that our method could be useful for accurate and automatic 3D building information extraction from GF-7 satellite images, which has potential for application in various fields. Our method is the first attempt to extract 3D building information in dense urban areas based on GF-7 satellite images, proving the ability of GF-7 satellite images to extract 3D information of buildings. Similarly, our future work will examine 3D modeling on urban buildings based on GF-7 satellite images.

Author Contributions: Conceptualization, J.W. and Q.M.; methodology, J.W.; software, J.W.; validation, J.W., Q.M. and X.H.; formal analysis, L.Z.; investigation, X.H.; resources, Q.M.; data curation, X.L.; writing—original draft preparation, J.W.; writing—review and editing, Q.M.; visualization, C.W.; supervision, M.Z.; project administration, Q.M.; funding acquisition, Q.M. All authors have read and agreed to the published version of the manuscript.

Funding: This research was funded by (the Major Projects of High Resolution Earth Observation Systems of National Science and Technology (05-Y30B01-9001-19/20-1)), (The National Key Research and Development Program of China (2020YFC0833100)).

Acknowledgments: Our gratitude to the Group of Photogrammetry and Computer Vision (GPCV), Wuhan University for providing WHU Building Dataset (https://study.rsgis.whu.edu.cn/pages/download/building_dataset.html).

Conflicts of Interest: The authors declare no conflict of interest.

References

1. Mahtta, R.; Mahendra, A.; Seto, K.C. Building up or spreading out? Typologies of urban growth across 478 cities of 1 million+. *Environ. Res. Lett.* **2019**, *14*, 124077. [CrossRef]
2. Seto, K.C.; Dhakal, S.; Bigio, A.; Blanco, H.; Delgado, G.C.; Dewar, D.; Huang, L.; Inaba, A.; Kansal, A.; Lwasa, S. Human settlements, infrastructure and spatial planning. In *Climate Change 2014: Mitigation of Climate Change. IPCC Working Group III Contribution to AR5*; Cambridge University Press: Cambridge, UK, 2014.
3. Berger, C.; Rosentreter, J.; Voltersen, M.; Baumgart, C.; Schmullius, C.; Hese, S. Spatio-temporal analysis of the relationship between 2D/3D urban site characteristics and land surface temperature. *Remote Sens. Environ.* **2017**, *193*, 225–243. [CrossRef]
4. Venter, Z.S.; Brousse, O.; Esau, I.; Meier, F. Hyperlocal mapping of urban air temperature using remote sensing and crowdsourced weather data. *Remote Sens. Environ.* **2020**, *242*, 111791. [CrossRef]
5. Huang, X.; Wen, D.; Li, J.; Qin, R. Multi-level monitoring of subtle urban changes for the megacities of China using high-resolution multi-view satellite imagery. *Remote Sens. Environ.* **2017**, *196*, 56–75. [CrossRef]
6. Xia, M.; Jia, K.; Zhao, W.; Liu, S.; Wei, X.; Wang, B. Spatio-temporal changes of ecological vulnerability across the Qinghai-Tibetan Plateau. *Ecol. Indic.* **2021**, *123*, 107274. [CrossRef]
7. Qin, R. Change detection on LOD 2 building models with very high resolution spaceborne stereo imagery. *ISPRS J. Photogramm. Remote Sens.* **2014**, *96*, 179–192. [CrossRef]
8. Hang, J.; Li, Y.; Sandberg, M.; Buccolieri, R.; Di Sabatino, S. The influence of building height variability on pollutant dispersion and pedestrian ventilation in idealized high-rise urban areas. *Build. Environ.* **2012**, *56*, 346–360. [CrossRef]
9. Facciolo, G.; De Franchis, C.; Meinhardt-Llopis, E. Automatic 3D reconstruction from multi-date satellite images. In Proceedings of the IEEE Conference on Computer Vision and Pattern Recognition, Honolulu, HI, USA, 21–26 July 2017; pp. 57–66.
10. Bullinger, S.; Bodensteiner, C.; Arens, M. 3D Surface Reconstruction From Multi-Date Satellite Images. *arXiv* **2021**, arXiv:2102.02502. [CrossRef]
11. Qin, R.; Song, S.; Ling, X.; Elhashash, M. 3D reconstruction through fusion of cross-view images. In *Recent Advances in Image Restoration with Applications to Real World Problems*; IntechOpen: London, UK, 2020; p. 123.
12. Gui, S.; Qin, R. Automated LoD-2 model reconstruction from very-high-resolution satellite-derived digital surface model and orthophoto. *ISPRS J. Photogramm. Remote Sens.* **2021**, *181*, 1–19. [CrossRef]
13. Huang, X.; Chen, H.; Gong, J. Angular difference feature extraction for urban scene classification using ZY-3 multi-angle high-resolution satellite imagery. *ISPRS J. Photogramm. Remote Sens.* **2018**, *135*, 127–141. [CrossRef]
14. Güneralp, B.; Zhou, Y.; Ürge-Vorsatz, D.; Gupta, M.; Yu, S.; Patel, P.L.; Fragkias, M.; Li, X.; Seto, K.C. Global scenarios of urban density and its impacts on building energy use through 2050. *Proc. Natl. Acad. Sci. USA* **2017**, *114*, 8945–8950. [CrossRef] [PubMed]
15. Xu, M.; Cao, C.; Jia, P. Mapping fine-scale urban spatial population distribution based on high-resolution stereo pair images, points of interest, and land cover data. *Remote Sens.* **2020**, *12*, 608. [CrossRef]
16. Tomás, L.; Fonseca, L.; Almeida, C.; Leonardi, F.; Pereira, M. Urban population estimation based on residential buildings volume using IKONOS-2 images and lidar data. *Int. J. Remote Sens.* **2016**, *37*, 1–28. [CrossRef]
17. Xie, Y.; Weng, A.; Weng, Q. Population estimation of urban residential communities using remotely sensed morphologic data. *IEEE Geosci. Remote Sens. Lett.* **2015**, *12*, 1111–1115.
18. Liu, P.; Liu, X.; Liu, M.; Shi, Q.; Yang, J.; Xu, X.; Zhang, Y. Building Footprint Extraction from High-Resolution Images via Spatial Residual Inception Convolutional Neural Network. *Remote Sens.* **2019**, *11*, 830. [CrossRef]
19. Huang, Z.; Cheng, G.; Wang, H.; Li, H.; Shi, L.; Pan, C. Building extraction from multi-source remote sensing images via deep deconvolution neural networks. In Proceedings of the 2016 IEEE International Geoscience and Remote Sensing Symposium (IGARSS), Beijing, China, 10–15 July 2016; pp. 1835–1838.
20. Lu, T.; Ming, D.; Lin, X.; Hong, Z.; Bai, X.; Fang, J. Detecting building edges from high spatial resolution remote sensing imagery using richer convolution features network. *Remote Sens.* **2018**, *10*, 1496. [CrossRef]
21. Li, Q.; Shi, Y.; Huang, X.; Zhu, X.X. Building Footprint Generation by Integrating Convolution Neural Network With Feature Pairwise Conditional Random Field (FPCRF). *IEEE Trans. Geosci. Remote Sens.* **2020**, *58*, 7502–7519. [CrossRef]
22. Zhang, Z.; Guo, W.; Li, M.; Yu, W. GIS-supervised building extraction with label noise-adaptive fully convolutional neural network. *IEEE Geosci. Remote Sens. Lett.* **2020**, *17*, 2135–2139. [CrossRef]
23. Wang, C.; Li, L. Multi-Scale Residual Deep Network for Semantic Segmentation of Buildings with Regularizer of Shape Representation. *Remote Sens.* **2020**, *12*, 2932. [CrossRef]

24. Zhu, Q.; Liao, C.; Hu, H.; Mei, X.; Li, H. MAP-Net: Multiple Attending Path Neural Network for Building Footprint Extraction From Remote Sensed Imagery. *IEEE Trans. Geosci. Remote Sens.* **2021**, *59*, 6169–6181. [CrossRef]
25. Liu, Y.; Chen, D.; Ma, A.; Zhong, Y.; Fang, F.; Xu, K. Multiscale U-Shaped CNN Building Instance Extraction Framework with Edge Constraint for High-Spatial-Resolution Remote Sensing Imagery. *IEEE Trans. Geosci. Remote Sens.* **2021**, *59*, 6106–6120. [CrossRef]
26. Cao, Y.; Huang, X. A deep learning method for building height estimation using high-resolution multi-view imagery over urban areas: A case study of 42 Chinese cities. *Remote Sens. Environ.* **2021**, *264*, 112590. [CrossRef]
27. Yang, X.; Wang, C.; Xi, X.; Wang, P.; Lei, Z.; Ma, W.; Nie, S. Extraction of multiple building heights using ICESat/GLAS full-waveform data assisted by optical imagery. *IEEE Geosci. Remote Sens. Lett.* **2019**, *16*, 1914–1918. [CrossRef]
28. Cheng, F.; Wang, C.; Wang, J.; Tang, F.; Xi, X. Trend analysis of building height and total floor space in Beijing, China using ICESat/GLAS data. *Int. J. Remote Sens.* **2011**, *32*, 8823–8835. [CrossRef]
29. Li, X.; Zhou, Y.; Gong, P.; Seto, K.C.; Clinton, N. Developing a method to estimate building height from Sentinel-1 data. *Remote Sens. Environ.* **2020**, *240*, 111705. [CrossRef]
30. Qi, F.; Zhai, J.Z.; Dang, G. Building height estimation using Google Earth. *Energy Build.* **2016**, *118*, 123–132. [CrossRef]
31. Liu, C.; Huang, X.; Wen, D.; Chen, H.; Gong, J. Assessing the quality of building height extraction from ZiYuan-3 multi-view imagery. *Remote Sens. Lett.* **2017**, *8*, 907–916. [CrossRef]
32. Xu, Y.; Ma, P.; Ng, E.; Lin, H. Fusion of worldview-2 stereo and multitemporal TerraSAR-X images for building height extraction in urban areas. *IEEE Geosci. Remote Sens. Lett.* **2015**, *12*, 1795–1799.
33. Chen, Q.; Wang, L.; Waslander, S.L.; Liu, X. An end-to-end shape modeling framework for vectorized building outline generation from aerial images. *ISPRS J. Photogramm. Remote Sens.* **2020**, *170*, 114–126. [CrossRef]
34. Zhang, W.; Qi, J.; Wan, P.; Wang, H.; Xie, D.; Wang, X.; Yan, G. An Easy-to-Use Airborne LiDAR Data Filtering Method Based on Cloth Simulation. *Remote Sens.* **2016**, *8*, 501. [CrossRef]
35. Ronneberger, O.; Fischer, P.; Brox, T. U-net: Convolutional networks for biomedical image segmentation. In Proceedings of the International Conference on Medical Image Computing and Computer-Assisted Intervention—MICCAI 2015, Munich, Germany, 5–9 October 2015; pp. 234–241.
36. He, K.; Zhang, X.; Ren, S.; Sun, J. Deep residual learning for image recognition. In Proceedings of the IEEE Conference on Computer Vision and Pattern Recognition (CVPR), Las Vegas, NV, USA, 27–30 June 2016; pp. 770–778.
37. Zhao, H.; Shi, J.; Qi, X.; Wang, X.; Jia, J. Pyramid Scene Parsing Network. In Proceedings of the 2017 IEEE Conference on Computer Vision and Pattern Recognition (CVPR), Honolulu, HI, USA, 21–26 July 2017; Volume 1, pp. 6230–6239.
38. Fu, J.; Liu, J.; Tian, H.; Li, Y.; Bao, Y.; Fang, Z.; Lu, H. Dual attention network for scene segmentation. In Proceedings of the IEEE/CVF Conference on Computer Vision and Pattern Recognition (CVPR), Long Beach, CA, USA, 16–20 June 2019; pp. 3146–3154.
39. Li, R.; Duan, C.; Zheng, S.; Zhang, C.; Atkinson, P.M. MACU-Net for Semantic Segmentation of Fine-Resolution Remotely Sensed Images. *IEEE Geosci. Remote Sens. Lett.* **2021**, 1–5. [CrossRef]
40. Wang, X.; Girshick, R.; Gupta, A.; He, K. Non-local neural networks. In Proceedings of the IEEE Conference on Computer Vision and Pattern Recognition, Salt Lake City, UT, USA, 18–22 June 2018; pp. 7794–7803.
41. Zhu, Z.; Xu, M.; Bai, S.; Huang, T.; Bai, X. Asymmetric non-local neural networks for semantic segmentation. In Proceedings of the IEEE/CVF International Conference on Computer Vision, Seoul, Korea, 27 October–2 November 2019; pp. 593–602.
42. Zhou, D.; Wang, G.; He, G.; Long, T.; Yin, R.; Zhang, Z.; Chen, S.; Luo, B. Robust Building Extraction for High Spatial Resolution Remote Sensing Images with Self-Attention Network. *Sensors* **2020**, *20*, 7241. [CrossRef] [PubMed]
43. Li, R.; Su, J.; Duan, C.; Zheng, S. Multistage Attention ResU-Net for Semantic Segmentation of Fine-Resolution Remote Sensing Images. *arXiv* **2020**, arXiv:2011.14302. [CrossRef]
44. Ji, S.; Wei, S.; Lu, M. Fully convolutional networks for multisource building extraction from an open aerial and satellite imagery data set. *IEEE Trans. Geosci. Remote Sens.* **2018**, *57*, 574–586. [CrossRef]
45. de Franchis, C.; Meinhardt-Llopis, E.; Michel, J.; Morel, J.M.; Facciolo, G. An automatic and modular stereo pipeline for pushbroom images. In *ISPRS Annals of the Photogrammetry, Remote Sensing and Spatial Information Sciences*; HAL: Bengaluru, India, 2014; Volume II-3, pp. 49–56. [CrossRef]
46. de Franchis, C.; Meinhardt-Llopis, E.; Michel, J.; Morel, J.-M.; Facciolo, G. Automatic sensor orientation refinement of Pléiades stereo images. In Proceedings of the 2014 IEEE Geoscience and Remote Sensing Symposium, Quebec City, QC, Canada, 13–18 July 2014; pp. 1639–1642.
47. de Franchis, C.; Meinhardt-Llopis, E.; Michel, J.; Morel, J.-M.; Facciolo, G. On stereo-rectification of pushbroom images. In Proceedings of the 2014 IEEE International Conference on Image Processing (ICIP), Paris, France, 27–30 October 2014; pp. 5447–5451.
48. Lowe, D.G. Distinctive image features from scale-invariant keypoints. *Int. J. Comput. Vis.* **2004**, *60*, 91–110. [CrossRef]
49. Hirschmuller, H. Stereo processing by semiglobal matching and mutual information. *IEEE Trans. Pattern Anal. Mach. Intell.* **2008**, *30*, 328–341. [CrossRef]
50. Zhang, K.; Chen, S.-C.; Whitman, D.; Shyu, M.-L.; Yan, J.; Zhang, C. A progressive morphological filter for removing nonground measurements from airborne LIDAR data. *IEEE Trans. Geosci. Remote Sens.* **2003**, *41*, 872–882. [CrossRef]
51. Long, J.; Shelhamer, E.; Darrell, T. Fully convolutional networks for semantic segmentation. In Proceedings of the IEEE Conference on Computer Vision and Pattern Recognition (CVRP), Boston, MA, USA, 7–12 June 2015; pp. 3431–3440.

Remote Sens. **2021**, *13*, 4532

52. Chen, L.-C.; Zhu, Y.; Papandreou, G.; Schroff, F.; Adam, H. Encoder-decoder with atrous separable convolution for semantic image segmentation. In Proceedings of the European Conference on Computer Vision (ECCV), Munich, Germany, 8–14 September 2018; pp. 801–818.
53. Badrinarayanan, V.; Kendall, A.; Cipolla, R. Segnet: A deep convolutional encoder-decoder architecture for image segmentation. *IEEE Trans. Pattern Anal. Mach. Intell.* **2017**, *39*, 2481–2495. [CrossRef]

Article

URNet: A U-Shaped Residual Network for Lightweight Image Super-Resolution

Yuntao Wang [1], Lin Zhao [2], Liman Liu [1,*], Huaifei Hu [1] and Wenbing Tao [2]

[1] School of Biomedical Engineering, South-Central University for Nationalities, Wuhan 430074, China; ytao-wang@scuec.edu.cn (Y.W.); huaifeihu@mail.scuec.edu.cn (H.H.)

[2] National Key Laboratory of Science and Technology on Multi-Spectral Information Processing, School of Artificial Intelligence and Automation, Huazhong University of Science and Technology, Wuhan 430074, China; linzhao@hust.edu.cn (L.Z.); wenbingtao@hust.edu.cn (W.T.)

* Correspondence: limanliu@mail.scuec.edu.cn

Citation: Wang, Y.; Zhao, L.; Liu, L.; Hu, H.; Tao, W. URNet: A U-Shaped Residual Network for Lightweight Image Super-Resolution. *Remote Sens.* **2021**, 13, 3848. https://doi.org/ 10.3390/rs13193848

Academic Editors: Wanshou Jiang, San Jiang and Xiongwu Xiao

Received: 9 July 2021
Accepted: 21 September 2021
Published: 26 September 2021

Publisher's Note: MDPI stays neutral with regard to jurisdictional claims in published maps and institutional affiliations.

Abstract: It is extremely important and necessary for low computing power or portable devices to design more lightweight algorithms for image super-resolution (SR). Recently, most SR methods have achieved outstanding performance by sacrificing computational cost and memory storage, or vice versa. To address this problem, we introduce a lightweight U-shaped residual network (URNet) for fast and accurate image SR. Specifically, we propose a more effective feature distillation pyramid residual group (FDPRG) to extract features from low-resolution images. The FDPRG can effectively reuse the learned features with dense shortcuts and capture multi-scale information with a cascaded feature pyramid block. Based on the U-shaped structure, we utilize a step-by-step fusion strategy to improve the performance of feature fusion of different blocks. This strategy is different from the general SR methods which only use a single *Concat* operation to fuse the features of all basic blocks. Moreover, a lightweight asymmetric residual non-local block is proposed to model the global context information and further improve the performance of SR. Finally, a high-frequency loss function is designed to alleviate smoothing image details caused by pixel-wise loss. Simultaneously, the proposed modules and high-frequency loss function can be easily plugged into multiple mature architectures to improve the performance of SR. Extensive experiments on multiple natural image datasets and remote sensing image datasets show the URNet achieves a better trade-off between image SR performance and model complexity against other state-of-the-art SR methods.

Keywords: single image super-resolution; lightweight image super-resolution; U-shaped residual network; dense shortcut; effective feature distillation; high-frequency loss

1. Introduction

Single image super-resolution (SISR) aims to reconstruct a high-resolution (HR) image from its low-resolution (LR) image. It has a wide range of applications in real scenes, such as medical imaging [1–3], video surveillance [4], remote sensing [5–7], high-definition display and imaging [8], super-resolution mapping [9], hyper-spectral images [10,11], iris recognition [12], and sign and number plate reading [13]. In general, this problem is inherently ill-posed because many HR images can be downsampled to an identical LR image. To address this problem, numerous super-resolution (SR) methods are proposed, including early traditional methods [14–17] and recent learning-based methods [18–20]. Traditional methods include interpolation-based methods and regularization-based methods. Early interpolation methods such as bicubic interpolation are based on sampling theory but often produce blurry results with aliasing artifacts in natural images. Therefore, some regularization-based algorithms use machine learning to improve the performance of SR, mainly including projection onto convex sets (POCS) methods and maximum a posteriori (MAP) methods. Patti and Altunbasak [15] consider a scheme to utilize a constraint to represent the prior belief about the structure of the recovered high-resolution image.

The POCS method assumes that each LR image imposes prior knowledge on the final solution. Later work by Hardie et al. [17] uses the L2 norm of a Laplacian-style filter over the super-resolution image to regularize their MAP reconstruction.

Recently, a great number of convolutional neural network-based methods have been proposed to address the image SR problem. As a pioneering work, Dong et al. [21,22] propose a three-layer network (SRCNN) to learn the mapping function from an LR image to an HR image. Some methods focus mainly on designing a deeper or wider model to further improve the performance of SR, e.g., VDSR [23], DRCN [24], EDSR [25], and RCAN [18]. Although these methods achieve satisfactory results, the increase in model size and computational complexity limits their applications in the real world.

To reduce the computational burden or memory consumption, CARN-M [26] proposes a cascading network architecture for mobile devices, but the performance of this method significantly drops. IDN [27] aggregates current information with partially retained local short-path information by an information distillation network. IMDN [19] designs an information multi-distillation block to further improve the performance of IDN. RFDN [28] proposes a more lightweight and flexible residual feature distillation network. However, these methods are not lightweight enough and the performance of image SR can still be further improved. To build a faster and more lightweight SR model, we first propose a lightweight feature distillation pyramid residual group (FDPRG). Based on the enhanced residual feature distillation block (E-RFDB) of E-RFDN [28], the FDPRG is designed by introducing a dense shortcut (DS) connection and a cascaded feature pyramid block (CFPB). Thus, the FDPRG can effectively reuse the learned feature with DS and capture multi-scale information with CFPB. Furthermore, we propose a lightweight asymmetric residual non-local block (ANRB) to capture the global context information and further improve the SISR performance. The ANRB is modified from ANB [29] by redesigning the convolution layers and adding a residual shortcut connection. It can not only capture non-local contextual information but also become a lightweight block benefitting from residual learning. Combined with the FDPRG, ANRB, and E-RFDB, we build a more powerful lightweight U-shaped residual network (URNet) for fast and accurate image SR by using a step-by-step fusion strategy.

In the image SR field, L1 loss (i.e., mean absolute error) and L2 loss (i.e., mean square error) are usually used to measure the pixel-wise difference between the super-resolved image and its ground truth. However, using only pixel-wise loss will often cause the results to lack high-frequency details and be perceptually unsatisfying with over-smooth textures, as depicted in Figure 1. Subsequently, content loss [30], texture loss [8], adversarial loss [31], and cycle consistency loss [32] are proposed to address this problem. In particular, the content loss transfers the learned knowledge of hierarchical image features from a classification network to the SR network. For the texture loss, it is still empirical to determine the patch size to match textures. For the adversarial loss and cycle consistency loss, the training process of generative adversarial nets (GANs) is still difficult and unstable. In this work, we propose a simple but effective high-frequency loss to alleviate the problem of over-smoothed super-resolved images. Specifically, we first extract the detailed information from the ground truth by using an edge detection algorithm (e.g., Canny). Our model also predicts a response map of detail texture. The mean square error between the response map and detail information is taken as our high-frequency loss, which makes our network pay more attention to detailed textures.

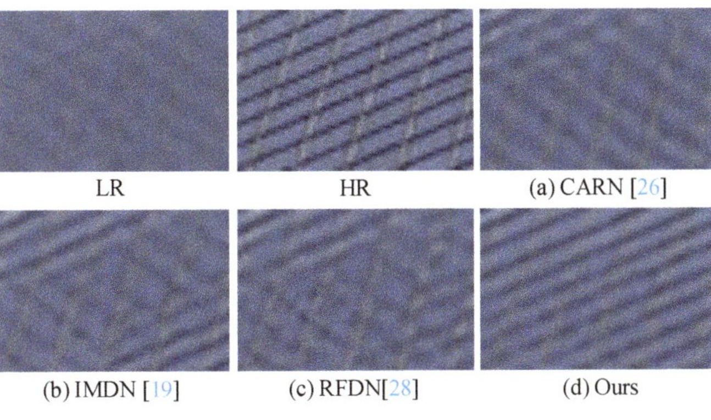

Figure 1. Visual results for ×3 SR on "img074" from Urban100. Our method obtains better visual quality than other SR methods.

The main contributions of this work can be summarized as follows:

(1) We propose a lightweight feature distillation pyramid residual group to better capture the multi-scale information and reconstruct the high-frequency detailed information of the image.

(2) We propose a lightweight asymmetric residual non-local block to capture the global contextual information and further improve the performance of SISR.

(3) We design a simple but effective high-frequency loss function to alleviate the problem of over-smoothed super-resolved images. Extensive experiments on multi-benchmark datasets demonstrate the superiority and effectiveness of our method in SISR tasks. It is worth mentioning that our designed modules and loss function can be combined with the numerous advancements in the image SR methods presented in the literature.

2. Related Work

In previous works, methods of image SR can be roughly divided into two categories: traditional methods [17,33,34] and deep learning-based methods [18,19,35,36]. Due to the limitation of space, we only briefly review the works related to deep learning networks for single image super-resolution, attention mechanism, and perceptual optimization.

2.1. Single Image Super-Resolution

The SRCNN [22] is one of the first pioneering works of directly applying deep learning to image SR. The SRCNN uses three convolution layers to map LR images to HR images. Inspired by this pioneering work, VDSR [23] and DRCN [24] stack more than 16 convolution layers based on residual learning to further improve the performance. To further unleash the power of the deep convolutional networks, EDSR [25] integrates the modified residual blocks into the SR framework to form a very deep and wide network. MemNet [37] and RDN [38] stack dense blocks to form a deep model and utilize all the hierarchical features from all the convolutional layers. SRFBN [39] proposes a feedback mechanism to generate effective high-level feature representations. EBRN [40] handles the texture SR with an incremental recovery process. Although these methods achieve significant performance, they are costly in memory consumption and computational complexity, limiting their applications in resource-constrained devices.

Recently, some fast and lightweight SISR architectures have been introduced to tackle image SR. These methods can be approximately divided into three categories: the knowledge distillation-based methods [19,27,28], the neural architecture search-based methods [41,42], and the model design-based methods [26,43]. Knowledge distillation aims to transfer the knowledge from a teacher network to a student network. IDN [27] proposes an information distillation network for better exploiting hierarchical features by separation processing of the current feature maps. Based on IDN, an information multi-distillation network (IMDN) [19] is proposed by constructing cascaded information multi-distillation blocks. RFDN [28] uses multiple feature distillation connections to learn more discriminative feature representations. FALSR [41] and MoreMNAS [42] apply neural architecture search to image SR. The performance of these methods is limited because of limitations in strategy. In addition, CARN [26] proposes a cascading mechanism based on a residual network to boost performance. LatticeNet [43] proposes a lattice block in which two butterfly structures are applied to combine two residual blocks. These works indicate that the lightweight SR networks can maintain a good trade-off between performance and model complexity.

2.2. Attention Mechanism

The attention mechanism is an important technique which has been widely used in various vision tasks (e.g., classification, object detection, and image segmentation). SENet [44] models channel-wise relationships to enhance the representational ability of the network. Non-Local [45] captures long-range dependencies by computing the response at a pixel position as a weighted sum of the features at all positions of an image. In the image SR domain, RCAN [18] and NLRN [46] improve the performance by considering attention mechanisms in the channel or the spatial dimension. SAN [35] proposes a second-order attention mechanism to enhance feature expression and correlation learning. CS-NL [47] proposes a cross-scale non-local attention module by exploring cross-scale feature correlations. HAN [48] models the holistic interdependencies among layers, channels, and positions. Due to the effectiveness of attention models, we also embed the attention mechanism into our framework to refine the high-level feature representations.

2.3. Perceptual Optimization

In the image SR field, the objective functions used to optimize models mostly contain a loss term with the pixel-wise distance between the prediction image and the ground truth image. However, researchers discovered that using this function alone leads to blurry and over-smoothed super-resolved images. Therefore, a variety of loss functions are proposed to guide the model optimization. Content loss [30] is introduced into SR to optimize the feature reconstruction error. EnhanceNet [8] uses a texture loss to produce visually more satisfactory results. MSDEPC [49] introduces an edge feature loss by using the phase congruency edge map to learn high-frequency image details. SRGAN [31] uses an adversarial loss to favor outputs residing on the manifold of natural images. CinCGAN [32] uses a cycle consistency loss to avoid the mode collapse issue of GAN and help minimize the distribution divergence.

3. U-Shaped Residual Network

In this section, we first describe the overall structure of our proposed network. Then, we elaborate on the feature distillation pyramid residual group and the asymmetric non-local residual block, respectively. Finally, we introduce the loss function of our network, including reconstruction loss and the proposed high-frequency loss.

3.1. Network Structure

As shown in Figure 2, our proposed U-shaped residual network (URNet) consists of three parts: the shallow feature extraction, the deep feature extraction, and the final image reconstruction.

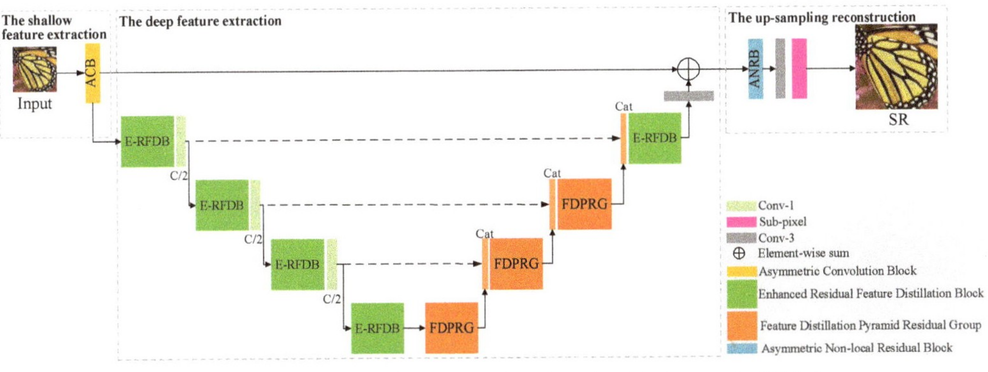

Figure 2. The architecture of the proposed U-shaped residual network (URNet).

Shallow Feature Extraction . Almost all previous works only used a 3×3 standard convolution as the first layer in their network to extract the shallow features from the input image. However, the extracted features are single scale and not rich enough. The importance of richer shallow features is ignored in subsequent deep learning methods. Inspired by the asymmetric convolution block (ACB) [50] for image classification, we adapt the ACB to SR domain to extract richer shallow features from the LR image. Specifically, 3×3, 1×3, and 3×1 convolution kernels are used to extract features from the input image in parallel. Then, the extracted features are fused by using an element-wise addition operation to generate richer shallow features. Compared with the standard convolution, the ACB can enrich the feature space and significantly improve the performance of SR with the addition of a few parameters and calculations.

Deep Feature Extraction. We use a U-shaped structure to extract deep features. In the downward flow of the U-shaped framework, we use the enhanced residual feature distillation block (E-RFDB) of E-RFDN [28] to extract features because the E-RFDN has shown its excellent performance in the super-resolution challenge of AIM 2020. In the early stage of deep feature extraction, there is no need for complex modules to extract features. Therefore, we only stack N E-RFDBs in the downward flow. The number of channels of the extracted feature map is halved by using a 1×1 convolution for each E-RFDB (except the last one).

Similarly, the upward flow of the U-shaped framework is composed of N basic blocks including $N - 1$ feature pyramid residual groups (FDPRG, see Section 3.2) and an E-RFDB. Based on the U-shaped structure, we utilize a step-by-step fusion strategy to fuse the features by using a *Concat* and FDPRG in the downward flow and upward flow. Specifically, the output features of each module in the downward flow are fused into the modules in the upward part in a back-to-front manner. This strategy transfers the information from a low level to a high level and allows the network to fuse the features of different receptive fields, resulting in effectively improving the performance of SR. The number of channels of the feature map increases with the use of the *Concat* operation. Especially for the last *Concat*, using the FDPRG will greatly increase the model complexity. Therefore, only one E-RFDB is used to extract features in the last upward flow.

Image Reconstruction. After the deep feature extraction stage, a simple 3×3 convolution is used to smooth the learned features. Then, the smoothed features are further fused with the shallow features (extracted by ACB) by an element-wise addition operation. In addition, the regression value of each pixel is closely related to the global context information in the image SR task. Therefore, we propose a lightweight asymmetric residual non-local block (ANRB, described in Section 3.3) to model the global context information and further refine the learned features. Finally, a learnable 3×3 convolution and a non-parametric sub-pixel [51] operation are used to reconstruct the HR image. Similar to [19,25,28], L1 loss is used to optimize our network. In particular, we propose a high-frequency loss function (see Section 3.4) to make our network pay more attention to learning high-frequency information.

3.2. Feature Distillation Pyramid Residual Group

In the upward flow of the U-shaped structure, we propose a more effective feature distillation pyramid residual group (FDPRG) to extract the deep features. As shown in Figure 3, the FDPRG consists of two main parts: a dense shortcut (DS) part based on three E-RFDBs and a cascaded feature pyramid block (CFPB). After the CFPB, a 3×3 convolution is used to refine the learned features.

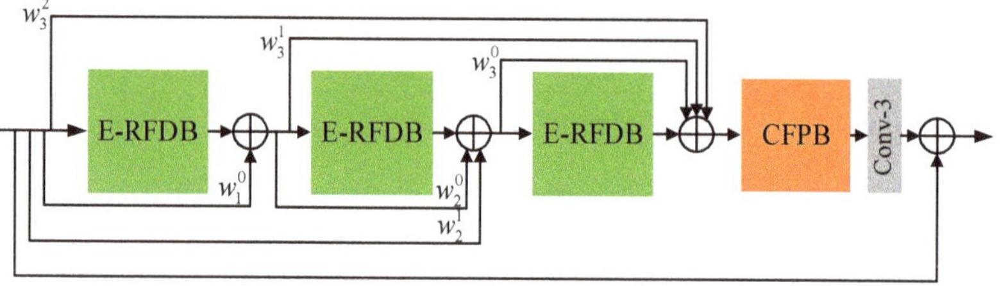

Figure 3. The feature distillation pyramid residual group (FDPRG). W_i^j is a learnable parameter.

Dense Shortcut. Residual shortcut (RS) connection is an important technique in various vision tasks. Benefitting from the RS, many SR methods have greatly improved the performance of image SR. RFDN also uses the RS between each RFDB. Although the RS can transfer the information from the input layer of the RFDB to the output layer of the RFDB, it lacks flexibility and simply adds the features of two layers. Later, we consider introducing a dense concatenation [52] to reuse the information of all previous layers. However, this dense connection is extremely GPU memory intensive. Inspired by the dense shortcut (DS) [53] for image classification, we adapt the DS to our SR model by removing the normalization in DS, because the DS has the efficiency of RS and the performance of the dense connection. As shown in Figure 3, the DS is used to connect the M E-RFDBs in a learnable manner for better feature extraction. In addition, the algorithm proves through experiments that the addition of DS reduces the memory and calculations, while slightly improving performance.

Cascaded Feature Pyramid Block. For the image SR task, the low-frequency information (e.g., simple texture) for an LR input image does not need to be reconstructed by a complex network, which allows more information in the low-level feature map. High-frequency information (e.g., edges or corners) needs to be reconstructed by a deeper network, so that the deep feature maps contain more high-frequency information. Hence, different scale features have different contributions to image SR reconstruction. Most previous methods do not utilize multi-scale information, which limits the improvement of image SR performance. Atrous spatial pyramid pooling (ASPP) [54] is an effective multi-scale feature extraction module, which adopts a parallel branch structure of convolutions with different dilation rates to extract multi-scale features, as shown in Figure 4a. However, the ASPP structure is more dependent on the setting of dilation rate parameters and each branch of ASPP is independent of the other.

Different from the ASPP, we propose a more effective multi-scale cascaded feature pyramid block (CFPB) to learn the different scale information, as shown in Figure 4b. The CFPB is designed by cascading multi-different scale convolution layers in a parallel manner. Then, the features of the different branches are fused by a *Concat* operation. The CFPB uses the idea of convolution cascading so that the next layer multi-scale features can be superimposed on the basis of the receptive field of the previous layer. Even if the dilation rate is small, it can still represent a larger receptive field. Additionally, in each parallel branch, the multi-scale features are no longer independent, which makes it easy for our network to learn multi-scale high-frequency information.

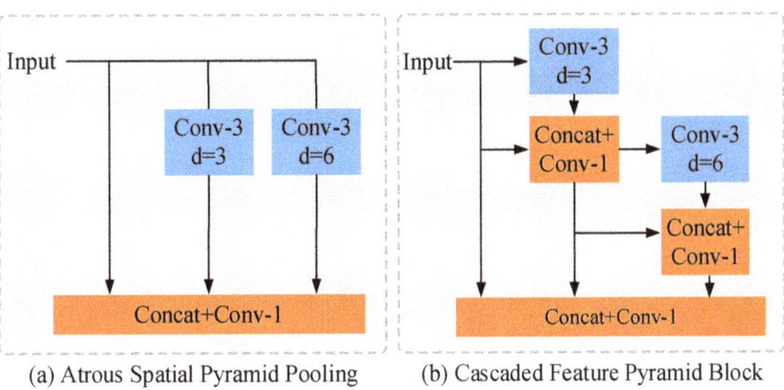

(a) Atrous Spatial Pyramid Pooling (b) Cascaded Feature Pyramid Block

Figure 4. CFPB (**b**) is an improment of ASPP (**a**).

3.3. Asymmetric Non-Local Residual Block

The non-local mechanism [45] is an attention model, which can effectively capture the long-range dependencies by modeling the connection relationship between a pixel position and all positions. In the image SR task, it is image-to-image learning. Most existing works only focus on learning detailed information while ignoring the long-range feature-wise similarities in natural images, which may produce incorrect textures globally. For the image "img092" (see Figure 8), other SR methods have learned the details of the texture (dark lines in the picture), but the direction of these lines is completely wrong in the global scope. The global texture learned by the proposed URNet after adding the non-local module is consistent with the GT image.

However, the classic Non-Local module has expensive calculation and memory consumption. It cannot be directly applied to the lightweight SR network. Inspired by the asymmetric non-local block (ANB) [29] for semantic segmentation, we propose a more lightweight asymmetric non-local residual block (ANRB, shown in Figure 5) for fast and lightweight image SR. Specifically, let $X \in R^{C \times H \times W}$ represent a feature map, where C and $H \times W$ are the numbers of channels and spatial size of X. We use three 1×1 convolutions to compress multi-channel features X into single-channel features $X_\phi, X_\theta, X_\gamma$, respectively. Afterwards, similar to the ANB, we use the pyramid pool sampling algorithm [55] to sample only $S(S \ll N = H \times W)$ representative feature points from the Key and Value branches. We perform four average pooling operations to obtain four feature maps with sizes of $1 \times 1, 3 \times 3, 6 \times 6, 8 \times 8$, respectively. Subsequently, we flatten and expand the four maps, then stitch them together to obtain a sampled feature map with a length of 110. Then, the non-local attention can be calculated as follows:

$$X_\phi = f_\phi(X), \quad X_\theta = f_\theta(X), \quad X_\gamma = f_\gamma(X), \tag{1}$$

$$\theta_P = P_\phi(X_\phi), \quad \gamma_P = P_\gamma(X_\gamma), \tag{2}$$

$$Y = Softmax(X_\phi^T \otimes \theta_P) \otimes \gamma_P, \tag{3}$$

where $f_\phi, f_\theta,$ and f_γ are 1×1 convolutions. P_ϕ and P_γ represent the pyramid pooling sampling for generating the sampled features θ_P and γ_P. $\otimes$ is matrix multiplication and Y is a feature map containing contextual information.

The last step of the attention mechanism generally uses dot multiplication to multiply the generated attention weight feature map Y with the original feature map to achieve the function of attention. However, the value of a large number of elements in Y, a matrix of $1 \times H \times W$, is close to zero due to the *Softmax* operation and the characteristics of the *Softmax* function itself: $\sum_i^H \sum_j^M (Softmax(y_{ij})) = 1$. If we directly use the operation of the dot multiplication for attention weighting, it will inevitably cause the value of the element in the weighted feature map to be too small, making the gradient disappear, which makes the gradient impossible to iterate.

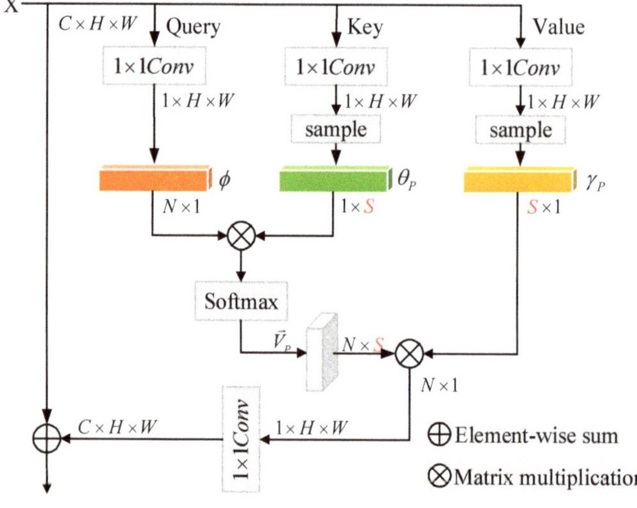

Figure 5. The asymmetric non-local residual block (ANRB).

In order to solve the above problems, we use the addition operation to generate the final attention weighted feature map $X_{weighted} = H_{1\times1}(Y) + X$, allowing the network to converge more easily, where $H_{1\times1}(\cdot)$ is a 1×1 convolution operation to convert the single-channel feature map Y into a C-channel feature map for the subsequent element-wise sum. Benefitting from the channel compression and the sampling operation, the ANRB is a lightweight non-local block. The ANRB is used to capture global context information for fast and accurate image SR.

3.4. Loss Function

In the SR domain, L1 loss (i.e., mean absolute error) and L2 loss (e.g., mean squared error) are the most frequently used loss functions for the image SR task. Similar to [18,19,25,51], we adopt L1 loss as the main reconstruction loss function to measure the differences between the SR images and the ground truth. Specifically, the L1 loss is defined as

$$\mathcal{L}_1 = \frac{1}{N} \sum_{i=1}^{N} \left\| I_{HR}^i - I_{SR}^i \right\|_1, \tag{4}$$

where I_{SR}^i, I_{HR}^i denote the i-th SR image generated by URNet and the corresponding i-th HR image used as ground truth. N is the total number of training samples.

For the image SR task, only using L1 loss or L2 loss will cause the super-resolved images to lack high-frequency details, presenting unsatisfying results with over-smooth textures. As depicted in Figure 6, comparing the natural image and the SR images generated by SR methods (e.g., RCAN [18] and IMDN [19]), we can see the reconstructed image is over-smooth in detailed texture areas. By applying edge detection algorithms to natural images and SR images, the difference is more obvious.

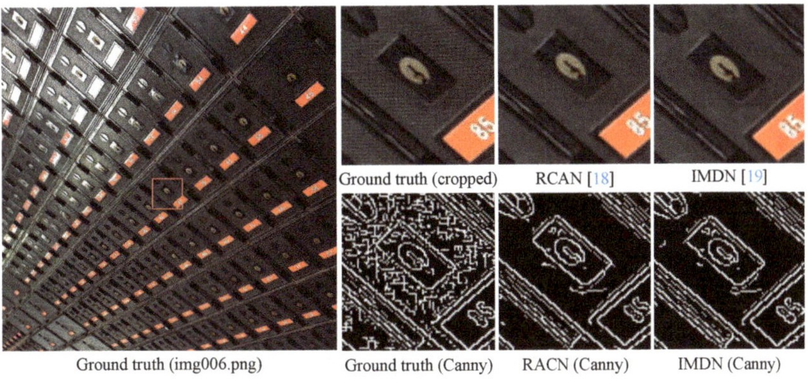

Ground truth (cropped) RCAN [18] IMDN [19]

Ground truth (img006.png) Ground truth (Canny) RACN (Canny) IMDN (Canny)

Figure 6. Ground truth/SR images and their edge images extracted by Canny operator.

Therefore, we propose a simple but effective high-frequency loss to alleviate this problem. Specifically, we first use the edge detection algorithm to extract the detailed texture maps of the HR and the SR images. Then, we adopt mean absolute error to measure the detailed differences between the SR image and the HR image. This process can be formulated as follows:

$$\mathcal{L}_{hf} = \frac{1}{N} \sum_{i=1}^{N} \left\| H_c(I_{HR}^i) - H_c(I_{SR}^i) \right\|_1,$$

(5)

where H_c denotes edge detection algorithm. In this work, we use Canny to extract detailed information from the SR images and the ground truth, respectively. Therefore, the training objective of our network is $\mathcal{L} = \alpha\mathcal{L}_{hf} + \beta\mathcal{L}_1$, where α and β are weights and used to adjust these two loss functions.

4. Experiments

4.1. Datasets and Metrics

DIV2K [56] is a high-quality image dataset, which contains 1000 DIVerse 2 K resolution RGB images including various scenes, such as animals, plants, and landscapes. The HR DIV2K is divided into 800 training images, 100 validation images, and 100 testing images. Similar to [19,27,28], we train all models with the DIV2K training images, and the corresponding LR images are generated by bicubic down-sampling the HR image with ×2, ×3, ×4 scale, respectively. To better evaluate the performance and generalization of our proposed URNet, we report the performance on four standard benchmark datasets including Set5 [57], Set14 [58], B100 [59], and Urban100 [16]. Following the previous works [19,26,28], the peak signal-to-noise ratio (PSNR) [60] and structural similarity index (SSIM) [61] are used to quantitatively evaluate our model on the Y channel in the YCbCr space converted from RGB space. PSNR is used to measure the differences between corresponding pixels of the super-resolved image and ground truth. SSIM is used to measure the structural similarity (e.g., luminance, contrast, and structures) between images.

4.2. Implementation Details

In order to clearly see the improvement effect of our method relative to RFDN, our model parameters and calculations are set as almost or less than RFDN's counterparts to exceed the performance of RFDN. The deeper or wider the convolutional network is, the better the performance is. Based on this, we tend to use as many modules as possible in the two flow branches. The number of channels, determining the width of the network, should not be too small. Therefore, we set $N = 4$, and the minimum number of channels to 8. Considering the complexity of the model, we use the most basic structure in [53], that is, setting $M = 3$. Then, considering the three-channel halving operations of the downward flow and the three *Concat* operations of the upward flow, we set the basic channel number of our URNet to 64. Specifically, for the four E-RFDBs in the downward flow (from top to bottom), the number of input channels is 64, 32, 16, and 8, respectively, while the number of input channels in the four modules in the upward flow (from bottom to top) is just the opposite.

Following the EDSR [25], the training data are augmented with random horizontal flips and 90 rotations. In the training phase, we randomly extract 32 LR RGB patches with the size of 64×64 from all the LR images in every batch. Our model is optimized by Adam with $\beta_1 = 0.9$, $\beta_2 = 0.999$, and $\epsilon = 10^{-8}$. The batch size is set to 32. The learning rate is initialized as 5×10^{-4} and halved for every 2×10^5 iterations for 1000 epochs. Each epoch has 1000 iterations of back-propagation. Similar to the IMDN [19], the hyper-parameter of Leaky ReLU is set as 0.05. The weight parameters of the loss function are set as $\alpha = 0.25$ and $\beta = 1.0$, respectively. The proposed method is implemented with PyTorch on a single GTX 1080Ti GPU.

4.3. Ablation Studies

To better validate the effectiveness of different blocks in our network, we conduct a series of ablation experiments on DIV2K. We first utilize the step-by-step fusion strategy to design a baseline model (denoted as URNet-B) based on the E-RFDB. Then, we gradually add different modules to the URNet-B. Detailed ablation experiment results are presented in Table 1. After adding the ACB into the URNet-B, the PSNR increases to 35.56 dB. Adding the DS and CFPB, we can see that the performance of image SR has increased from 35.56 dB to 35.59 dB. After adding all the blocks into the URNet-B, the PSNR increases to 35.62 dB. This is mainly because our model can consistently accumulate the hierarchical features to form more representative features and it is well focused on spatial context information. These results demonstrate the effectiveness of our ACB, FDPRG (including DS and CFPB), and ANRB.

Table 1. Ablation experiment results of different blocks on DIV2K val. **Bold** indicates the best performance ($\times 2$ SR).

URNet-B	✓	✓	✓	✓	✓
ACB		✓	✓	✓	✓
FDPRG/DS			✓	✓	✓
FDPRG/CFPB				✓	✓
ANRB					✓
PSNR (dB)	35.54	35.56	35.58	35.59	**35.62**

Afterwards, we conduct ablation experiments on the four benchmark datasets on $\times 2$ scale SR to validate the effectiveness of our proposed high-frequency loss $\mathcal{L}_{hf}$ against other loss functions widely used in the field of SR (see Section 2.3). For the adversarial loss and the cyclic consistency loss, these two loss functions are suitable for the GAN, but not for our proposed URNet. Therefore, we only report the comparison results with the other five loss functions (see Table 2). For the content loss (denoted as $\mathcal{L}_c$) and the texture loss (denoted as $\mathcal{L}_t$), we use the same configuration with SRResNet [31] and EnhanceNet [8], respectively. We observe a trend that using content loss or texture loss yields worse performance. In practice, these two loss functions are used in combination with the adversarial loss in the GAN of SR.

Table 2. Performance of different loss functions. Best results are **bolded** ($\times 2$ SR).

		Set5	Set14	B100	Urban100
$\mathcal{L}_1$	PSNR	38.020	33.685	32.228	32.356
	SSIM	0.9606	0.9184	0.9003	0.9303
$\mathcal{L}_2$	PSNR	37.999	**33.692**	32.181	32.184
	SSIM	0.9605	**0.9191**	0.8998	0.9291
$\mathcal{L}_c$	PSNR	35.823	31.776	30.283	30.145
	SSIM	0.9350	0.8763	0.8439	0.8822
$\mathcal{L}_t$	PSNR	35.267	31.230	29.870	29.587
	SSIM	0.9328	0.8747	0.8518	0.8900
$\mathcal{L}_1 + \mathcal{L}_{hf}$	PSNR	**38.063**	33.684	**32.240**	**32.415**
	SSIM	**0.9608**	0.9187	**0.9005**	**0.9310**

As shown in Figure 7, we visualize the performance difference for the other three loss functions (including $\mathcal{L}_1$, $\mathcal{L}_2$, and $\mathcal{L}_1 + \mathcal{L}_{hf}$). Compared with $\mathcal{L}_1$ and $\mathcal{L}_1 + \mathcal{L}_{hf}$, the performance of $\mathcal{L}_2$ on the four datasets is generally lower, especially on Urban100 with richer texture details. This is because the $\mathcal{L}_2$ loss uses the square of the pixel value error, so high-value differences are more important than low-value differences, resulting in too smooth results (in the case of minimum error values). Therefore, the $\mathcal{L}_1$ loss function is more widely used than the $\mathcal{L}_2$ loss in the image super-resolution [25,62]. After adding the high-frequency loss $\mathcal{L}_{hf}$ to the total loss function, the performance of image SR achieves significant improvement on both Set5 and Urban100. Compared with only using $\mathcal{L}_1$ loss, our high-frequency loss also achieves comparable PSNR and SSIM scores on the Set14 and B100 datasets. Our high-frequency loss performs especially well on Urban100 because the dataset has richer structured texture information. The high-frequency loss makes our network more focused on the texture structure of images.

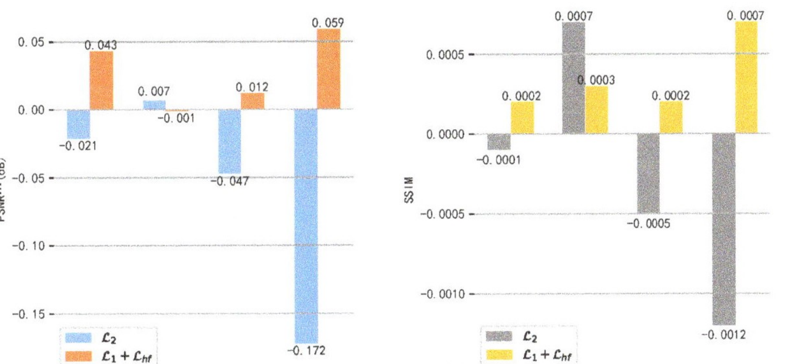

Figure 7. Comparison results of the performance difference between the three loss functions. We take PSNR/SSIM scores of $\mathcal{L}_1$ as a baseline and the PSNR/SSIM scores of $\mathcal{L}_2$ and the proposed $\mathcal{L}_1 + \mathcal{L}_{hf}$ are subtracted from it, respectively.

In order to further gain a clearer insight on the improvements of the step-by-step fusion strategy based on the U-shaped structure, we conduct experiments to compare this strategy and the general *Concat* operation to fuse the features of all blocks. Specially, we train the URNet-B and E-RFDN from scratch with the same experiment configurations to validate the effectiveness of this fusion strategy, because these two models are built based on the E-RFDB and using different fusion strategies. The experiment results are presented in Table 3.

We can see that the URNet-B not only achieves significant performance improvements on the four benchmark datasets, especially in Urban100 (PSNR: **+0.11 dB**), but also has fewer parameters (URNet-B: **567.6 K** vs. E-RFDN: 663.9 K) and calculations (FLOPs: **35.9 G** vs. 41.3 G). These results demonstrate that the step-by-step fusion strategy can not only reduce model complexity but also effectively preserve the hierarchical information to facilitate subsequent feature extraction.

Table 3. The comparison of different fusion strategies (the step-by-step and *Concat* the features of all blocks). URNet-B achieves the best PSNR (dB) scores on the four benchmark datasets ($\times$2 SR).

Method	Set5	Set14	B100	Urban100	Params	FLOPs
E-RFDN [28]	37.99	**33.56**	32.19	32.16	663.9 K	41.3 G
URNet-B	**38.03**	**33.56**	**32.20**	**32.27**	**567.6 K**	**35.9 G**

4.4. Comparison with State-of-the-Art Methods

In this section, numerous experiments are described on the four public SR benchmark datasets mentioned above. We extensively compare our proposed method with various state-of-the-art lightweight SISR methods, including Bicubic, SRCNN [21], FSRCNN [63], VDSR [23], DRCN [24], LapSRN [64], DRRN [65], MemNet [37], IDN [27], SRMDNF [66], CARN [26], IMDN [19], and RFDN-L [28]. Similar to [18,25], we also introduce a self-ensemble strategy to improve our URNet and denote the self-ensembled one as URNet+.

Quantitative Results by PSNR/SSIM. Table 4 presents quantitative comparisons for $\times$2, $\times$3, and $\times$4 SR. For a clearer and fairer comparison, we re-train the RFDN-L [28] by using the same experimental configurations as in their paper. We test the IMDN [19] (using the official pre-trained models (https://github.com/Zheng222/IMDN, accessed on 15 September 2021)), RFDN-L, and our URNet with the same environment. The results of other methods come from their papers. Compared with all the aforementioned approaches, our URNet performs the best in almost all cases. For all scaling factors, the proposed method achieves obvious improvement in the Urban100 dataset. These results indicate that our algorithm could successfully reconstruct satisfactory results for images with rich and detailed structures.

Qualitative Results. The qualitative results are illustrated in Figure 8. For challenging details in images "img006", "img067", and "img092" of the Urban100 [16] dataset, we observe that most of the compared methods would suffer from blurring edges and noticeable artifacts. IMDN [19] and RFDN-L [28] can alleviate blurred edges and recover more details (e.g., "img006" and "img067") but produce different degrees of the fake information (e.g., "img092"). In contrast, our URNet gains much better results in recovering sharper and more precise edges, more faithful to the ground truth. Especially for the image "img092" on the $\times$4 SR, the texture direction of the reconstructed edges from all compared methods is completely wrong. The URNet can make full use of the learned features and obtain clearer contours without serious artifacts. These comparisons indicate that the URNet can better recover more informative components in HR images and show satisfactory image SR results than other methods.

Figure 8. Visual qualitative comparisons of the state-of-the-art lightweight methods and our URNet on Urban100 dataset for ×2, ×3, and ×4 SR. Zoom in for best view.

Model Parameters. For the lightweight image SR, the number of model parameters is a key factor to take into account. Table 4 depicts the comparison of image SR performance and model parameters on the four benchmark datasets with scale factor ×2, ×3, and ×4, respectively. To obtain a more comprehensive understanding of the model complexity, the comparisons of the model parameters and performance are visualized in Figure 9. We can see that the proposed URNet achieves a better trade-off between the performance of image SR and model complexity than other state-of-the-art lightweight models.

Table 4. The average performance of the state-of-the-art methods for scale factor ×2, ×3, and ×4 on the four benchmark datasets Set5, Set14, B100, and Urban100. Best and second best results are **bolded** and underlined.

Method	Scale	Params	Set5 PSNR/SSIM	Set14 PSNR/SSIM	B100 PSNR/SSIM	Uban100 PSNR/SSIM
Bicubic		-	33.66/0.9299	30.24/0.8688	29.56/0.8431	26.88/0.8403
SRCNN [21]		8 K	36.66/0.9542	32.45/0.9067	31.36/0.8879	29.50/0.8946
FSRCNN [63]		13 K	37.00/0.9558	32.63/0.9088	31.53/0.8920	29.88/0.9020
VDSR [23]		666 K	37.53/0.9587	33.03/0.9124	31.90/0.8960	30.76/0.9140
DRCN [24]		1774 K	37.63/0.9588	33.04/0.9118	31.85/0.8942	30.75/0.9133
LapSRN [64]		251 K	37.52/0.9591	32.99/0.9124	31.80/0.8952	30.41/0.9103
DRRN [65]	×2	298 K	37.74/0.9591	33.23/0.9136	32.05/0.8973	31.23/0.9188
MemNet [37]		678 K	37.78/0.9597	33.28/0.9142	32.08/0.8978	31.31/0.9195
IDN [27]		553 K	37.83/0.9600	33.30/0.9148	32.08/0.8985	31.27/0.9196
SRMDNF [66]		1511 K	37.79/0.9601	33.32/0.9159	32.05/0.8985	31.33/0.9204
CARN [26]		1592 K	37.76/0.9590	33.52/0.9166	32.09/0.8978	31.92/0.9256
IMDN [19]		694 K	38.00/0.9605	33.63/0.9177	32.18/0.8996	32.17/0.9283
RFDN-L [28]		626 K	38.03/0.9606	33.65/0.9183	32.17/0.8996	32.16/0.9282
URNet (ours)		612 K	38.06/0.9608	33.68/0.9187	32.24/0.9005	32.42/0.9310
URNet+ (ours)		612 K	**38.14/0.9611**	**33.70/0.9190**	**32.29/0.9009**	**32.61/0.9325**
Bicubic		-	30.39/0.8682	27.55/0.7742	27.21/0.7385	24.46/0.7349
SRCNN [21]		8 K	32.75/0.9090	29.30/0.8215	28.41/0.7863	26.24/0.7989
FSRCNN [63]		13 K	33.18/0.9140	29.37/0.8240	28.53/0.7910	26.43/0.8080
VDSR [23]		666 K	33.66/0.9213	29.77/0.8314	28.82/0.7976	27.14/0.8279
DRCN [24]		1774 K	33.82/0.9226	29.76/0.8311	28.80/0.7963	27.15/0.8276
LapSRN [64]		502 K	33.81/0.9220	29.79/0.8325	28.82/0.7980	27.07/0.8275
DRRN [65]	×3	298 K	34.03/0.9244	29.96/0.8349	28.95/0.8004	27.53/0.8378
MemNet [37]		678 K	34.09/0.9248	30.00/0.8350	28.96/0.8001	27.56/0.8376
IDN [27]		553 K	34.11/0.9253	29.99/0.8354	28.95/0.8013	27.42/0.8359
SRMDNF [66]		1528K	34.12/0.9254	30.04/0.8382	28.97/0.8025	27.57/0.8398
CARN [26]		1592 K	34.29/0.9255	30.29/0.8407	29.06/0.8034	28.06/0.8493
IMDN [19]		703 K	34.36/0.9270	30.32/0.8417	29.09/0.8047	28.16/0.8519
RFDN-L [28]		633 K	34.39/0.9271	30.35/0.8419	29.11/0.8054	28.24/0.8534
URNet (ours)		621 K	34.51/0.9281	30.40/0.8433	29.14/0.8061	28.40/0.8574
URNet+ (ours)		621 K	**34.60/0.9288**	**30.48/0.8444**	**29.19/0.8072**	**28.57/0.8599**
Bicubic		-	28.42/0.8104	26.00/0.7027	25.96/0.6675	23.14/0.6577
SRCNN [21]		8 K	30.48/0.8626	27.50/0.7513	26.90/0.7101	24.52/0.7221
FSRCNN [63]		13 K	30.72/0.8660	27.61/0.7550	26.98/0.7150	24.62/0.7280
VDSR [23]		666 K	31.35/0.8838	28.01/0.7674	27.29/0.7251	25.18/0.7524
DRCN [24]		1774 K	31.53/0.8854	28.02/0.7670	27.23/0.7233	25.14/0.7510
LapSRN [64]		251 K	31.54/0.8852	28.09/0.7700	27.32/0.7275	25.21/0.7562
DRRN [65]	×4	298 K	31.68/0.8888	28.21/0.7720	27.38/0.7284	25.44/0.7638
MemNet [37]		678 K	31.74/0.8893	28.26/0.7723	27.40/0.7281	25.50/0.7630
IDN [27]		553 K	31.82/0.8903	28.25/0.7730	27.41/0.7297	25.41/0.7632
SRMDNF [66]		1552 K	31.96/0.8925	28.35/0.7787	27.49/0.7337	25.68/0.7731
CARN [26]		1592 K	32.13/0.8937	28.60/0.7806	27.58/0.7349	26.07/0.7837
IMDN [19]		715 K	32.21/0.8948	28.58/0.7810	27.55/0.7353	26.04/0.7838
RFDN-L [28]		643 K	32.23/0.8953	28.59/0.7814	27.56/0.7362	26.14/0.7871
URNet (ours)		633K	32.20/0.8952	28.63/0.7826	27.60/0.7369	26.23/0.7905
URNet+ (ours)		633K	**32.35/0.8969**	**28.71/0.7840**	**27.66/0.7383**	**26.41/0.7945**

4.5. Model Anaysis

Model Calculations. It is not enough to measure the weight of the model only by the model parameters. Calculation consumption is also an important metric. In Table 5, we report the comparison of URNet and other state-of-the-art algorithms (e.g., CARN [26],

IMDN [19], and RFDN-L [28]) in terms of FLOPs (using a single image with the size 256×256) and PSNR/SSIM (using the Set14 dataset with the $\times 4$ scale factor). As we can see, our URNet achieves higher PSNR/SSIM than other methods while using fewer calculations. These results demonstrate that our method can balance the calculation costs and the performance of image reconstruction well.

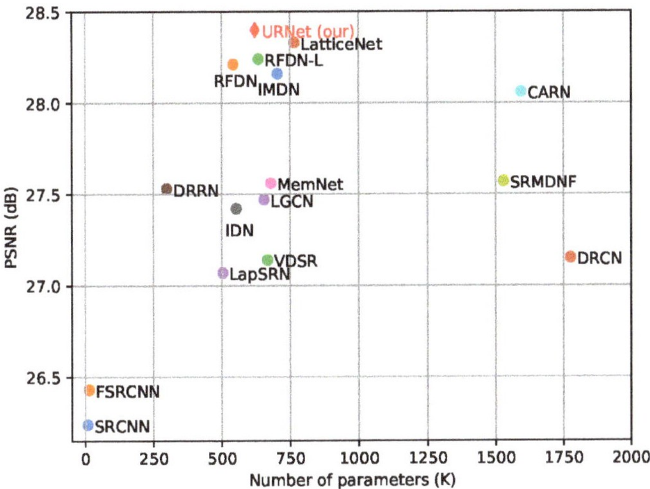

Figure 9. PSNR vs. the number of parameters. The comparison is conducted on Urban100 with the $\times 3$ scale factor.

Table 5. PSNR/SSIM vs. FLOPs on Set14 ($\times 4$).

	CARN [26]	IMDN [19]	RFDN-L [28]	URNet (ours)
SSIM	0.7806	0.7810	0.7814	**0.7826**
PSNR	28.60	28.58	28.59	**28.63**
FLOPs (G)	103.58	46.60	41.54	**39.51**

Lightweight Analyses. We also choose two non-lightweight methods and one SOTA lightweight SISR method, i.e., EDSR [25], RCAN [18], and IMDN [19], for comparison. We use official codes (https://github.com/cszn/KAIR, accessed on 15 September 2021) (AIM 2020 efficient super-resolution challenge (https://data.vision.ee.ethz.ch/cvl/aim20/, accessed on 15 September 2021)) to test the running time of these methods in a feed-forward process on the B100 ($\times 4$) dataset. The results are reported in Table 6. We can observe that both methods, EDSR and RCAN, outperform our URNet. This is a reasonable result since they have a deeper and wider network structure that contains large quantities of convolutional layers and parameters. Actually, the parameters of EDSR and RCAN are 40 M and 16 M, while that of ours is only 0.6 M. However, compared with other methods, URNet runs the fastest inference speed. Simultaneously, our URNet achieves dominant performance in terms of parameter usage and time consumption, compared to IMDN. These comparison results show that our method can obtain fast and accurate image SR.

Table 6. Comparison with non-lightweight and SOTA lightweight methods.

	Scale	EDSR [25]	RCAN [18]	IMDN [19]	URNet (ours)
Set5	2	38.11/0.9602	38.27/0.9614	38.00/0.9605	38.06/0.9608
	3	34.65/0.9280	34.74/0.9299	34.36/0.9270	34.51/0.9281
	4	32.46/0.8968	32.63/0.9002	32.21/0.8948	32.20/0.8952
Set14	2	33.92/0.9195	34.12/0.9216	33.63/0.9177	33.68/0.9187
	3	30.52/0.8462	30.65/0.8482	30.32/0.8417	30.40/0.8433
	4	28.80/0.7876	28.87/0.7889	28.58/0.7810	28.63/0.7826
B100	2	32.32/0.9013	32.41/0.9027	32.18/0.8996	32.24/0.9005
	3	29.25/0.8093	29.32/0.8111	29.09/0.8047	29.14/0.8061
	4	27.71/0.7420	27.77/0.7436	27.55/0.7353	27.60/0.7369
Urban100	2	32.93/0.9351	33.24/0.9384	32.17/0.9283	32.42/0.9310
	3	28.80/0.8653	29.09/0.8702	28.16/0.8519	28.40/0.8574
	4	26.64/0.8033	26.82/0.8087	26.04/0.7838	26.23/0.7905
Parameters (K)		43,090	15,592	715	633
FLOPs (G)		3293.9	1044.0	46.6	39.5
Running Time (Sec.)		0.2178	0.2596	0.0939	0.0310

4.6. Remote Sensing Image Super-Resolution

To better evaluate the generalization of our method, we also conduct experiments on the remote sensing datasets. The natural image SR and remote sensing image SR belong to different image domains but the same task. Consequently, we can use the URNet trained on the natural image dataset (i.e., DIV2K) as a pre-trained model and fine-tune the model on the remote sensing dataset. By transferring the external knowledge from the natural image domain to the remote sensing domain, our proposed URNet achieves a better performance on the remote sensing image SR task.

Following most remote sensing image SR methods [67–71], we conduct experiments on the UC Merced [72] land-use dataset. The UC Merced dataset is one of the most popular image collections in the remote sensing community, which contains 21 classes of land-use scenes in total with 100 aerial images per class. These images have a high spatial resolution (0.3 m/pixel). We randomly select 840 images (40 images per class) from the UC Merced as the training set, and we randomly select 40 images from the training set as a validation set. Moreover, we construct a testing set named UCTest by randomly choosing 120 images from the remaining images of the UC Merced dataset. The LR-HR image pair acquisition operation and implementation details are the same as for experiments on the DIV2K dataset. The model is trained for 100 epochs with an initial learning rate of 0.0001 and the input patch size set to 16×16. Similarly, we also re-train RFDN-L [28] by using the same training strategies. MPSR [68] randomly selects 800 images from the UC Merced dataset as the training samples. For a fair and convincing comparison, we re-train the MPSR by using the same experimental configurations as in their paper and the same dataset as this paper.

The NWPU-RESISC45 [73] dataset is a public benchmark with spatial resolution varying from 30 m to 0.2 m per pixel. We also randomly select 180 images from the NWPU-RESISC45 dataset as a testing set (named RESISCTest) to validate the robustness of our model.

Table 7 shows the quantitative results of the state-of-the-art SR methods on remote sensing datasets UCTest and RESISCTest for scale factor ×4. We can see that our proposed URNet and URNet-T (using the pre-trained model) achieve the highest PSNR and SSIM scores on these two datasets. The methods could gain better performance by using the strategy of the pre-trained model, which means that this strategy allows low-level feature information from DIV2K to be shared to another dataset, achieving better performance on super-resolving remote sensing images. The performance of MPSR is further improved on UCTest by using the same strategy but fails on RESISCTest because the MPSR-T is a non-lightweight model (MPSR-T: 12.3 M vs. URNet-T: 633 K, and MPSR-T: 835.5 G vs. URNet-T: 39.5 G, in terms of parameters and FLOPs) and more likely to overfit on the training set.

Table 7. The PSNR/SSIM of UCTest and RESISCTest with a scale factor of ×4. (*-T denotes using the pre-trained model.)

		RFDN-L [28]	MPSR [68]	URNet (ours)	RFDN-L-T	MPSR-T	URNet-T (ours)
UCTest	PSNR	29.03	29.09	**29.15**	29.37	29.34	**29.58**
	SSIM	0.7940	0.7953	**0.7968**	0.8047	0.8060	**0.8102**
RESISCTest	PSNR	29.06	29.09	**29.13**	29.09	29.01	**29.19**
	SSIM	0.7710	0.7718	**0.7730**	0.7721	0.7706	**0.7750**

To fully demonstrate the effectiveness of our method, we also show the ×4 SR visual results from UCTest's "agricultural81" in Figure 10 and RESISCTest's "harbor_450" in Figure 11. We can see that our proposed URN-T shows significant improvements, reducing aliasing, blur artifacts, and better reconstructing high-fidelity image details.

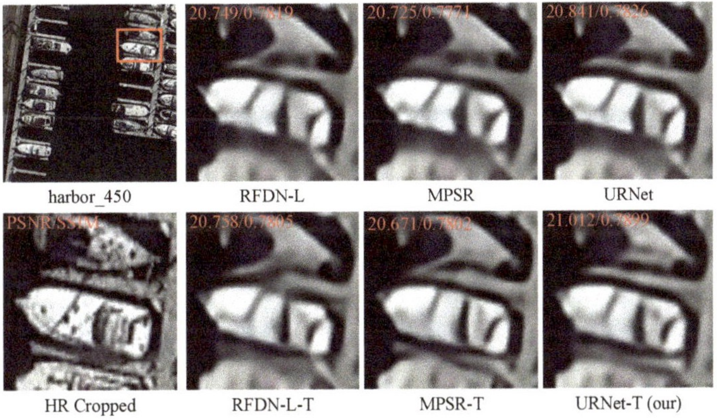

Figure 10. Comparison of reconstructed HR images of "agricultural81" obtained from UCTest dataset with 256 × 256 pixel images using different methods with a scale factor of ×4.

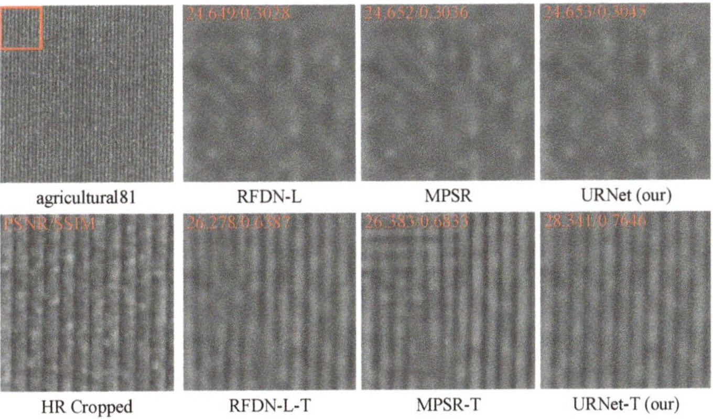

Figure 11. Comparison of reconstructed HR images of "harbor_450" obtained from RESISCTest dataset with 256 × 256 pixel images using different methods with a scale factor of ×4.

5. Conclusions

In this paper, we introduce a novel lightweight U-shaped residual network (URNet) for fast and accurate image SR. Specifically, we design an effective feature distillation pyramid residual group (FDPRG) to extract deep features from an LR image based on the E-RFDB. The FDPRG can effectively reuse the shallow features with dense shortcut connections and capture multi-scale information with a cascaded feature pyramid block. Based on the U-shaped structure, we utilize a step-by-step fusion strategy to fuse the features of different blocks and further refine the learned features. In addition, we introduce a lightweight asymmetric non-local residual block to capture the global context information and further improve the performance of image SR. In particular, to alleviate the problem of smoothing image details caused by pixel-wise loss, we design a simple but effective high-frequency loss to help optimize our model. Extensive experiments indicate the URNet achieves a better trade-off between image SR performance and model complexity against other state-of-the-art SR methods. In the future, our method will be applied to super-resolution images with fuzzy or even real degradation models. At the same time, we will also consider deep separable convolutions or other lightweight convolutions as an alternative to standard convolutions to further reduce the number of parameters and calculations.

Author Contributions: Y.W. and L.Z. have equal contribution to this work and are co-first authors. Conceptualization, Y.W. and L.L.; methodology, Y.W. and L.Z.; software, Y.W.; validation, L.L., H.H., and W.T.; writing—original draft preparation, Y.W. and L.Z.; writing—review and editing, Y.W., L.Z., and L.L.; supervision, W.T.; project administration, L.L.; funding acquisition, L.L., H.H., and W.T. All authors have read and agreed to the published version of the manuscript.

Funding: This work was supported in part by the National Natural Science Foundation of China (Grant 61976227, 62176096, and 62076257) and in part by the Natural Science Foundation of Hubei Province under Grant 2019CFB622.

Data Availability Statement: Code is available at https://github.com/ytao-wang/URNet, accessed on 15 September 2021.

Acknowledgments: The authors are grateful to the Editor and reviewers for their constructive comments, which significantly improved this work.

Conflicts of Interest: The authors declare no conflict of interest.

References

1. Isaac, J.S.; Kulkarni, R. Super resolution techniques for medical image processing. In Proceedings of the 2015 International Conference on Technologies for Sustainable Development, Mumbai, India, 4–6 February 2015; pp. 1–6.
2. Liu, H.; Xu, J.; Wu, Y.; Guo, Q.; Ibragimov, B.; Xing, L. Learning deconvolutional deep neural network for high resolution medical image reconstruction. *Inf. Sci.* **2018**, *468*, 142–154. [CrossRef]
3. Yamashita, K.; Markov, K. Medical Image Enhancement Using Super Resolution Methods. In *International Conference on Computational Science*; Springer: Berlin/Heidelberg, Germany, 2020; pp. 496–508.
4. Rasti, P.; Uiboupin, T.; Escalera, S.; Anbarjafari, G. Convolutional neural network super resolution for face recognition in surveillance monitoring. In *International Conference on Articulated Motion and Deformable Objects*; Springer: Berlin/Heidelberg, Germany, 2016; pp. 175–184.
5. Xu, W.; Guangluan, X.; Wang, Y.; Sun, X.; Lin, D.; Yirong, W. High quality remote sensing image super-resolution using deep memory connected network. In Proceedings of the IGARSS 2018-2018 IEEE International Geoscience and Remote Sensing Symposium, Valencia, Spain, 22–27 July 2018; pp. 8889–8892.
6. Ma, W.; Pan, Z.; Yuan, F.; Lei, B. Super-resolution of remote sensing images via a dense residual generative adversarial network. *Remote Sens.* **2019**, *11*, 2578. [CrossRef]
7. Gong, Y.; Liao, P.; Zhang, X.; Zhang, L.; Chen, G.; Zhu, K.; Tan, X.; Lv, Z. Enlighten-GAN for Super Resolution Reconstruction in Mid-Resolution Remote Sensing Images. *Remote Sens.* **2021**, *13*, 1104. [CrossRef]
8. Sajjadi, M.S.M.; Schölkopf, B.; Hirsch, M. EnhanceNet: Single Image Super-Resolution through Automated Texture Synthesis. In Proceedings of the IEEE International Conference on Computer Vision, Venice, Italy, 22–29 October 2017; pp. 4501–4510.
9. Wang, P.; Wang, L.; Leung, H.; Zhang, G. Super-Resolution Mapping Based on Spatial–Spectral Correlation for Spectral Imagery. *IEEE Trans. Geosci. Remote Sens.* **2020**, *59*, 2256–2268. [CrossRef]
10. Wan, W.; Guo, W.; Huang, H.; Liu, J. Nonnegative and nonlocal sparse tensor factorization-based hyperspectral image super-resolution. *IEEE Trans. Geosci. Remote Sens.* **2020**, *58*, 8384–8394. [CrossRef]

Remote Sens. **2021**, *13*, 3848

11. Li, J.; Cui, R.; Li, B.; Song, R.; Li, Y.; Du, Q. Hyperspectral image super-resolution with 1D–2D attentional convolutional neural network. *Remote Sens.* **2019**, *11*, 2859. [CrossRef]
12. Nguyen, K.; Sridharan, S.; Denman, S.; Fookes, C. Feature-domain super-resolution framework for Gabor-based face and iris recognition. In Proceedings of the 2012 IEEE Conference on Computer Vision and Pattern Recognition, Providence, RI, USA, 16–21 June 2012; pp. 2642–2649.
13. Zhou, F.; Yang, W.; Liao, Q. A coarse-to-fine subpixel registration method to recover local perspective deformation in the application of image super-resolution. *IEEE Trans. Image Process.* **2011**, *21*, 53–66. [CrossRef] [PubMed]
14. Stark, H.; Oskoui, P. High-resolution image recovery from image-plane arrays, using convex projections. *JOSA A* **1989**, *6*, 1715–1726. [CrossRef] [PubMed]
15. Patti, A.J.; Altunbasak, Y. Artifact reduction for set theoretic super resolution image reconstruction with edge adaptive constraints and higher-order interpolants. *IEEE Trans. Image Process.* **2001**, *10*, 179–186. [CrossRef]
16. Huang, J.B.; Singh, A.; Ahuja, N. Single Image Super-Resolution From Transformed Self-Exemplars. In Proceedings of the IEEE Conference on Computer Vision and Pattern Recognition, Boston, MA, USA, 7–12 June 2015; pp. 5197–5206.
17. Hardie, R.C.; Barnard, K.J.; Armstrong, E.E. Joint MAP registration and high-resolution image estimation using a sequence of undersampled images. *IEEE Trans. Image Process.* **1997**, *6*, 1621–1633. [CrossRef]
18. Zhang, Y.; Li, K.; Li, K.; Wang, L.; Zhong, B.; Fu, Y. Image Super-Resolution Using Very Deep Residual Channel Attention Networks. In Proceedings of the European Conference on Computer Vision, Munich, Germany, 8–14 September 2018; pp. 294–310.
19. Hui, Z.; Gao, X.; Yang, Y.; Wang, X. Lightweight Image Super-Resolution with Information Multi-distillation Network. In Proceedings of the 27th ACM International Conference on Multimedia, Nice, France, 21–25 October 2019; pp. 2024–2032.
20. Feng, X.; Zhang, W.; Su, X.; Xu, Z. Optical Remote Sensing Image Denoising and Super-Resolution Reconstructing Using Optimized Generative Network in Wavelet Transform Domain. *Remote Sens.* **2021**, *13*, 1858. [CrossRef]
21. Dong, C.; Loy, C.C.; He, K.; Tang, X. Learning a deep convolutional network for image super-resolution. In Proceedings of the European Conference on Computer Vision, Zurich, Switzerland, 6–12 September 2014; pp. 184–199.
22. Dong, C.; Loy, C.C.; He, K.; Tang, X. Image super-resolution using deep convolutional networks. *IEEE Trans. Pattern Anal. Mach. Intell.* **2015**, *38*, 295–307. [CrossRef] [PubMed]
23. Kim, J.; Kwon Lee, J.; Mu Lee, K. Accurate image super-resolution using very deep convolutional networks. In Proceedings of the IEEE Conference on Computer Vision and Pattern Recognition, Las Vegas, NV, USA, 27–30 June 2016; pp. 1646–1654.
24. Kim, J.; Lee, J.K.; Lee, K.M. Deeply-recursive convolutional network for image super-resolution. In Proceedings of the IEEE Conference on Computer Vision and Pattern Recognition, 2016; pp. 1637–1645.
25. Lim, B.; Son, S.; Kim, H.; Nah, S.; Lee, K.M. Enhanced Deep Residual Networks for Single Image Super-Resolution. In Proceedings of the IEEE Conference on Computer Vision and Pattern Recognition Workshops, Honolulu, HI, USA, 21–26 July 2017; pp. 1132–1140.
26. Ahn, N.; Kang, B.; Sohn, K.A. Fast, Accurate, and Lightweight Super-Resolution with Cascading Residual Network. In Proceedings of the European Conference on Computer Vision, Munich, Germany, 8–14 September 2018; pp. 252–268.
27. Hui, Z.; Wang, X.; Gao, X. Fast and Accurate Single Image Super-Resolution via Information Distillation Network. In Proceedings of the IEEE Conference on Computer Vision and Pattern Recognition, Salt Lake City, UT, USA, 18–23 June 2018; pp. 723–731.
28. Liu, J.; Tang, J.; Wu, G. Residual Feature Distillation Network for Lightweight Image Super-Resolution. In Proceedings of the European Conference on Computer Vision AIM Workshops, Glasgow, UK, 23–28 August 2020.
29. Zhu, Z.; Xu, M.; Bai, S.; Huang, T.; Bai, X. Asymmetric non-local neural networks for semantic segmentation. In Proceedings of the IEEE International Conference on Computer Vision, Seoul, Korea, 27 October 2019; pp. 593–602.
30. Justin, J.; Alexandre, A.; Li, F.-F. Perceptual Losses for Real-Time Style Transfer and Super-Resolution. In Proceedings of the European Conference on Computer Vision;Springer: Berlin, Germany, 2016; pp. 694–711.
31. Ledig, C.; Theis, L.; Huszar, F.; Caballero, J.; Cunningham, A.; Acosta, A.; Aitken, A.; Tejani, A.; Totz, J.; Wang, Z.; et al. Photo-Realistic Single Image Super-Resolution Using a Generative Adversarial Network. In Proceedings of the IEEE Conference on Computer Vision and Pattern Recognition, Honolulu, HI, USA, 21–26 July 2017; pp. 5892–5900.
32. Yuan, Y.; Liu, S.; Zhang, J.; Zhang, Y.; Dong, C.; Lin, L. Unsupervised Image Super-Resolution using Cycle-in-Cycle Generative Adversarial Networks. In Proceedings of the IEEE Conference on Computer Vision and Pattern Recognition Workshops, Salt Lake City, UT, USA, 18–22 June 2018; pp. 701–710.
33. Zhang, H.; Yang, Z.; Zhang, L.; Shen, H. Super-resolution reconstruction for multi-angle remote sensing images considering resolution differences. *Remote Sens.* **2014**, *6*, 637–657. [CrossRef]
34. Chantas, G.K.; Galatsanos, N.P.; Woods, N.A. Super-resolution based on fast registration and maximum a posteriori reconstruction. *IEEE Trans. Image Process.* **2007**, *16*, 1821–1830. [CrossRef] [PubMed]
35. Dai, T.; Cai, J.; Zhang, Y.; Xia, S.T.; Zhang, L. Second-order Attention Network for Single Image Super-Resolution. In Proceedings of the IEEE Conference on Computer Vision and Pattern Recognition, Long Beach, CA, USA, 15–20 June 2019; pp. 11065–11074.
36. Feng, X.; Su, X.; Shen, J.; Jin, H. Single space object image denoising and super-resolution reconstructing using deep convolutional networks. *Remote Sens.* **2019**, *11*, 1910. [CrossRef]
37. Tai, Y.; Yang, J.; Liu, X.; Xu, C. MemNet: A Persistent Memory Network for Image Restoration. In Proceedings of the IEEE International Conference on Computer Vision, Venice, Italy, 22–29 October 2017; pp. 4539–4547.

38. Zhang, Y.; Tian, Y.; Kong, Y.; Zhong, B.; Fu, Y. Residual Dense Network for Image Super-Resolution. In Proceedings of the IEEE Conference on Computer Vision and Pattern Recognition, Salt Lake City, UT, USA, 18–23 June 2018; pp. 2472–2481.
39. Li, Z.; Yang, J.; Liu, Z.; Yang, X.; Jeon, G.; Wu, W. Feedback Network for Image Super-Resolution. In Proceedings of the IEEE Conference on Computer Vision and Pattern Recognition, Long Beach, CA, USA, 16–20 June 2019; pp. 3862–3871.
40. Qiu, Y.; Wang, R.; Tao, D.; Cheng, J. Embedded Block Residual Network: A Recursive Restoration Model for Single-Image Super-Resolution. In Proceedings of the IEEE International Conference on Computer Vision, Seoul, Korea, 27 October–2 November 2019; pp. 4180–4189.
41. Chu, X.; Zhang, B.; Ma, H.; Xu, R.; Li, J.; Li, Q. Fast, accurate and lightweight super-resolution with neural architecture search. *arXiv* **2019**, arXiv:1901.07261.
42. Chu, X.; Zhang, B.; Xu, R.; Ma, H. Multi-objective reinforced evolution in mobile neural architecture search. *arXiv* **2019**, arXiv:1901.01074.
43. Luo, X.; Xie, Y.; Zhang, Y.; Qu, Y.; Li, C.; Fu, Y. LatticeNet: Towards Lightweight Image Super-resolution with Lattice Block. In Proceedings of the European Conference on Computer Vision, Glasgow, UK, 23-28 August 2020.
44. Hu, J.; Shen, L.; Sun, G. Squeeze-and-Excitation Networks. In Proceedings of the IEEE Conference on Computer Vision and Pattern Recognition, Salt Lake City, UT, USA, 18–23 June 2018; pp. 7132–7174.
45. Wang, X.; Girshick, R.; Gupta, A.; He, K. Non-Local Neural Networks. In Proceedings of the IEEE Conference on Computer Vision and Pattern Recognition, Salt Lake City, UT, USA, 18–23 June 2018; pp. 7794–7803.
46. Liu, D.; Wen, B.; Fan, Y.; Loy, C.C.; Huang, T.S. Non-Local Recurrent Network for Image Restoration. In *Advances in Neural Information Processing Systems*; MIT Press: Cambridge, MA, USA, 2018; pp. 1680–1689.
47. Mei, Y.; Fan, Y.; Zhou, Y.; Huang, L.; Huang, T.S.; Shi, H. Image Super-Resolution With Cross-Scale Non-Local Attention and Exhaustive Self-Exemplars Mining. In Proceedings of the IEEE Conference on Computer Vision and Pattern Recognition, Seattle, WA, USA, 13–19 June 2020.
48. Niu, B.; Wen, W.; Ren, W.; Zhang, X.; Yang, L.; Wang, S.; Zhang, K.; Cao, X.; Shen, H. Single Image Super-Resolution via a Holistic Attention Network. In Proceedings of the European Conference on Computer Vision, Glasgow, UK, 23–28 August 2020; pp. 191–207.
49. Liu, H.; Fu, Z.; Han, J.; Shao, L.; Hou, S.; Chu, Y. Single image super-resolution using multi-scale deep encoder–decoder with phase congruency edge map guidance. *Inf. Sci.* **2019**, *473*, 44–58. [CrossRef]
50. Ding, X.; Guo, Y.; Ding, G.; Han, J. Acnet: Strengthening the kernel skeletons for powerful cnn via asymmetric convolution blocks. In Proceedings of the IEEE International Conference on Computer Vision, Seoul, Korea, 27 October–2 November 2019; pp. 1911–1920.
51. Wang, Z.; Liu, D.; Yang, J.; Han, W.; Huang, T. Deep networks for image super-resolution with sparse prior. In Proceedings of the IEEE International Conference on Computer Vision, Santiago, Chile, 7–13 December 2015; pp. 370–378.
52. Huang, G.; Liu, Z.; Van Der Maaten, L.; Weinberger, K.Q. Densely connected convolutional networks. In Proceedings of the IEEE Conference on Computer Vision and Pattern Recognition, Honolulu, HI, USA, 22–25 July 2017; pp. 4700–4708.
53. Zhang, C.; Benz, P.; Argaw, D.M.; Lee, S.; Kim, J.; Rameau, F.; Bazin, J.C.; Kweon, I.S. Resnet or densenet? introducing dense shortcuts to resnet. In Proceedings of the IEEE Winter Conference on Applications of Computer Vision, Waikoloa, Hawaii, US, 5–9 January 2021 ; pp. 3550–3559.
54. Chen, L.C.; Papandreou, G.; Kokkinos, I.; Murphy, K.; Yuille, A.L. Deeplab: Semantic image segmentation with deep convolutional nets, atrous convolution, and fully connected crfs. *IEEE Trans. Pattern Anal. Mach. Intell.* **2017**, *40*, 834–848. [CrossRef]
55. Zhao, H.; Shi, J.; Qi, X.; Wang, X.; Jia, J. Pyramid scene parsing network. In Proceedings of the IEEE Conference on Computer Vision and Pattern Recognition, Honolulu, HI, USA, 22–25 July 2017; pp. 2881–2890.
56. Timofte, R.; Agustsson, E.; Van Gool, L.; Yang, M.H.; Zhang, L. Ntire 2017 challenge on single image super-resolution: Methods and results. In Proceedings of the IEEE Conference on Computer Vision and Pattern Recognition Workshops, Honolulu, HI, USA, 21–26 July 2017; pp. 114–125.
57. Bevilacqua, M.; Roumy, A.; Guillemot, C.; Alberi-Morel, M.L. Low-complexity single-image super-resolution based on non-negative neighbor embedding. In Proceedings of the 2012 British Machine Vision Conference, Surrey, UK, 3–7 September 2012.
58. Zeyde, R.; Elad, M.; Protter, M. On single image scale-up using sparse-representations. In *International Conference on Curves and Surfaces*; Springer: Berlin/Heidelberg, Germany, 2010; pp. 711–730.
59. Arbelaez, P.; Maire, M.; Fowlkes, C.; Malik, J. Contour detection and hierarchical image segmentation. *IEEE Trans. Pattern Anal. Mach. Intell.* **2010**, *33*, 898–916. [CrossRef]
60. Gao, X.; Lu, W.; Tao, D.; Li, X. Image quality assessment based on multiscale geometric analysis. *IEEE Trans. Image Process.* **2009**, *18*, 1409–1423.
61. Wang, Z.; Bovik, A.C.; Sheikh, H.R.; Simoncelli, E.P. Image quality assessment: From error visibility to structural similarity. *IEEE Trans. Image Process.* **2004**, *13*, 600–612. [CrossRef]
62. Zhao, H.; Gallo, O.; Frosio, I.; Kautz, J. Loss functions for image restoration with neural networks. *IEEE Trans. Comput. Imaging* **2016**, *3*, 47–57. [CrossRef]
63. Dong, C.; Loy, C.C.; Tang, X. Accelerating the super-resolution convolutional neural network. In Proceedings of the European Conference on Computer Vision, Amsterdam, The Netherlands, 11–14 October 2016; pp. 391–407.

64. Lai, W.S.; Huang, J.B.; Ahuja, N.; Yang, M.H. Deep laplacian pyramid networks for fast and accurate super-resolution. In Proceedings of the IEEE Conference on Computer Vision and Pattern Recognition, Honolulu, HI, USA, 21–26 July 2017; pp. 624–632.
65. Tai, Y.; Yang, J.; Liu, X. Image super-resolution via deep recursive residual network. In Proceedings of the IEEE Conference on Computer Vision and Pattern Recognition, Honolulu, HI, USA, 21–26 July 2017; pp. 3147–3155.
66. Zhang, K.; Zuo, W.; Zhang, L. Learning a single convolutional super-resolution network for multiple degradations. In Proceedings of the IEEE Conference on Computer Vision and Pattern Recognition, Salt Lake City, UT, USA, 18–23 June 2018; pp. 3262–3271.
67. Lei, S.; Shi, Z.; Zou, Z. Super-resolution for remote sensing images via local–global combined network. *IEEE Geosci. Remote Sens. Lett.* **2017**, *14*, 1243–1247. [CrossRef]
68. Dong, X.; Xi, Z.; Sun, X.; Gao, L. Transferred multi-perception attention networks for remote sensing image super-resolution. *Remote Sens.* **2019**, *11*, 2857. [CrossRef]
69. Dong, X.; Sun, X.; Jia, X.; Xi, Z.; Gao, L.; Zhang, B. Remote sensing image super-resolution using novel dense-sampling networks. *IEEE Trans. Geosci. Remote Sens.* **2020**, *59*, 1618–1633. [CrossRef]
70. Ma, Y.; Lv, P.; Liu, H.; Sun, X.; Zhong, Y. Remote Sensing Image Super-Resolution Based on Dense Channel Attention Network. *Remote Sens.* **2021**, *13*, 2966. [CrossRef]
71. Dharejo, F.A.; Deeba, F.; Zhou, Y.; Das, B.; Jatoi, M.A.; Zawish, M.; Du, Y.; Wang, X. TWIST-GAN: Towards Wavelet Transform and Transferred GAN for Spatio-Temporal Single Image Super Resolution. *arXiv* **2021**, arXiv:2104.10268.
72. Yang, Y.; Newsam, S. Bag-of-visual-words and spatial extensions for land-use classification. In Proceedings of the 18th SIGSPATIAL international conference on advances in geographic information systems, San Jose, CA, USA, 2–5 November 2010; pp. 270–279.
73. Cheng, G.; Han, J.; Lu, X. Remote sensing image scene classification: Benchmark and state of the art. *Proc. IEEE* **2017**, *105*, 1865–1883. [CrossRef]

remote sensing

Article

Semantic Segmentation of 3D Point Cloud Based on Spatial Eight-Quadrant Kernel Convolution

Liman Liu [1], Jinjin Yu [1], Longyu Tan [1], Wanjuan Su [2,*], Lin Zhao [2] and Wenbing Tao [2]

[1] School of Biomedical Engineering, South-Central University for Nationalities, Wuhan 430074, China; limanliu@mail.scuec.edu.cn (L.L.); 2020110498@mail.scuec.edu.cn (J.Y.); 2018110451@mail.scuec.edu.cn (L.T.)
[2] National Key Laboratory of Science and Technology on Multi-Spectral Information Processing, School of Artificial Intelligence and Automation, Huazhong University of Science and Technology, Wuhan 430074, China; linzhao@hust.edu.cn (L.Z.); wenbingtao@hust.edu.cn (W.T.)
* Correspondence: suwanjuan@hust.edu.cn

Abstract: In order to deal with the problem that some existing semantic segmentation networks for 3D point clouds generally have poor performance on small objects, a Spatial Eight-Quadrant Kernel Convolution (SEQKC) algorithm is proposed to enhance the ability of the network for extracting fine-grained features from 3D point clouds. As a result, the semantic segmentation accuracy of small objects in indoor scenes can be improved. To be specific, in the spherical space of the point cloud neighborhoods, a kernel point with attached weights is constructed in each octant, the distances between the kernel point and the points in its neighborhood are calculated, and the distance and the kernel points' weights are used together to weight the point cloud features in the neighborhood space. In this case, the relationship between points are modeled, so that the local fine-grained features of the point clouds can be extracted by the SEQKC. Based on the SEQKC, we design a downsampling module for point clouds, and embed it into classical semantic segmentation networks (PointNet++, PointSIFT and PointConv) for semantic segmentation. Experimental results on benchmark dataset ScanNet V2 show that SEQKC-based PointNet++, PointSIFT and PointConv outperform the original networks about 1.35–2.12% in terms of MIoU, and they effectively improve the semantic segmentation performance of the networks for small objects of indoor scenes, e.g., the segmentation accuracy of small object "picture" is improved from 0.70% of PointNet++ to 10.37% of SEQKC-PointNet++.

Keywords: spatial eight-quadrant kernel convolution; 3D point cloud; semantic segmentation; indoor scene

Citation: Liu, L.; Yu, J.; Tan, L.; Su, W.; Zhao, L.; Tao, W. Semantic Segmentation of 3D Point Cloud Based on Spatial Eight-Quadrant Kernel Convolution. *Remote Sens.* **2021**, *13*, 3140. https://doi.org/10.3390/rs13163140

Academic Editors: Wanshou Jiang, San Jiang and Xiongwu Xiao

Received: 24 June 2021
Accepted: 4 August 2021
Published: 8 August 2021

Publisher's Note: MDPI stays neutral with regard to jurisdictional claims in published maps and institutional affiliations.

1. Introduction

Since the semantic understanding and analysis of a 3D point cloud is the basis for realizing scene understanding [1,2], the application of semantic segmentation of 3D point cloud has been more and more extensive in recent years [3–5], such as augmented/virtual reality [6] and intelligent robot [7]. Moreover, in the field of self-driving, the accurate perception of the environment based on LIDAR point cloud data is the key to realize information decision-making and driving safely in the complex dynamic environment. Particularly, accurate segmentation of small objects can help self-driving vehicles make correct decisions in time in some cases. Semantic segmentation of 3D point clouds aims to predict the label of each point in the point clouds, making different classes of objects with corresponding labels. The performance of semantic segmentation based on deep neural networks depends on the strength of the feature extraction ability of the network [8–10], especially for the small objects in the scene, which requires the network to be able to extract more fine-grained local semantic information.

PointNet [11] was proposed to use raw point clouds as the input of the network at the first time, which uses Multi-Layer Perception (MLP) [12] to learn features from the point cloud adaptively. The core concept of PointNet is to approximate a general function defined on a point set by applying a symmetric function, so it can efficiently extract information from the unordered point cloud. However, it is difficult for PointNet to learn local features and the relationship between points in the point cloud. To address this problem, PointNet++ [13] uses a sampling and grouping strategy to divide the point cloud into several small local regions, so it can leverage the PointNet to extract local features. Jiang et al. [14] proposed the Orientation-Encoding and Scale-Awareness module in PointSIFT network to extract features. The orientation-encoding first integrates the feature information of the points in each of the eight spatial directions in the point cloud space, then performs a three-stage ordered convolution to encode these feature information. Meanwhile, the network connects multi-scale features by stacking orientation encoding in the process of propagation. For local regions of point clouds, Wu et al. [15] uses inverse density-weighted convolution to capture local features. The method weights the features of the points by inverse density and weight function which is generated by kernel density estimation method [16] and the coordinates of the points. PointConv gains great improvement in terms of the semantic segmentation, but the network tends to lose the information of large object edges and small objects in the sampling process.

The semantic segmentation performance of the above methods for small objects in 3D scenes is generally poor. All objects in the point cloud are composed of points connected with each other, so that the semantic segmentation of each point depends on the relationship of points. Using the MLP to extract features will treat all the points equally, and the result may be biased toward the categories that account for a larger proportion of the point cloud data. In contrast, objects with a smaller percentage of points in the scene need to be distinguished using point-to-point associations. As mentioned above, the MLP treats each point in the point cloud equally, neglecting the connection between points, and thus the extracted features of the point cloud are not distinguishable enough, resulting in low accuracy of the final 3D point cloud semantic segmentation results, especially for small objects.

Different from previous methods that use the MLP to extract features directly, the proposed Spatial Eight-Quadrant Kernel Convolution (SEQKC) algorithm generates eight kernel points with coordinates and shared weights in the neighborhood space of point cloud at first; Then, the distance between each kernel point and its neighbors are calculated, and the inverse distance is used as the coefficient of the weight; Finally, the kernel point weights are used to weight the features of each point, and all the weighted features are aggregated and input into the MLP. As a result, the proposed method can effectively extract the local fine-grained information of point clouds by modeling the relationship between points in space, which can improve the semantic segmentation accuracy of small objects in 3D scenes. Furthermore, we design a downsampling module based on the SEQKC which is combined with the existing methods (e.g., PointNet++, PointSIFT and PointConv networks) to do feature extraction for point clouds. In order to save computational time and memory, the downsampling module consists of SEQKC module and set sbstraction module. Moreover, to learn both global and local features well at the same time, the downsampling module is similar to a residual network structure, so that multi-scale semantic information of the point cloud can be extracted.

In summary, our contributions are:

- To capture point-to-point connections in point clouds for better local feature extraction, the spatial eight-quadrant kernel convolution is proposed in this paper.
- A downsampling module based on spatial eight-quadrant kernel convolution is designed, which can be combined with existing methods and further improve the semantic segmentation performance.
- Extensive experiments have been conducted on ScanNet V2 dataset [17], the experimental results show that our method can help the network effectively extract local semantic information of point clouds and improve the semantic segmentation accuracy of small objects.

2. Literature Review

In this section, we briefly review the approaches of semantic segmentation for 3D point clouds using deep learning network.

2.1. Projection-Based Semantic Segmentation Methods for Point Clouds

Some networks project 3D point clouds into 2D images, so they can use 2D convolution to process the point clouds. Tatarchenko et al. [18] proposed the tangent convolution, where a plane tangent to the local surface of each point is constructed, then the local surface is projected onto the tangent plane and convolved on the projection plane. Lawin et al. [19] observe 3D point clouds from different views, then integrate and input multiple projections to the FCN network, and synthetic image per-pixel evaluation scores are output, finally the evaluation scores are mapped back to individual views to obtain semantic labels for each point. However, the performance of this method is greatly influenced by the viewpoint selection, and it is difficult to handle the occluded point clouds, and the projection is highly prone to lose point cloud information, so these methods are not suitable for dealing with point clouds that possess complex geometric structures. Additionally, in the self-driving scenario, poor synchronization of LIDAR and camera may lead to bad point cloud projection in the image, resulting in 3D points with an erroneous semantic class [20].

2.2. Voxelization-Based Semantic Segmentation Method for Point Clouds

Since point cloud data have irregular structure in 3D space, early point clouds were often processed by voxelization methods to enable them to be processed using standard 3D convolution. Huang et al. [21] voxelized point clouds and input them into 3D convolutional neural networks for semantic segmentation, the predicted semantic labels for each point are output and compared with the real semantic labels for back-propagation learning. Due to the sparsity of point clouds, this method can only cope with some more regular point clouds. Tchapmi et al. [22] proposed SEGCloud for feature extraction in the fine-grained space of point clouds and performing semantic segmentation that encompasses the whole point cloud.

The discretization method can maintain the geometric structure of the point cloud, and the standard 3D convolution can be well adapted to this format. However, the voxelization method will inevitably make some points shift the original position and lead to discretization artifacts, and there is also the problem of information loss in the process. Since point cloud data contain variable types of objects, it is difficult to choose a suitable grid resolution, and a high resolution will make the network training slow and computationally expensive, while a low one will lose important information and lead to wrong results.

2.3. Point-Based Semantic Segmentation Method for Point Clouds

Point-based methods can directly use raw point clouds. PointNet is a pioneer work which used the point cloud data directly, but the operation of global pooling makes the network lose local information of the point cloud, which cannot meet the needs of point cloud semantic segmentation for small objects. In order to learn local features of the point cloud, subsequent work mainly uses hierarchical networks or feature weighting methods. PointNet++ is a representative work for hierarchical networks, which uses a sampling and grouping strategy to extract the point cloud local features, the iteration of downsampling to expand the receive field of the network, and feature interpolation to finally achieve point cloud semantic segmentation. In [23], a large outdoor public dataset for 3D semantic segmentation (PC-Urban) is proposed and baseline semantic segmentation results on PC-Urban are produced by PointNet++ and PointConv. Unal et al. [24] proposed a detection aware 3D semantic segmentation method which leverages localization features from an auxiliary 3D object detection task. By utilizing multitask training, the shared feature representation of the network is guided to be aware of per class detection features that aid tackling the differentiation of geometrically similar classes. Hua et al. [25] proposed a method to integrate adjacent points in the local region space of the point cloud and then convolve the integrated point cloud features using kernel point weights. Thomas et al. [26] proposed a kernel point convolution operator to construct kernel points in the 3D point cloud space, and weighted the features by calculating the Euclidean distance from the point to the kernel point. Ye et al. [27] proposed to extract multi-scale features of point clouds using a pointwise pyramid structure and apply Recurrent Neural Networks (RNNs) to achieve end-to-end learning. In order to extract the contextual features of each local region of the point cloud during the point cloud feature propagation, Engelmann et al. [28] applying RNNs to do point cloud semantic segmentation, it encodes features that contain different scales of the point cloud by [28] using merged units or recursive merged units to extract the detailed point cloud feature information.

3. Methods

3.1. Spatial Kernel Convolution Algorithm

Previous works directly use shared MLPs separately on each point, followed by the operation of global max-pooling. The shared MLP acts as a set of learning spatial encodings, and the global characteristics of the point cloud are calculated as the maximum response between all points for each of these encodings. Although the kernel of point convolution can be implemented by MLPs in this way, local spatial relationships in the data have not been considered, and it makes the convolution operator more complex and the convergence of the network harder. To this end, we propose the idea of using spatial kernel convolution, like image convolutions, whose weights are directly learned without the intermediate representation of a MLP. Furthermore, local relationships between points can be modeled by distance weighting, so that the local fine-grained feature can be extracted by the proposed method.

Specifically, multiple kernel points with shared weights are generated in the point cloud neighborhood space, and each kernel point is accompanied with coordinates. Before aggregating the neighborhood points features into the MLP, the distances between the kernel points and several points in their neighborhood are calculated, then features of each point are weighted by the distances and the kernel point weights, finally all the weighted features are aggregated and input into the MLP network.

As shown in Figure 1, the $p_1, p_2 \ldots p_n$ represent n neighborhood points, $w_1, w_2 \ldots w_k$ denote the weights of the k kernel points, and the features of each point in the neighborhood are multiplied with the weights $w_1, w_2 \ldots w_k$. Unlike the conventional fully connected method, features of each point are only connected to some kernel points. The dashed lines in Figure 1 which connect the features and weights indicate that the features of each point are weighted using the reverse distance from each point in the neighborhood to the kernel point, so the closer the point is to the kernel point, the greater the weight is. The method that uses inverse distance weighting features can not only well distinguish the strength of the relationship between each point and the centroid, but also extract fine-grained features of the point cloud more effectively. The features of each point are weighted by multiple spatial kernels and then combined again into new features $m_1, m_2 \ldots m_n$. Then we input the new features that contain tight connections into MLP, as a result, the local point cloud features will be effectively extracted.

We study the influence of different kernel points number on ability of feature extraction. We change the number of kernel points in point cloud semantic segmentation experiments. The experimental results show that the best number of spatial kernel points is 8. Because the number of feature points in the neighborhood space is limited, and the feature information extraction of each kernel point is accompanied by distance weighting. When the number of kernel points increases, the point with the same feature may be similar to several kernel points at the same time, and the information extracted from the feature point by these kernel points may be repetitions, resulting in the entire spatial kernel convolution module doing redundant and meaningless work, which increases the network computation in vain. Since the eight kernel points are uniformly distributed in the eight quadrants of the neighborhood space, we call our point cloud feature extraction method as Spatial Eight-Quadrant Kernel Convolution (SEQKC).

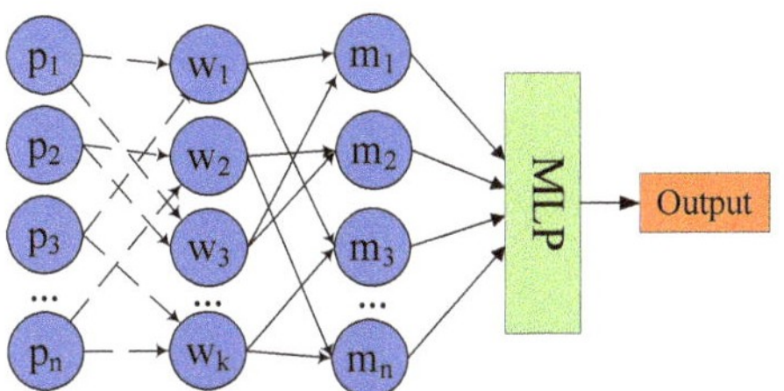

Figure 1. The process of spatial kernel convolution, where p_n denotes n neighborhood points, w_k denotes weights of the k kernel points, dashed lines denote features of each point are weighted using the reverse distance from each point in the neighborhood, m_n denotes the new feature that is combined with features weighted by multiple spatial kernels.

3.2. Spatial Eight-Quadrant Kernel Convolution Algorithm

In this section we describe the details of the SEQKC. As shown in Figure 2, a spherical neighborhood space is constructed, where the center point is p and the radius is r. The SEQKC takes p as the origin and unfolds the spatial eight quadrants. In each quadrant space, there is a point with the weight W_k, and the spatial coordinate of the kernel point is $[r/2, r/2, r/2]$. The dimension of the weight c is matched with the dimension of current point cloud feature. The SEQKC module g is used to convolute all the point features F in the neighborhood, which can be defined as:

$$(\mathcal{F} * g) = \sum_{p_i \in N_p} g(p_i, K) f_i \tag{1}$$

where p_i denotes the points in the neighborhood, the f_i is the feature of point p_i, K denotes the eight quadrant kernel points. We define the kernel function g for neighborhood points as:

$$g(p_i, K) = \sum_{k_i \in K} W_{k_j} / dist(p_i, k_j) \tag{2}$$

where $dist(p_i, k_j)$ is the Euclidean distance between each point in the neighborhood and the quadrant kernel point. We use it restricts the influence of kernel point to each neighborhood points, so that the weights of the points that closer to the quadrant kernel point are larger, while the points far from the quadrant kernel point are small and less affected by the weight. Using the distance-constrained weights in eight directions to dot product the feature of each point, the algorithm can aggregate the global feature from all neighborhood.

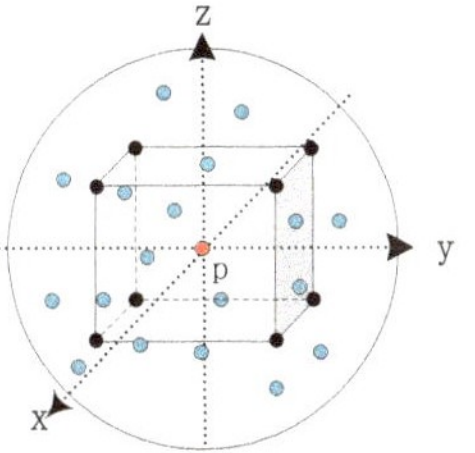

Figure 2. The structure of spatial eight-quadrant kernel in point cloud neighborhoods. The red point is the center point p in the constructed spherical neighborhood space, the black points are kernel points of eight quadrants in the constructed spherical neighborhood space, and the blue points are neighborhood points of the center point.

3.3. Spatial Eight-Quadrant Kernel Convolution Module

Based on the proposed SEQKC, we design a feature encoder for 3D point clouds, namely SEQKC module. The structure of SEQKC module is shown in Figure 3. The input data of the module are divided into point cloud coordinates $(b, n1, 3)$ and point clouds feature $(b, n1, c_{in})$, which b denotes the batch size of input data, $n1$ is the number of points for each batch, 3 indicates the X, Y, Z coordinates of the point cloud, and c_{in} is the dimension of the current point cloud feature. In addition, we employed the sampling and grouping strategy of PointNet++ to our encoder. That is to say, we use the farthest point sampling to find $n2$ points, which are used to be the centroid to construct the spherical neighborhood with the radius r. In every neighborhood, we select k support points, if there are not enough points, the center point is repeatedly used instead. We combine the coordinates and features of the points in each neighborhood, so the data sizes are $(b, n2, k, 3)$ and $(b, n2, k, c_{in})$ respectively. The neighborhood points need to be decentered before computing with each quadrant kernel point, namely, the coordinates of each point in the neighborhood need to be subtracted from the coordinates of corresponding centroids. After that, the eight-quadrant kernel points are constructed with coordinate size $(8, 3)$ and the weight dimension $(8, c_{in})$. We compute the distances from each point in the neighborhood to the eight-quadrant kernel points , then use them as weight coefficients for quadrant kernel point. Although the convolution kernel has only eight kernel points with eight different weights, when each weight is given a different distance weighting, it disguisedly increases the kernel point weights to the same number of points as the neighborhood points. After distance weighting, the weight dimension is raised to $(b, n2, k, 8, c_{in})$, which is the same as point cloud feature dimension. They are multiplied with the neighboring point cloud features, added to the bias, and followed by batch normalization layer and ReLu activation to batch normalize the features and remove the data less than zero in the features. Finally, the feature information of point cloud is extracted by MLP, and the feature information is output after dimensionality upgrading.

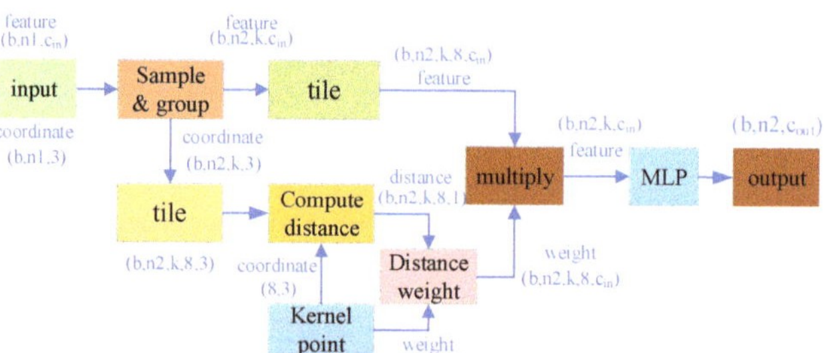

Figure 3. The encoder module of spatial eight-quadrant kernel convolution. First, sampling and grouping the input point clouds to form the neighborhood space. Then, using the spatial eight-quadrant kernel convolution algorithm to weight the features in neighborhoods. Finally, the weighted features are aggregated and input to the MLP.

4. Spatial Eight-Quadrant Kernel Convolution Algorithm-Based Networks

We apply the proposed SEQKC module based on the spatial eight-quadrant kernel convolution algorithm to PointNet++, PointSIFT, PointConv networks as a downsampling module together with the Set Abstraction Module (SA) for feature extraction of the point cloud. The SA module consists of a sampling layer, a grouping layer, and a feature extraction layer. The sampling layer samples the point clouds globally by the farthest point sampling method; the grouping layer constructs multiple local regions by the K-nearest neighbor algorithm [29] or by specifying the radius of the sampled points; the feature extraction layer extracts the sampled and grouped point clouds using MLP to increase the feature dimension.

To save computational time and memory, we fused the SEQKC module with the SA module, which is shown in Figure 4. The SEQKC fusion module is similar to a residual network structure. The cascaded features contain multi-scale semantic information about the point cloud, it can allow the deep neural network to learn the global features of the grouped neighborhood as well as the shape features of small local regions.

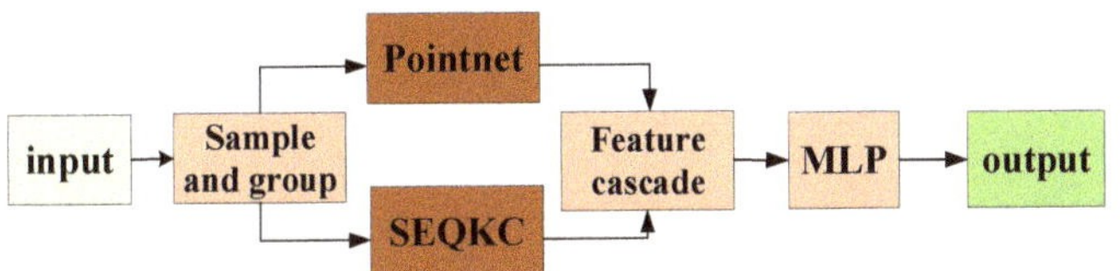

Figure 4. The SEQKC fusion network. After sampling and grouping the point cloud, the SA module of PointNet and SEQKC module are cascaded to extract features.

4.1. Seqkc-Based Pointnnet++ Network

PointNet++ is a hierarchical neural network, by iterating the farthest point sampling to downsample point clouds, the network is able to learn features of point clouds from local to global. The network uses SA modules to extract point cloud features. Each SA module extracts the local information of the spherical neighborhood. By stacking SA modules, the number of point clouds decreases, the local information converges to global information, and the network's receptive field changes from small to large. Subsequently, the feature propagation (FP) module is used, and the input number of points is linearly interpolated from N to N'. Finally, the number of point clouds is restored to the original number of point clouds while keeping the feature dimension unchanged, and the semantic segmentation is achieved.

The SA module of the PointNet++ network directly inputs the sampled and grouped features into the MLP, which leads to the network learning more about the global shape information of each grouping neighborhood, lacking the processing of local area information in the neighborhood. As a result, the network is unable to segment small objects surrounded by large objects, and the robustness of the semantic segmentation for point cloud is poor. We incorporated the SEQKC module into the PointNet++ network, and the improved network structure is shown in Figure 5.

Figure 5. The structure of SEQKC-PointNet++ nerwork. The basic structure is the same as Point-Net++. Fusion module denotes the SEQKC module with the SA module, and FP module is a feature propagation module which uses linear interpolation weighted by distances to upsample the point cloud.

We directly replace the original SA module in the encoding layers of PointNet++ with SEQKC module, so that the network can extract richer features that contain more detailed contextual relationship between points. For complex environments, the network can propagate richer semantic information for the prediction of semantic labels. At the same time, the improved network is more accurate in terms of segmentation results of small objects with more natural and realistic boundary transitions due to the extracted semantic features contain more fine-grained information.

4.2. Seqkc-Based Pointsift Network

The PointSIFT network is improved based on PointNet++, which combines point cloud Orientation-Encoding and Scale-Awareness unit. In the past, point cloud local descriptors were usually unordered operations, while ordered operations may provide more information. With this in mind, PointSIFT uses Orientation-Encoding, which is a three-stage operator that convolves the $2 \times 2 \times 2$ cube along X, Y, and Z axes successively. The PointSIFT module is used before the SA module in the network to integrate the features of point cloud. Such an approach enhances the ability of the network to extract distinguished features, so that the network has a stronger semantic segmentation capability.

Similarly, we replace the SA module of the PointSIFT network for point cloud feature downsampling with the proposed SEQKC module. Through the interaction between the small neighborhood information contained by the features of each point and other points in the spherical neighborhood, the point cloud features output by SEQKC fusion module contain the relationship between the points. The structure of the improved PointSIFT network based on the SEQKC feature fusion module is shown in Figure 6.

With the PointSIFT module for point cloud feature preprocessing, the features obtained by the SEQKC fusion module contain more fine-grained multi-scale information about the local neighborhood of the point cloud. After three times of downsampling by the SEQKC fusion module, the acquired global features can be upsampled using the original feature decoding module of the PointSIFT network. The features obtained by downsampling with the SEQKC fusion module contain richer and more detailed semantic information than that of the original network, so the PointSIFT network with the improved downsampling module has higher semantic segmentation accuracy compared with the original PointSIFT network.

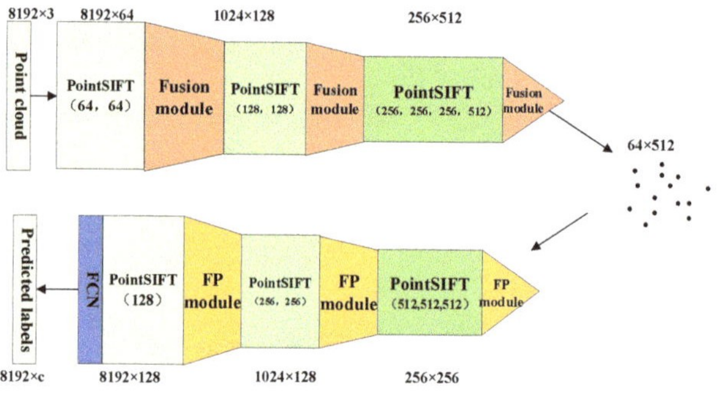

Figure 6. The structure of SEQKC-PointSIFT network. The structure of SEQKC-PointSIFT is the same as that of PointSIFT, except that SA module is replaced with Fusion module.

4.3. Seqkc-Based Pointconv Network

Inspired by 2D image convolution, PointConv approximates the continuous convolution operation with a discrete function that performs the convolution operation on a non-uniformly sampled point cloud. In the original PointConv network, the inverse density weighting method is used to weight the point features in the local region, and the local coordinates of the points in the region are used to construct the weights which is multiplied by the point feature. In the network based on SEQKC, the inverse distance weighting method is used to weight the point features in the local region, and the weight of the constructed kernel points is multiplied by the point features.

Since both points in sparse region and dense region are important to the whole point cloud, and the number of points in the dense region accounts for a relatively large number, so directly feeding the whole points into the network will make the network learn more information from the dense points and neglect the sparse points. To avoid this problem, the SEQKC algorithm extracts the features of the points in the region by eight spatial kernel points from the viewpoint of distance. The spatial kernel points are uniformly distributed in the local region of the point cloud, and in this region, both the points in the dense and sparse places will have their features weighted by their nearest kernel points, so that the influence of each point on the region is balanced.

These two methods obtain balanced local space features from the viewpoint of density and distance, and have different scale point cloud semantic information. Considering the differences between these two methods, we combine them to construct a new downsampling module called the point cloud double convolution (DC) module, which is shown in Figure 7. For the input point cloud features, the improved point cloud DC module uses a sampling and grouping strategy to obtain individual local features of the point clouds.

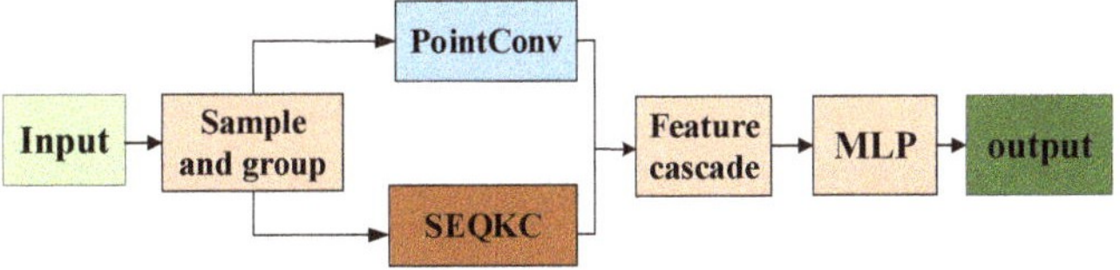

Figure 7. The double convolution module. After sampling and grouping the point cloud, the SEQKC with PointConv convolution are cascaded to extract features.

The improved PointConv network based on the DC module is shown in Figure 8. It uses the PointConv convolution method with the spatial eight-quadrant kernel convolution method to extract features from each local region of the point cloud, then cascades the two features which contains semantic information at different scales. Finally, the cascaded new features are fed into the MLP for training, and the point cloud features are extracted.

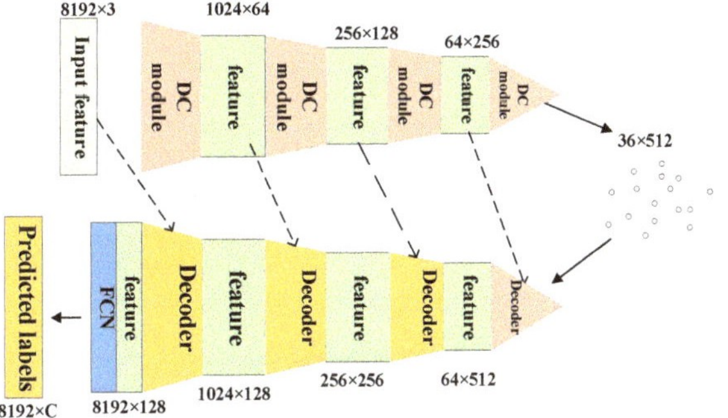

Figure 8. The structure of SEQKC-PointConv network. The network consists of downsampling (DC module) and upsampling (Decoder module) procedures. The DC module is double convolution module mentioned above, and Decoder is the PointDeconv module in [15].

5. Experiment

To ensure the comparability of the experimental results, we evaluate our networks on the benchmark dataset ScanNet V2 [17] (Richly annotated 3D Reconstructions of Indoor Scenes). This dataset contains a large number of indoor scenes, obtained by camera scanning at different viewpoints and 3D reconstruction. The indoor scenes are rich in types and sizes, including not only large-scale indoor scenes such as apartments and libraries, but also many small indoor scenes such as storage rooms and bathrooms. Each scene may contain 19 different categories of objects such as doors, windows, chairs, tables, etc., and one unknown type. We used 1201 indoor scenes for training and the remaining 312 indoor scenes for testing. For a fair comparison, we follow we follow the PointNet++ [13], PointSIFT [14] and PointConv [15] to divide ScanNet dataset into the training set and the test set in the corresponding experiments. Note that the scenes of PointConv [15] in the training set and the test set are different from that of PointNet++ [13] and PointSIF [14], but the amount of data in the training set and the test set is the same. In all experiments, we implement the models with Tensorflow on a GTX 1080Ti GPU.

5.1. Hyperparameter Setting

Each training session uses non-uniform sampling to collect input points in each point cloud scene, and the input points size of the network is set to 8192. If the training platform does not have enough memory, the number of input points can be reduced to 4096, and the batch size is 12. The network uses Adam optimizer, the learning rate is set by exponential decay method, the initial learning rate is set to 0.001, decay step is 200,000 and decay rate is 0.7, max epoch is 1000.

5.2. Loss Function

To measure the degree of inconsistency between the predicted semantic labels of the model and the true semantic labels , we use the sparse cross entropy loss [30] function, and the formulas are shown in Equations (3) and (4).

$$p = softmax(logits) = \frac{e^{logits_j}}{\sum_{j=0}^{K} e^{logits_j}} \tag{3}$$

$$loss = -\sum_{j=0}^{K} y * \ln P_i \tag{4}$$

where *logits* denotes the semantic label predicted by the network for any point, which is the probability score of each category, *logits$_j$* denotes the probability of the point on the *j*th category. The formula shows that P also has $K + 1$ dimension, the natural logarithm of each dimension on p is obtained and multiplied with the actual semantic label y. The negative sum of all the dimensions is the desired loss.

5.3. Evaluation Criteria

In the experiments, we use the following evaluation metrics.

(1) Point calibrated average accuracy (caliacc). MPA treats each class equally and takes the average of all the accuracy of the classes. In fact, the proportion of each class in the point cloud is different, so that this method is flawed. As shown in Equation (5), the caliacc uses the proportion of each class in the point cloud points to weight the accuracy of each class, and sums them.

$$caliacc = \sum_{i=0}^{K} W_i \frac{P_{ii}}{\sum_{j=0}^{K} P_{ij}} \tag{5}$$

(2) Mean Intersection over Union (MIoU). The IoU refers to the ratio of intersection and union of two sets. For the prediction results of point cloud semantic labels, the more points that are predicted to be the right semantic labels, and the fewer points whose semantic labels are not of that class are predicted to be of that class. The IoU is more convincing than judging whether the semantic labels of only one class of points are predicted accurately without considering the PA values of other classes. As shown in Equation (6), the MIoU is obtained by summing the IoU for each class and taking the mean value.

$$MIoU = \frac{1}{K+1} \sum_{i=0}^{K} \frac{P_{ii}}{\sum_{j=0}^{K} P_{ij} + \sum_{j=0}^{K} P_{ij} - P_{ii}} \tag{6}$$

5.4. The Experiments to Verify the Number of Kernel Point

In order to find the optimal number of kernel points for the spatial kernel convolution algorithm, we distribute different numbers of kernel points uniformly in the spherical neighborhood, then conduct point cloud semantic segmentation experiments separately.

The distribution of kernel points in the spherical neighborhood of the point cloud under each number is shown in Figure 9. The experiments are conducted based on the PointNet++ network, and the point cloud downsampling module of the PointNet++ network is replaced with SEQKC module, while the rest of the network structure remains unchanged. A comparison of the specific results is shown in Table 1, where $k = 0$ denotes the original network.

Table 1. The comparison results for different number of kernel point.

k	MIoU(%)	Caliacc (%)
0	40.56	83.71
6	41.94	84.35
8	**42.32**	**84.85**
12	41.58	84.76
8	41.87	84.73
8	41.41	84.74

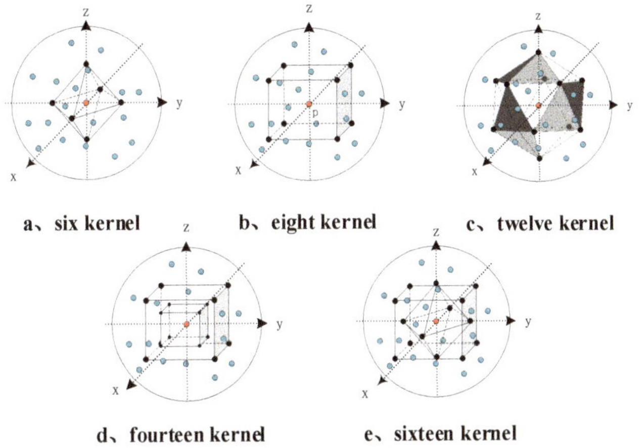

a、 six kernel   b、 eight kernel   c、 twelve kernel

d、 fourteen kernel   e、 sixteen kernel

Figure 9. The distribution of kernel in spatial for different kernel number.

It can be seen that all PointNet++ networks using SEQKC module obtain an improvement in both MIoU and caliacc compared to the original network. As the number of kernel points increases, it reaches a maximum at 8 kernel points, and the network's caliacc finally remains at about 84.74%. MIoU also reaches a maximum at 8 kernel points and then starts to decrease, so the spatial kernel convolution has the best semantic segmentation performance for 3D point clouds when 8 kernel points are used.

5.5. The Experiment to Verify Cascaded Seqkc Structure

Our proposed SA module with SEQKC fusion module is actually a residual structure of SEQKC module. In order to verify whether this residual structure is valid, experiments were conducted on the uncascaded SEQKC module and the cascaded SEQKC module with sampling and grouping operations, respectively. The experiments were based on PointNet++ network, and the results are shown in Table 2.

Table 2. Experimental results of cascaded SEQKC module.

Method	MIoU (%)	Caliacc (%)
Pointnet++	40.56	83.71
SEQKC	42.13	83.56
SEQKC (Cascaded)	42.32	84.85

From Table 2, it can be seen that the SEQKC module can effectively improve the MIoU of point cloud semantic segmentation results with or without cascade, and the cascaded SEQKC module can improve the semantic segmentation caliacc of points. This indicates that the features extracted by the SEQKC method contain more detailed spatial scale and can imply the point-to-point connection information to improve the integrity of the object semantic segmentation results, thus improving the MIoU values.

In addition, in order to verify the importance of the kernel point coordinates, we cancel kernel point coordinates in SEQKC, and apply the eight weights without distance weighting to the input point cloud features. The results are shown in Table 3. The eight kernel points that lose the coordinates are convolved with the point cloud features without the distance weighting, and the network segmentation results are slightly lower in terms of caliacc than the network with the kernel point coordinates, and their MIoU is 2.52% lower than the network with the kernel point coordinates. This indicates that the coordinate positions of the kernel points can indeed help the network analyze the connection between points and help the network find the accurate location of the points in the final semantic segmentation results.

Table 3. The comparison results for SEQKC with and without kernel point coordinates.

Method	MIoU (%)	Caliacc (%)
SEQKC (without coordinate)	39.79	84.40
SEQKC	42.32	84.85

5.6. The Experiment of Semantic Segmentation-Based on Enhanced Networks

We embed the SEQKC module into classical semantic segmentation networks to evaluate the performance of the algorithm, and for the fairness of comparison, we ensure that the network parameters are the same as the original network except for the added module.

In order to better show the effectiveness and stability of our method, we did three repeated experiments on SEQKC-PointNet++, SEQKC-PointSIFT and SEQKC-PointConv with the same setting, and the results are shown in Table 4. It shows that the proposed method has stable improvement, although there is randomness in the training process. We compared the MIoU of the semantic segmentation results across all networks. From Table 4, we can see that the networks with the SEQKC module improve the MIoU compared to the original networsk, with a minimum improvement of 1.35%, indicating that our module can more accurately identify the points of small objects in indoor scenes, and has a stronger performance of the segmentation in visual.

Table 4. The comparison results of the enhanced networks for semantic segmentation.

Method	MIoU (%)	Caliacc (%)
PointNet++ [13]	40.56	83.71
SEQKC-PointNet++	42.68 $\pm$ 0.42	84.71 $\pm$ 0.75
PointSIFT [14]	42.47	85.04
SEQKC-PointSIFT	43.82 $\pm$ 0.22	85.66 $\pm$ 0.81
PointConv [15]	48.28	-
SEQKC-PointConv	50.12 $\pm$ 0.33	-

In order to analyze it in detail, Tables 5 and 6 show the segmentation results of each category on the Scannet dataset under different networks. To make the comparison of experimental results more intuitive, we highlight the small objects in Tables 5 and 6. As can be seen from the table, the semantic segmentation results of large objects are almost the same. However, the segmentation performance of our network is significantly better than that of the original network in terms of small objects, such as chair, shower curtain, sink, toilet, picture and so on.

Table 5. Experimental results of all categories on Scannet dataset in terms of Caliacc (%).

Method	Wall	Floor	Chair	Table	Desk	Bed	Bookshelf	Sofa	Sink	Bathtub	Toilet	Curtain	Counter	Door	Window	Shower Curtain	Refridgerator	Picture	Cabinet	Otherfurniture
PointNet++	91.17	98.45	79.03	66.44	43.72	74.86	73.45	74.78	38.34	80.46	63.10	44.09	39.40	15.80	22.60	54.05	46.19	0.70	36.33	27.67
SEQKC-PointNet++	90.46	97.80	**83.09**	65.64	**48.00**	72.71	60.13	**79.75**	**56.38**	**87.93**	**77.03**	**51.84**	**38.21**	**29.97**	**26.57**	**82.30**	**54.24**	**10.37**	**39.89**	**37.75**
PointSIFT	90.36	98.24	81.65	60.35	45.39	78.04	78.52	80.85	59.57	82.33	75.69	50.92	32.89	37.83	33.76	57.19	58.22	1.93	42.71	28.07
SEQKC-PointSIFT	**92.16**	**98.51**	**87.19**	**63.85**	**48.72**	76.97	74.62	74.72	52.43	**87.69**	**83.81**	**64.12**	**35.43**	27.73	28.38	50.93	43.33	**4.66**	36.47	25.60

Table 6. IoU (%) comparison of various semantic segmentation results based on PointConv.

Method	Wall	Floor	Chair	Table	Desk	Bed	Bookshelf	Sofa	Sink	Bathtub	Toilet	Curtain	Counter	Door	Window	Shower Curtain	Refridgerator	Picture	Cabinet	Otherfurniture
PointConv	65.84	92.11	35.53	56.32	73.69	55.85	54.64	29.13	35.74	37.99	12.58	42.30	36.97	41.15	27.75	40.90	75.03	56.10	66.93	29.03
SEQKC-PointConv	64.53	85.61	32.53	**58.77**	**74.65**	**60.85**	**58.17**	**32.22**	**38.90**	37.52	10.82	40.85	**38.97**	**52.43**	**31.75**	**49.88**	**76.09**	50.72	**67.28**	**30.25**

Compared to PiontNet++, the semantic segmentation accuracy of small objects (chair, desk, sink, bathtub, toilet, counter, shower curtain and picture) are improved by SEQKC-PointNet++. Especially, the semantic segmentation accuracy of SEQKC-PointNet++ significantly improved by 28.25% for shower curtains, 18.04% for sinks, 14.17% for doors, 13.93% for toilets, and 9.67% for picture. The reason why the "picture" is so difficult to segment is that most of them are hung on the wall, and they are almost integrated with the wall in the point cloud space, and the percentage of points in the whole Scannet dataset is only 0.04%. Therefore, the semantic segmentation of pictures requires the network to be able to extract fine-grained and discriminative features. It can be seen that our improved PointNet++ network based on SEQKC accomplishes this task well.

SEQKC-PointSIFT also has improved the semantic segmentation accuracy of small objects, such as chair, desk, bathtub, toilet, counter, curtain and picture. Among them, the semantic segmentation accuracy of curtain is improved by 14.80% and toilet is improved by 12.12%, and other objects also have small improvements, these objects have rich geometric structure, thanks to the SEQKC can carefully handle the relationship between the points in the point cloud space, the semantic segmentation accuracy of the small objects with the smallest percentage of points has improved.

Since the improvement of PointConv network for 3D point cloud semantic segmentation is more on the IoU, we compared the IoU for each category in the semantic segmentation results as shown in Table 6. As we can see from Table 6, the improved network with the SEQKC module has improved the IoU in 13 categories compared to the original network. The largest improvement is for doors, with a 11.38% increase in the IoU, the semantic segmentation IoU of shower curtain has also increased by 8.98%. This indicates that the embedded modules in the PointConv network substantially help the network to obtain more useful local features of the point cloud and strengthen the network's ability to identify the structure of small scale objects in the point cloud space.

The results of the semantic segmentation were visualized using Meshlab software, and the results are shown in Figure 10. As shown in Figure 10, when the original network segment small objects in the 3D scene, the network sometimes could not recognize small objects surrounded by large objects and often confused them with the background or other large objects. Otherwise, the network was insensitive to the boundary information and produced irregular object boundaries after segmentation. The network combined with the SEQKC module is able to extract richer local semantic features, better segmentation of small objects and clearer segmentation boundaries due to the enhanced relationship between local points of the point cloud. The results show that the SEQKC algorithm can correctly analyze the detailed information of the local region of the point cloud, and using SEQKC module can effectively help the network extract more local feature information of the point cloud, improve the semantic segmentation accuracy of the network.

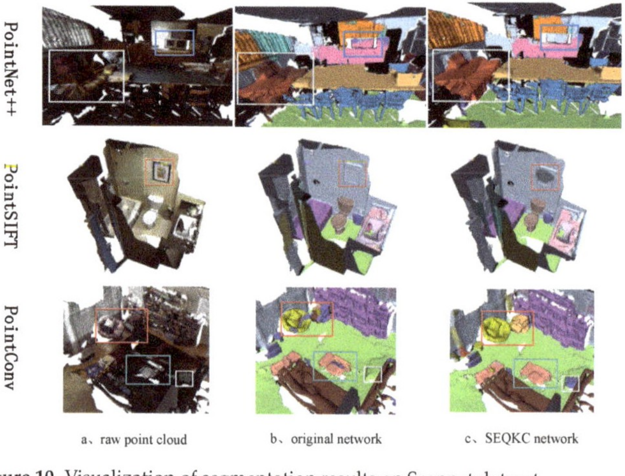

Figure 10. Visualization of segmentation results on Scannet dataset.

5.7. Running Time

Table 7 illustrates a comparison of running time. For a fair comparison, we conduct all experiments on a single GTX 1080Ti GPU with the same environment. The training time is for training one epoch, and the test time is for evaluating 4×8192 points. As we can see, compare with the original networks, when segmenting the same scenes, our algorithms take about the same time, but our algorithms improve the segmentation accuracy, especially for small objects in the scenes.

Table 7. The comparison results of running time.

Method	Training Time (s)	Test Time (s)
PointNet++	98	0.14
SEQKC-PointNet++	105	0.19
PointSIFT	160	0.25
SEQKC-PointSIFT	170	0.26
PointConv	200	0.30
SEQKC-PointConv	230	0.32

6. Discussion

Each point in the point cloud space does not exist in isolation, the relationship of points between different objects or different parts of the same object are different. Such point-to-point relationships are important for the point cloud semantic segmentation task. To capture such relationships, we propose the spatial eight-quadrant kernel convolution algorithm, which captures point-to-point connections by constructing kernel points and weighting the neighborhood points with kernel point weights and distances, so that local fine-grained information of the point cloud can be extracted.

The proposed algorithm can be added to previous point cloud semantic segmentation network to improve the semantic segmentation performance. After incorporating the spatial eight-quadrant kernel convolution algorithm, the network is more sensitive to small objects and boundary features in point clouds, and the segmentation accuracy of the network is significantly improved.

7. Conclusions

In this paper, we propose a point cloud feature extraction algorithm, which called spatial eight-quadrant kernel convolution. The proposed algorithm models the relationship between points and extracts local fine-grained features from point clouds, so that it can improve the performance of semantic segmentation for small objects. We modified the downsampling module with the spatial eight-quadrant kernel convolution algorithm and apply it to classical point cloud semantic segmentation networks. The results show that small-scale objects in 3D point clouds are easily affected by large-scale objects, resulting in wrong semantic categories or partial boundary erosion by large-scale objects. The proposed SEQKC module can help the network to extract fine-grained feature information from the point cloud and improve the semantic segmentation ability of small-scale objects in complex and variable environments. Extensive experimental results on ScanNet v2 dataset show that The proposed model can help the network to improve the semantic segmentation.

Author Contributions: Methodology, L.L.; Software, J.Y. and L.T.; Validation, L.L., J.Y. and L.T.; Writing—original draft preparation, J.Y.; Writing—review and editing, W.S. and L.Z.; Visualization, J.Y. and L.T.; Supervision, W.T.; Funding acquisition, L.L. All authors have read and agreed to the published version of the manuscript.

Funding: This research was funded by the National Natural Science Foundation of China under Grant (61976227,61772213) and in part by the Natural Science Foundation of Hubei Province under Grant 2019CFB622.

Data Availability Statement: The ScanNet V2 Dataset used for this study can be accessed at http://www.scan-net.org/ accessed on 11 June 2018.

Acknowledgments: We would like to thank our colleagues for their helpful suggestions during the experiment, and thank the editor and the anonymous reviewers for their valuable comments and suggestions that greatly improve the quality of this paper.

Conflicts of Interest: The authors declare no conflict of interest.

References

1. Lamas, D.; Soilán, M.; Grandío, J.; Riveiro, B. Automatic Point Cloud Semantic Segmentation of Complex Railway Environments. *Remote Sens.* **2021**, *13*, 2332. [CrossRef]
2. Luo, N.; Yu, H.; Huo, Z.; Liu, J.; Wang, Q.; Xu, Y.; Gao, Y. KVGCN: A KNN Searching and VLAD Combined Graph Convolutional Network for Point Cloud Segmentation. *Remote Sens.* **2021**, *13*, 1003. [CrossRef]
3. Rusu, R.B.; Cousins, S. 3D is here: Point Cloud Library (PCL). In Proceedings of the 2011 IEEE International Conference on Robotics and Automation, Shanghai, China, 9–13 May 2011; IEEE: Piscataway, NJ, USA, 2011; pp. 1–4.
4. Wang, J.; Zhu, M.; Wang, B.; Sun, D.; Wei, H.; Liu, C.; Nie, H. KDA3D: Key-Point Densification and Multi-Attention Guidance for 3D Object Detection. *Remote Sens.* **2020**, *12*, 1895. [CrossRef]
5. Platt, C.; Young, S.; Austin, R.; Patterson, G.R.; Mitchell, D.; Miller, S.D. LIRAD Observations of Tropical Cirrus Clouds in MCTEX. Part I: Optical Properties and Detection of Small Particles in Cold Cirrus. *J. Atmos. Sci.* **2002**, *59*, 3145–3162. [CrossRef]
6. Carmigniani, J.; Furht, B.; Anisetti, M.; Ceravolo, P.; Damiani, E.; Ivkovic, M. Augmented reality technologies, systems and applications. *Multimed. Tools Appl.* **2010**, *51*, 341–377. [CrossRef]
7. Laskey, M.; Lee, J.; Chuck, C.; Gealy, D.V.; Hsieh, W.Y.S.; Pokorny, F.T.; Dragan, A.; Goldberg, K. Robot grasping in clutter: Using a hierarchy of supervisors for learning from demonstrations. In Proceedings of the 2016 IEEE International Conference on Automation Science and Engineering (CASE), Fort Worth, TX, USA, 21–25 August 2016; IEEE: Piscataway, NJ, USA, 2016; pp. 827–834.
8. Geng, X.; Ji, S.; Lu, M.; Zhao, L. Multi-Scale Attentive Aggregation for LiDAR Point Cloud Segmentation. *Remote Sens.* **2021**, *13*, 691. [CrossRef]

9. Pierdicca, R.; Paolanti, M.; Matrone, F.; Martini, M.; Morbidoni, C.; Malinverni, E.S.; Frontoni, E.; Lingua, A.M. Point Cloud Semantic Segmentation Using a Deep Learning Framework for Cultural Heritage. *Remote Sens.* **2020**, *12*, 1005. [CrossRef]
10. Kwak, J.; Sung, Y. DeepLabV3-Refiner-Based Semantic Segmentation Model for Dense 3D Point Clouds. *Remote Sens.* **2021**, *13*, 1565. [CrossRef]
11. Charles, R.Q.; Su, H.; Kaichun, M.; Guibas, L.J. PointNet: Deep Learning on Point Sets for 3D Classification and Segmentation. In Proceedings of the 2017 IEEE Conference on Computer Vision and Pattern Recognition (CVPR), Honolulu, HI, USA, 21–26 July 2017; pp. 77–85.
12. Gardner, M.; Dorling, S. Artificial neural networks (the multilayer perceptron)—A review of applications in the atmospheric sciences. *Atmos. Environ.* **1998**, *32*, 2627–2636. [CrossRef]
13. Qi, C.R.; Yi, L.; Su, H.; Guibas, L.J. PointNet++: Deep Hierarchical Feature Learning on Point Sets in a Metric Space. In Proceedings of the Advances in Neural Information Processing Systems, Long Beach, CA, USA, 4–9 December 2017; pp. 5105–5114.
14. Jiang, M.; Wu, Y.; Lu, C. PointSIFT: A SIFT-like Network Module for 3D Point Cloud Semantic Segmentation. *arXiv* **2018**, arXiv:1807.00652.
15. Wu, W.; Qi, Z.; Li, F. PointConv: Deep Convolutional Networks on 3D Point Clouds. In Proceedings of the 2019 IEEE/CVF Conference on Computer Vision and Pattern Recognition (CVPR), Long Beach, CA, USA, 15–20 June 2019; pp. 9613–9622.
16. Sheather, S.; Jones, M. A reliable data-based bandwidth selection method for kernel density estimation. *J. R. Stat. Soc. Ser. B-Methodol.* **1991**, *53*, 683–690. [CrossRef]
17. Dai, A.; Chang, A.X.; Savva, M.; Halber, M.; Funkhouser, T.; Nießner, M. ScanNet: Richly-Annotated 3D Reconstructions of Indoor Scenes. In Proceedings of the 2017 IEEE Conference on Computer Vision and Pattern Recognition (CVPR), Honolulu, I II, USA, 21–26 July 2017; pp. 2432–2443.
18. Tatarchenko, M.; Park, J.; Koltun, V.; Zhou, Q.Y. Tangent Convolutions for Dense Prediction in 3D. In Proceedings of the 2018 IEEE/CVF Conference on Computer Vision and Pattern Recognition, Salt Lake City, UT, USA, 18–22 June 2018; pp. 3887–3896.
19. Lawin, F.J.; Danelljan, M.; Tosteberg, P.; Bhat, G.; Khan, F.S.; Felsberg, M. Deep projective 3D semantic segmentation. In Proceedings of the International Conference on Computer Analysis of Images and Patterns, Ystad, Sweden, 22–24 August 2017; pp. 95–107.
20. Muresan, M.P.; Nedevschi, S. Multi-object tracking of 3D cuboids using aggregated features. In Proceedings of the 2019 IEEE 15th International Conference on Intelligent Computer Communication and Processing (ICCP), Cluj-Napoca, Romania, 5–7 September 2019; pp. 11–18.
21. Huang, J.; You, S. Point cloud labeling using 3D Convolutional Neural Network. In Proceedings of the 2016 23rd International Conference on Pattern Recognition (ICPR), Cancun, Mexico, 4–8 December 2016; pp. 2670–2675.
22. Tchapmi, L.; Choy, C.; Armeni, I.; Gwak, J.; Savarese, S. SEGCloud: Semantic Segmentation of 3D Point Clouds. In Proceedings of the 2017 International Conference on 3D Vision (3DV , Qingdao, China, 10–12 October 2017; pp. 537–547.
23. Ibrahim, M.; Akhtar, N.; Wise, M.; Mian, A. Annotation Tool and Urban Dataset for 3D Point Cloud Semantic Segmentation. *IEEE Access* **2021**, *9*, 35984–35996. [CrossRef]
24. Unal, O.; Van Gool, L.; Dai, D. Improving Point Cloud Semantic Segmentation by Learning 3D Object Detection. In Proceedings of the IEEE/CVF Winter Conference on Applications of Computer Vision, Waikoloa, HI, USA, 3–8 January 2021; pp. 2950–2959.
25. Hua, B.S.; Tran, M.K.; Yeung, S.K. Pointwise Convolutional Neural Networks. In Proceedings of the 2018 IEEE/CVF Conference on Computer Vision and Pattern Recognition, Salt Lake City, UT, USA, 18–22 June 2018; pp. 984–993.
26. Thomas, H.; Qi, C.R.; Deschaud, J.E.; Marcotegui, B.; Goulette, F.; Guibas, L. KPConv: Flexible and Deformable Convolution for Point Clouds. In Proceedings of the 2019 IEEE/CVF International Conference on Computer Vision (ICCV), Seoul, Korea, 27–28 October 2019; pp. 6410–6419.
27. Ye, X.; Li, J.; Huang, H.; Du, L.; Zhang, X. 3D Recurrent Neural Networks with Context Fusion for Point Cloud Semantic Segmentation. In Proceedings of the European Conference on Computer Vision (ECCV), Munich, Germany, 8–14 September 2018; pp. 403–417.
28. Engelmann, F.; Kontogianni, T.; Hermans, A.; Leibe, B. Exploring Spatial Context for 3D Semantic Segmentation of Point Clouds. In Proceedings of the 2017 IEEE International Conference on Computer Vision Workshops (ICCVW), Venice, Italy, 22–29 October 2017; pp. 716–724.
29. Raktrakulthum, P.A.; Netramai, C. Vehicle classification in congested traffic based on 3D point cloud using SVM and KNN. In Proceedings of the 2017 9th International Conference on Information Technology and Electrical Engineering (ICITEE), Phuket, Thailand, 12–13 October 2017; pp. 1–6.
30. Rubinstein, R. Optimization of computer simulation models with rare events. *Eur. J. Oper. Res.* **1997**, *99*, 89–112. [CrossRef]

remote sensing

Article

Building Multi-Feature Fusion Refined Network for Building Extraction from High-Resolution Remote Sensing Images

Shuhao Ran [1], **Xianjun Gao** [1], **Yuanwei Yang** [1,2,3,*], **Shaohua Li** [1], **Guangbin Zhang** [1] and **Ping Wang** [4,5]

1   School of Geosciences, Yangtze University, Wuhan 430100, China; 201500880@yangtzeu.edu.cn (S.R.); junxgao@yangtzeu.edu.cn (X.G.); lish@yangtzeu.edu.cn (S.L.); 202072509@yangtzeu.edu.cn (G.Z.)
2   Beijing Key Laboratory of Urban Spatial Information Engineering, Beijing 100045, China
3   Hunan Provincial Key Laboratory of Geo-Information Engineering in Surveying, Mapping and Remote Sensing, Hunan University of Science and Technology, Xiangtan 411201, China
4   Aerospace Information Research Institute, Chinese Academy of Sciences, Beijing 100094, China; wangping@aircas.ac.cn
5   Key Laboratory of Earth Observation of Hainan Province, Sanya 572029, China
*   Correspondence: 516042@yangtzeu.edu.cn; Tel.: +86-156-2354-2326

Abstract: Deep learning approaches have been widely used in building automatic extraction tasks and have made great progress in recent years. However, the missing detection and wrong detection causing by spectrum confusion is still a great challenge. The existing fully convolutional networks (FCNs) cannot effectively distinguish whether the feature differences are from one building or the building and its adjacent non-building objects. In order to overcome the limitations, a building multi-feature fusion refined network (BMFR-Net) was presented in this paper to extract buildings accurately and completely. BMFR-Net is based on an encoding and decoding structure, mainly consisting of two parts: the continuous atrous convolution pyramid (CACP) module and the multiscale output fusion constraint (MOFC) structure. The CACP module is positioned at the end of the contracting path and it effectively minimizes the loss of effective information in multiscale feature extraction and fusion by using parallel continuous small-scale atrous convolution. To improve the ability to aggregate semantic information from the context, the MOFC structure performs predictive output at each stage of the expanding path and integrates the results into the network. Furthermore, the multilevel joint weighted loss function effectively updates parameters well away from the output layer, enhancing the learning capacity of the network for low-level abstract features. The experimental results demonstrate that the proposed BMFR-Net outperforms the other five state-of-the-art approaches in both visual interpretation and quantitative evaluation.

Keywords: high-resolution remote sensing images; building extraction; multiscale features; aggregate semantic information; feature pyramid

Citation: Ran, S.; Gao, X.; Yang, Y.; Li, S.; Zhang, G.; Wang, P. Building Multi-Feature Fusion Refined Network for Building Extraction from High-Resolution Remote Sensing Images. *Remote Sens.* **2021**, *13*, 2794. https://doi.org/10.3390/rs13142794

Academic Editor: Mohammad Awrangjeb

Received: 24 May 2021
Accepted: 14 July 2021
Published: 16 July 2021

Publisher's Note: MDPI stays neutral with regard to jurisdictional claims in published maps and institutional affiliations.

1. Introduction

The building is one of the most important artificial objects. Accurately and automatically extracting buildings from high-resolution remote sensing images is of great significance in many aspects, such as urban planning, map data updating, emergency response, etc. [1–3]. In recent years, with the rapid development of sensor technology and unmanned aerial vehicle (UAV) technology, many high-resolution remote sensing images have been produced widely. The high-resolution remote sensing images can provide more fine detail features, increasing the challenge of building extraction. On the one hand, the diverse roof materials of buildings are represented in detail, leading to undetected building results. On the other hand, the similar difference between a building and its adjacent non-building objects results in some wrong detection. These difficulties are the primary factor influencing the building results that can be used in realistic applications. As a result, accurately and automatically extracting buildings from high-resolution remote sensing images is a challenging but crucial task [4].

Currently, the conventional method of feature extraction based on artificial design has gradually given way to the neural network method based on deep learning technology for extracting buildings from VHR images. The traditional extraction methods mainly use the subjective experience to design and extract typical building features in remote sensing images [5,6] and then combine with some ways of image processing and analysis to improve the accuracy of building extraction [7,8]. However, their performance is still severely limited by the capability of feature representation. Therefore, using the neural network to automatically extract high and low dimension image features and identify them at the pixel level is one of the most famous building extraction approaches.

With the development of computer vision technology based on deep learning, the convolutional neural networks (CNNs) are gradually applied to remote sensing image classification [9–11] or ground object detection [12,13]. Long et al. [14] first designed the fully convolutional network (FCN) in 2015, using the convolution layer to replace the fully connected layer of the CNNs. It has an end-to-end pixel-level recognition capability and makes the semantic segmentation process easier to complete. Since then, the emphasis of research on building extraction from remote sensing images using deep learning technologies has shifted from CNNs to FCN [15–18]. In the optimization research of the building extraction method based on FCN, in order to improve the accuracy and integrity of building detection, the work mainly focuses on three aspects. The first is model improvement, which mainly focuses on optimizing the internal structure of an existing classic network to increase the performance of the model [19,20]. The second is data improvement, which consists of establishing a high-quality and high-precision sample set, increasing sample data in the study field, and realizing sample data improvement by fusing multisource data like DSM [21,22]. The third is classifier synthesis, which introduces the conditional random field and attention mechanism to improve classification accuracy [23,24]. While the encoding and decoding structure in FCN can realize an end-to-end network structure, the original image information lost during the encoding phase is difficult to recover during the decoding phase, resulting in the fuzzy edges of building extraction results, a loss of building details, and building details a reduction in extraction accuracy. As a result, the improvement of FCN models (Supplementary Materials) primarily focuses on two types: improving the internal structure of FCN to make full use of multiscale features and optimizing the upsampling stage of FCN.

In the category of improving the internal structure of FCN, many researchers enhance the network performance by enhancing the multiscale feature extraction fusion ability of the model. Researchers realized the detection of objects of various scales in the early days by constructing the inception structure [25]. However, since it uses the common convolution operation, multiple branch convolution can result in a lot of extra calculations. Zhao et al. [26] introduced a pyramid pooling module (PPM) to achieve multiscale feature fusion. They first used the multiscale pooling operation to get the pooling results of various sizes, then channel reduction and upsampling were made, and the feature maps were finally concatenated. Yu [27] et al. proposed atrous convolution by padding zero in the adjacent weights of the ordinary convolution kernel and expanding the size of the equivalent convolution kernel. The dilation rate is defined as the gap between two adjacent effective weights. Chen et al. [28] first proposed the atrous spatial pyramid pooling (ASPP) module in DeepLabv2 based on the idea of atrous convolution. The ASPP module captures the features of different scale objects by parallel multiple atrous convolutions with different dilation rates and realizes the fusion by connecting them. Subsequently, the PPM and ASPP module are widely used in building extraction tasks [29–31]. However, the aforementioned multiscale feature extraction fusion methods generally use large-scale dilation rates and pooling windows to acquire a large range of image information, which typically results in the loss of effective information and decreases the completeness of buildings with variable spectral characteristics. The buildings extracted by MAP-Net [31] and BRRNet [32] cannot obtain a complete detection result under the condition of building with complex spectral characteristics, as shown in the first image in Figure 1.

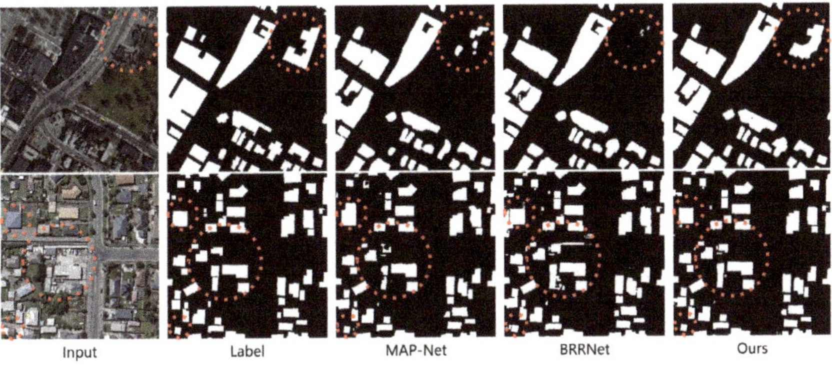

Figure 1. Typical building extraction results by some existing methods in the spectrum confusion area.

In terms of the category of optimizing the upsampling stage of FCN, they provide more semantic information of a multiscale context for the upsampling stage, allowing it to recover part of the semantic information and improve the segmentation accuracy. SegNet [33] records the location information of max values in the MaxPooling operation by using the pooling indices structure and recovers it in the upsampling stage, which improves the segmentation accuracy. By fusing low-level detail information in the encoding stage with high-level semantic information in the decoding stage, Ronneberger et al. [34] proposed a U-Net network model based on the FCN structure, which enhanced the accuracy of building extraction. Since then, multiple building extraction networks based on U-Net have been created, such as ResUNet-a [35], MA-FCN [36], and U-Net-Modified [37]. Nonetheless, these networks that improve the upsampling stage usually only predict via the last layer of the network. It fails to use feature information from other levels fully. For example, multiscale semantic information from the context, including color and edge from high-level and low-level output results, cannot be aggregated. Thus, buildings with similar spectral characteristics to nearby ground objects cannot be detected accurately. Although the MA-FCN outputs at each level of the expanding path and fuses multiple output results at the end of network, the large-scale upsampling operation is not precise enough, which will integrate too much invalid information and reduce the network performance. Moreover, the existing FCNs always have a large number of parameters and a deep structure. Suppose the network is constrained only by the results of the last layer. In that case, the update range of the parameters faring away from the output layer will be significantly attenuated due to the distance, thereby weakening the semantic information of the abstract features and reducing the performance of the network. As the second image in Figure 1 shows, the spectral characteristics of a building and its adjacent ground objects are similar. The existing methods cannot effectively distinguish them and get a false detection.

Given the issues mentioned above, this paper proposes a building multi-feature fusion refined network (BMFR-Net). It takes U-Net as the main backbone, mainly including the continuous atrous convolution pyramid (CACP) module and multiscale output fusion constraint (MOFC) structure. The CACP module takes the end feature maps of the contracting path as input and realizes multiscale feature extraction and fusion by parallel continuous small scale atrous convolution, then feeds the fusion results into the subsequent expanding path. In the expansion path, the MOFC structure enhances the ability of the network to aggregate multiscale semantic information from the context by integrating the multilevel output results into the network. It constructs the multilevel joint loss constraint to update the network parameters effectively. Finally, the accurate and complete extraction of buildings is realized at the end of the network.

The main contributions of this paper include the following aspects:

(1)   The BMFR-Net is proposed to extract buildings from high-resolution remote sensing images accurately and completely. Experimental results on the Massachusetts Building Dataset [12] and WHU Building Dataset [38] shows that the BMFR-Net outperforms the other five state-of-the-art (SOTA) methods in both visual interpretation and quantitative evaluations

(2)   This paper designed a new multiscale feature extraction and fusion module named CACP. By paralleling the continuous small-scale atrous convolution in line with HDC constraints for multiscale feature extraction at the end of the contracting path, which can reduce the loss of effective information and enhance the continuity between local information.

(3)   The MOFC structure is explored in this paper, which can enhance the ability of the network to aggregate multiscale semantic information from the context by integrating each layer output results into the expanding path. In addition, we use the multilevel output results to construct the multilevel joint weighted loss function and determine the best combination of weights to effectively update network parameters.

The rest of this paper is arranged as follows. In Section 2, the BMFR-Net is introduced in detail. The experimental conditions, results, and analysis are given in Section 3. The influence of each module or structure on the network performance is discussed in Section 4. Finally, Section 5 concludes the whole paper.

2. Methodology

This section mainly describes the method proposed in this paper. Firstly, we overview the overall framework of BMFR-Net in Section 2.1. Then, the CACP module and the MOFC structure in BMFR-Net are described in detail in Sections 2.2 and 2.3. Finally, in Section 2.4, the multilevel joint weighted loss function is introduced.

2.1. Overall Framework

To better address the problem of missing detection and incorrect detection of buildings extracted from high-resolution remote sensing images due to spectrum uncertainty, we proposed an end-to-end deep learning neural network named BMFR-Net, as shown in Figure 2.

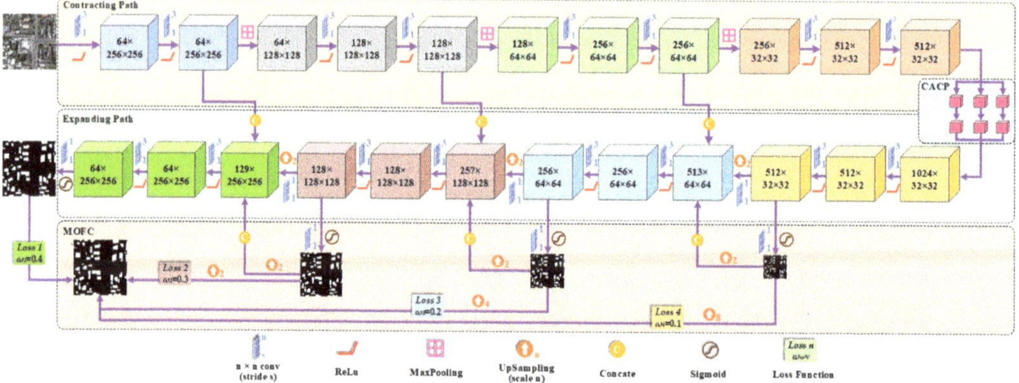

Figure 2. The overall structure of the proposed building multi-feature fusion refined network (BMFR-Net). The upper part is the contraction path, the middle part is the expanding path, the bottom part is the MOFC structure, and the right part is the CACP module.

The BMFR-Net mainly comprises the CACP module and the MOFC structure and uses the U-Net as the main backbone after the last stage is removed. At the end of the contracting path, the CACP module is fused. It can effectively reduce the loss of effective information in multiscale feature extraction and fusion by parallel continuous small-scale atrous convolution. Then the MOFC structure outputs at each level of the expanding path. It reversely integrates the output results into the network to enhance the ability to aggregate multiscale semantic information from the context. Besides, the MOFC structure realizes the joint constraints on the network by combining the multilevel loss functions. It can effectively update the network parameters in the contracting path in BMFR-Net, which are located far away from the output layer and enhance the learning capacity of the network for shallow features.

2.2. Continuous Atrous Convolution Pyramid Module

To alleviate the information loss in the multiscale feature extraction process, we proposed the CACP module in this section. Buildings are often densely spaced in high-resolution remote sensing images of urban scenes, as is well recognized, and the size difference is obvious. Therefore, it is necessary to obtain multiscale features to extract different scale buildings completely. We propose CACP, a new multiscale feature extraction and fusion module inspired by hybrid dilated convolution (HDC) [39], as shown in Figure 3.

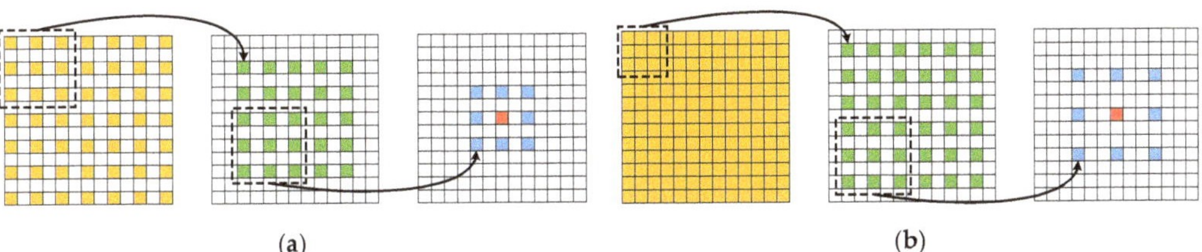

(a)   (b)

Figure 3. Illustration of the HDC. All the atrous convolution layers with a kernel size of 3 × 3: (**a**) from left to right, continuous atrous convolution with the dilation rate of 2, the red pixel can only get information from the input feature map in a checkerboard fashion, and most of the information is lost; (**b**) from left to right, continuous atrous convolution with the dilation rates of 1, 2, and 3, respectively, the receptive field of the red pixel covers the whole input feature map without any holes or edge loss.

As shown in Figure 4, the CACP module is made up of three small blocks: feature map channel reduction, multiscale feature extraction, and multiscale feature fusion. To begin, in the block of feature map channel reduction, the input channel number of the feature map is reduced by half to reduce the calculation amount. Following that, the reduced feature maps are fed into the multiscale feature extraction block, which extracts multiscale features through five parallel branches. The first three branches are continuous small-scale atrous convolution branches. In this paper, the dilation rates of the three branches are (1,2,3), (1,3,5), and (1,3,9). The gridding phenomenon is alleviated and local information such as texture and geometry loss is effectively minimized by placing HDC constraints on the dilation rate of continuous atrous convolution. The fourth is the global average pooling branch, which is used to obtain image-level features. The fifth branch is designed as a residual [40] branch to integrate the original information and facilitate the error backpropagation to the shallow network. Besides, the batch normalization and ReLu activation functions are performed after the atrous convolution process. Finally, the extracted features are fused by pixel addition in the multiscale features fusion block and the channel number of the feature map is restored to its target number.

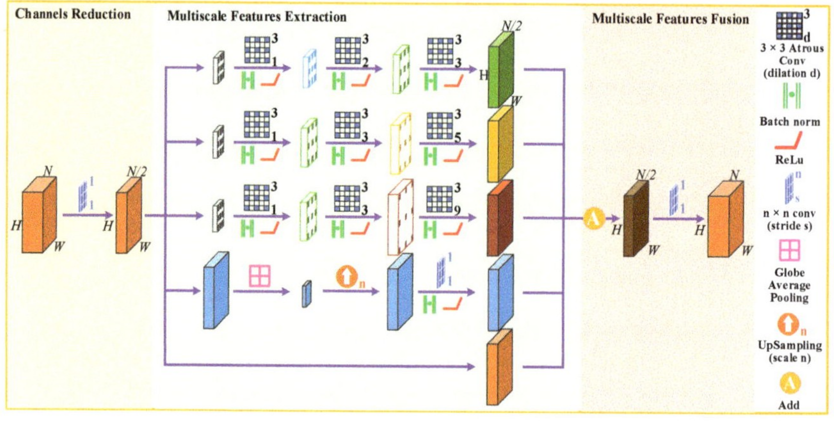

Figure 4. The architecture of the continuous atrous convolution pyramid (CACP) module.

In comparison to the ASPP module, the CACP module replaces the single-layer large-scale atrous convolution in the ASPP module with continuous small-scale atrous convolution. The CACP module can enhance the relevance of local information such as texture and geometry and slow down the loss of high-level semantic information that helps target extraction in the atrous convolution process to improve the completeness of buildings with variable spectral characteristics. The CACP module can also be easily incorporated into other networks to enhance multiscale feature extraction and fusion.

2.3. Multiscale Output Fusion Constraint Structure

This section designs a multiscale output fusion constraint structure in order to increase the ability to aggregate multiscale semantic information from the context and reduce the difficulty of updating parameters in the contracting path in BMFR-Net, which are located far away from the output layer. At present, the U-Net and other networks for building extraction from remote sensing images usually only generate results at the last layer. The network is insufficient to aggregate multiscale semantic information from the context since these frameworks fail to make full use of feature information from other levels. Additionally, most of the existing networks usually have more deep layers. Due to single-level network constraints, it is difficult to efficiently change parameters far away from the output layer. As a consequence, the precision of the building extraction results is insufficient for practical applications.

Inspired by FPN [41], the MOFC structure is designed to solve the above problems, and its structure is shown in Figure 5. In this paper, we took U-Net as the main backbone. Firstly, the MOFC structure uses a convolution layer with kernel size 1×1 and the sigmoid activation function for prediction production at the end of each level of the expanding path, as shown by the purple arrow in Figure 5. Next, the predicted results except the last level are upsampled twice. Then, as shown by the red arrow in Figure 5, the upsampling feature map is connected with the feature map of the adjacent level that has the skip connection. Moreover, except for the last level, the output results are upsampled to the size of the input image and evaluated with the ground truth to construct the multilevel joint weighted loss function, as shown in the orange arrow in Figure 5. In the end, the building extraction result is generated at the end of the network.

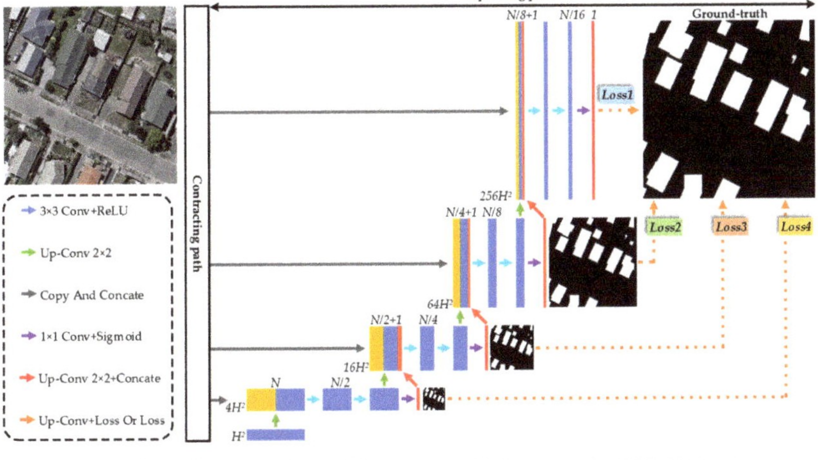

Figure 5. The architecture of the multiscale output fusion constraint (MOFC) structure.

Since the MOFC integrates the predicted results of different levels in the expanding path into the network and constructs a multilevel loss function to constrain the network jointly, the proposed network with the MOFC structure can obtain the unique high-level semantic information about buildings and low-level semantic information such as color and edge from high-level and low-level output results, respectively, to provide more multiscale semantic information from the context for the upsampling process. Furthermore, it can more efficiently update parameters in the contracting path that is far away from the output layer than the current network, extracting buildings with identical spectral features to be accurate.

2.4. Multilevel Joint Weighted Loss Function

The loss function was used to calculate the difference between expected and actual outcomes and it is extremely significant in neural network training. Building extraction is a two-class semantic segmentation task in which loss functions such as binary cross entropy loss (BCE loss) [42] and dice loss [43] are widely used. The basic expressions of BCE loss and dice loss are shown in Equations (1) and (2):

$$l_{BCE} = -\frac{1}{N} \sum_{i=1}^{N} (g_i \times \log p_i + (1 - g_i) \times \log(1 - p_i)) \tag{1}$$

$$l_{Dice} = 1 - \frac{2 \times \sum_{i=1}^{N}(g_i \times p_i)}{\sum_{i=1}^{N} g_i + \sum_{i=1}^{N} p_i} \tag{2}$$

where l_{BCE} is BCE loss, l_{Dice} is dice loss, N denotes the total number of pixels in the image, and g_i denotes whether the ith pixel in the ground truth belongs to a building. If it belongs to a building, $g_i = 1$, otherwise $g_i = 0$. p_i denotes the probability that the ith pixel in the predicted result is a building.

Since BMFR-Net adopts a multiscale output fusion constraint structure, it has predicted results at every level of the expanding path, so it is necessary to weight all loss functions of predicted results to obtain the final loss function. The $loss_{BMFR-Net}$ is expressed in Equation (3):

$$loss_{BMFR-Net} = \sum_{n=1}^{4} \omega_n C_n \tag{3}$$

where C_n denotes the nth output restriction (loss function) in BMFR-Net from the end of the network to the beginning of extending path. For example, C_1 represents the output constraint at the end of the network, C_4 represents the output constraint at the beginning of the expanding path. ω_n denotes the weight value of the nth output constraint.

3. Experiments and Results

In this section, the experimental evaluation of the effectiveness of the proposed BMFR-Net is presented and compared with the other five SOTA methods. Section 3.1 illustrates the open-source data set used in the experiment. Section 3.2 describes the parameter setting details and environment conditions of the experiment. Section 3.3 presents the evaluation metrics. Section 3.4.1 shows the comparative experiment results with analysis.

3.1. Dataset

1. WHU Building Dataset

The aerial imagery dataset of the WHU Building Dataset was published by Ji et al. [38] in 2018. The entire aerial image data set covers an area of about 450 km^2 in Christchurch, New Zealand. The dataset contains 8189 images with 0.3 m spatial resolution, all of which are 512 pixels $\times$ 512 pixels. The dataset was divided into the training set, validation set, and test set. Due to the limited GPU memory, it is difficult to achieve direct training of such a large range of images, so we resized all the images to 256 pixels $\times$ 256 pixels. Finally, the training set contained 18,944 images, the validation set contained 4144 images, and the test set contained 9664 images. The partially cropped images and the corresponding building labels are shown in Figure 6a.

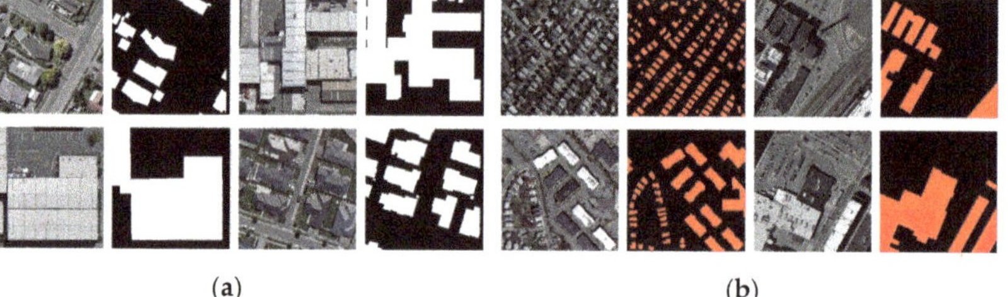

(a)   (b)

Figure 6. Two aerial datasets image and corresponding building label images: (**a**) the white area in the label map of WHU Building Dataset represents the building; (**b**) the red area in the label map of Massachusetts Building Dataset represents buildings.

2. Massachusetts Building Dataset

The Massachusetts Building Dataset was open-sourced by Mnih [12] in 2013, which contains a total of 155 aerial images and building label images of the Boston area. The spatial resolution of the images is 1 m and the size of each image is 1500 pixels $\times$ 1500 pixels. The dataset was divided into three parts: the training set contained 137 images, the validation set contained four images, and the test set contained ten images. Due to the limitation of GPU memory, we also trimmed all images to 256 pixels $\times$ 256 pixels. We cropped the original image in the form of a sliding window, starting from the top left corner, from left to right, and then from top to bottom. The remaining part less than 256 was expanded to 256 $\times$ 256. Some incomplete images were eliminated and the final training set included 4392 images, the validation set included 144 images, and the test set included 360 images. The partially cropped images and the corresponding building labels are shown in Figure 6b.

3.2. Experiment Settings

All of the experiments in this paper were performed on the workstation running a 64-bit version of Windows 10. The workstation is equipped with Intel(R) Core (TM) i7-9700 K CPU @ 3.60 GHz, 32 GB memory, and a GPU of NVIDIA GeForce RTX 2080 Ti with an 11 GB RAM. All the networks were implemented on TensorFlow1.14 [44] and Keras 2.2.4 [45].

The image with a size of 256 pixels × 256 pixels was the input for all networks. The 'the_normal' distribution initialization method was chosen to initialize the parameters of the convolution kernel during the network training stage. In addition, Adam [46] was used as the model optimizer, with a learning rate of 0.0001 and a mini-batch size of 6. All networks used dice loss as the loss function. Due to the difference in image data quantity, resolution, and label accuracy, the network was trained with 200 epochs for the Massachusetts Building Dataset and 50 epochs for the WHU Building Dataset.

3.3. Evaluation Metrics

In order to evaluate the performance of the network proposed in this paper accurately, we selected five evaluation metrics commonly used in semantic segmentation tasks to evaluate the experimental results, including 'overall accuracy (OA)', 'Precision', 'Recall, 'F$_1$-Score', and 'intersection over union (IoU)'. The OA refers to the ratio of all pixels correctly classified to all pixels participating in the evaluation calculation and its calculation formula shows in Equation (4). The precision refers to the proportion of pixels classified as positive categories in all pixels classified as positive categories, as shown in Equation (5). The recall refers to the proportion of pixels correctly classified as positive categories in all pixels of positive categories, as shown in Equation (6). The F$_1$-Score is the harmonic mean of precision and recall, which is a comprehensive evaluation index, as shown in Equation (7). The IoU is the intersection ratio of all predicted positive class pixels and real positive class pixels over their union, as shown in Equation (8):

$$OA = \frac{TP + TN}{TP + TN + FP + FN} \tag{4}$$

$$P = \frac{TP}{TP + FP} \tag{5}$$

$$R = \frac{TP}{TP + FN} \tag{6}$$

$$F_1 - Score = \frac{2 \times P \times R}{P + R} \tag{7}$$

$$IoU = \frac{TP}{TP + FP + FN} \tag{8}$$

where TP (true-positive) is the number of correctly identified building pixels; FP (false positive) is the number of wrongly classified background pixels; FN (false negative) is the number of improperly classified building pixels; TN (true-negative) is the number of correctly classified background pixels.

We employed the object-based evaluation approach [47] in addition to the pixel-based evaluation method to evaluate network performance. Object-based evaluation is based on a single building area: if the ratio of a single extracted result and the ground-truth intersection region to the ground-truth is 0, (0, 0.6), and [0.6, 1.0], it will be recorded as FP, FN, and TP, respectively.

3.4. Comparisons and Analysis

Several comparative experiments were carried out on the selected dataset to evaluate the effectiveness of the BMFR-Net proposed in this paper. First, we tested the performance of BMFR-Net under different loss functions. Then, BMFR-Net is compared with the other five SOTA methods in accuracy and training efficiency.

3.4.1. Comparative Experiments of Different Loss Functions

We used the BCE loss and dice loss to train BMFR-Net, respectively, to verify the influence of different loss functions on the performance of BMFR-Net and the effectiveness of dice loss. The experimental details were given in Section 3.2. The experimental results and some building extraction results are shown in Table 1 and Figure 7.

Table 1. Quantitative evaluation (%) of BMFR-Net with different loss functions. The best metric value is highlighted in bold and underline.

Datasets	BCE Loss	Dice Loss	Pixel-Based Performance Parameter					Object-Based Performance Parameter			
			OA	Precision	Recall	IoU	F_1-Score	Precision	Recall	IoU	F_1-Score
WHU Building Dataset	√		98.68	93.93	94.27	88.85	94.10	91.38	**89.64**	86.56	90.01
		√	**98.74**	**94.31**	**94.42**	**89.32**	**94.36**	**91.67**	89.61	**86.68**	**90.12**
Massachusetts Building Dataset	√		94.38	**86.92**	82.29	73.22	84.54	88.12	72.13	67.65	78.28
		√	**94.46**	85.39	**84.89**	**74.12**	**85.14**	**90.49**	**79.78**	**75.56**	**84.14**

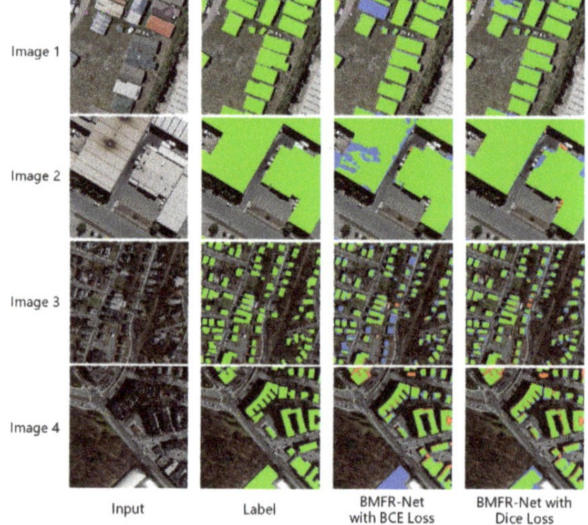

Figure 7. Typical building extraction results of BMFR-Net with different loss functions. Images 1 and 2 belong to the WHU Building Dataset and images 3 and 4 belong to the Massachusetts Building Dataset. In the graph, green represents TP, red represents FP, and blue represents FN.

According to the above results, the BMFR-Net improves the pixel-based IoU and F_1-Score by 0.47% and 0.26% and 0.9% and 0.6% on the two separate datasets when using dice loss, respectively. Additionally the integrity of building results was improved. Additionally, from the perspective of object-based, the recall of building results on the Massachusetts Building Dataset was significantly enhanced by 7.65% after using the dice loss function. That is because dice loss can solve the problem caused by the data imbalance between the number of background pixels and the number of building pixels and avoid falling into the local optimum. Unlike BCE loss, which treats all pixels equally, dice loss prioritizes the foreground detail. The ground truth usually has only two kinds of values in the binary classification task: 0 and 1. Only the foreground (building) pixels can be activated during the dice coefficient calculation using dice loss, while the background pixels are cleared. Thus, dice loss is adopted as the loss function of BMFR-Net.

3.4.2. Comparative Experiments with SOTA Methods

We compared BMFR-Net to the other five SOTA approaches, including U-Net [34], SegNet [33], DeepLabV3+ [48], MAP-Net [31], and BRRNet [32], to further assess the efficacy of the introduced network in this paper. We chose U-Net as one of the comparison methods since BMFR-Net uses U-Net as its main backbone. The SegNet was selected as the comparison method since it has the same encoding and decoding structure as U-Net and has a unique MaxPooling indices structure. Besides, the DeepLabV3+ is the latest structure of the DeepLab series network, which has a codec structure and includes an improved Xception structure and an ASPP module. Considering that the residual structure and atrous convolution have a profound impact on the development of neural networks, we selected BRRNet, a building extraction network based on U-Net and the integrating residual structure and atrous convolution. Moreover, we also used MAP-Net as a comparison method, which is an advanced network for building extraction.

To ensure the fairness of the comparative experiment, we reduced the number of parameters of SegNet and DeeplabV3+, which are used for multiclass segmentation of natural images. The last encoding and first decoding stage of SegNet was removed and the number of repetitions with the middle flow in the DeepLabV3+ was changed to the same eight times as the original Xception.

1.   The comparative experiments on the WHU Building Dataset

The quantitative evaluation results of building extraction on the WHU Building Dataset are shown in Table 2. Our proposed BMFR-Net got higher scores in all evaluation metrics than other methods. As compared to BRRNet, the second-best performance, BMFR-Net, was 3.13% and 1.14% higher in pixel-based and object-based IoU, respectively, and 1.78% and 1.03% higher in the pixel-based and object-based F_1-score, respectively.

Table 2. Quantitative evaluation (%) of several SOTA methods on the WHU Building Dataset. The best metric value is highlighted in bold and underline.

Methods	Pixel-Based Performance Parameter					Object-Based Performance Parameter			
	OA	Precision	Recall	IoU	F_1-Score	Precision	Recall	IoU	F_1-Score
U-Net [34]	98.20	90.25	94.00	85.34	92.09	90.01	88.60	85.46	88.85
SegNet [33]	98.03	89.43	93.36	84.08	91.35	88.98	88.34	84.38	88.13
DeepLabV3+ [48]	98.28	91.80	92.84	85.73	92.32	89.22	87.44	83.84	87.74
MAP-Net [31]	98.10	91.30	91.61	84.26	91.46	88.83	88.68	84.04	88.10
BRRNet [32]	98.33	91.52	93.68	86.19	92.58	90.31	88.86	85.54	89.09
BMFR-Net (ours)	**98.74**	**94.31**	**94.42**	**89.32**	**94.36**	**91.67**	**89.61**	**86.68**	**90.12**

Extensive area building extraction examples by different methods are shown in Figures 8 and 9. According to the typical building extraction results are shown in Figure 10, we can see that the BMFR-Net results are the most accurate and complete with the fewest FP and FN. When the spectral characteristics of a building and its adjacent ground objects are similar, as shown in images 1, 2, and 3 in Figure 10, other approaches cannot distinguish effectively. In contrast, BMFR-Net obtains accurate building extraction results by fusing the MOFC structure in the expanding path. On the one hand, the MOFC structure in BMFR-Net enhances the network of the ability to aggregate multiscale semantic information from the context and provides more effective information for the discrimination of pixels at each level. On the other hand, the MOFC structure realizes effective updating of parameters in the contracting path in BMFR-Net, which are located far away from the output layer, making the semantic information contained in the low-level abstract features richer and more accurate.

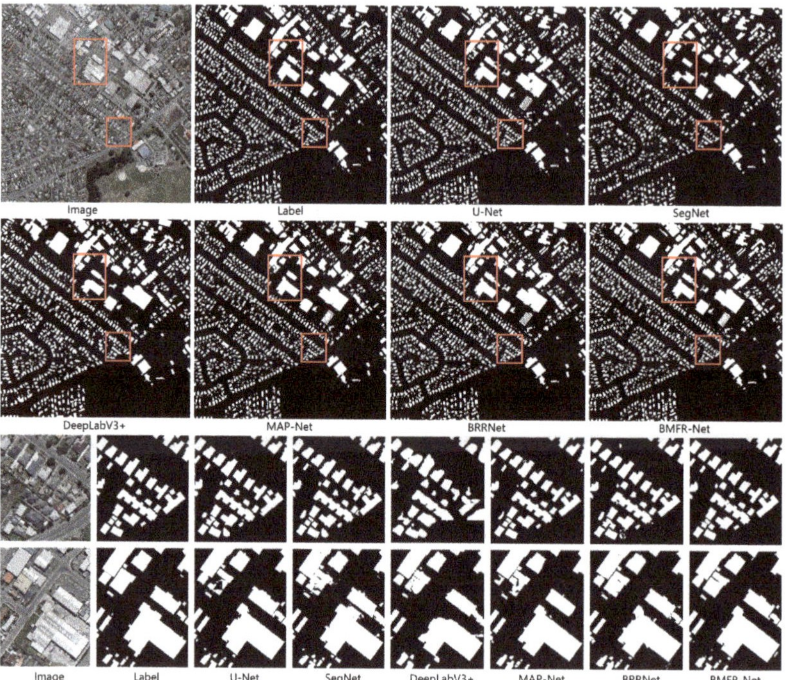

Figure 8. Extensive area building extraction results by different methods on the WHU Buildings Dataset. The bottom column represents the corresponding partial details.

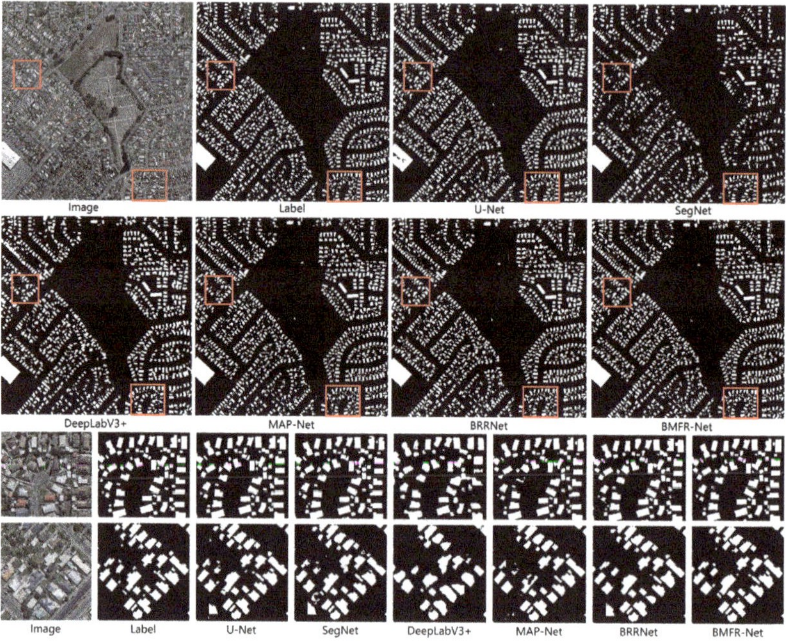

Figure 9. Extensive area building extraction results by different methods on the WHU Buildings Dataset. The bottom column represents the corresponding partial details.

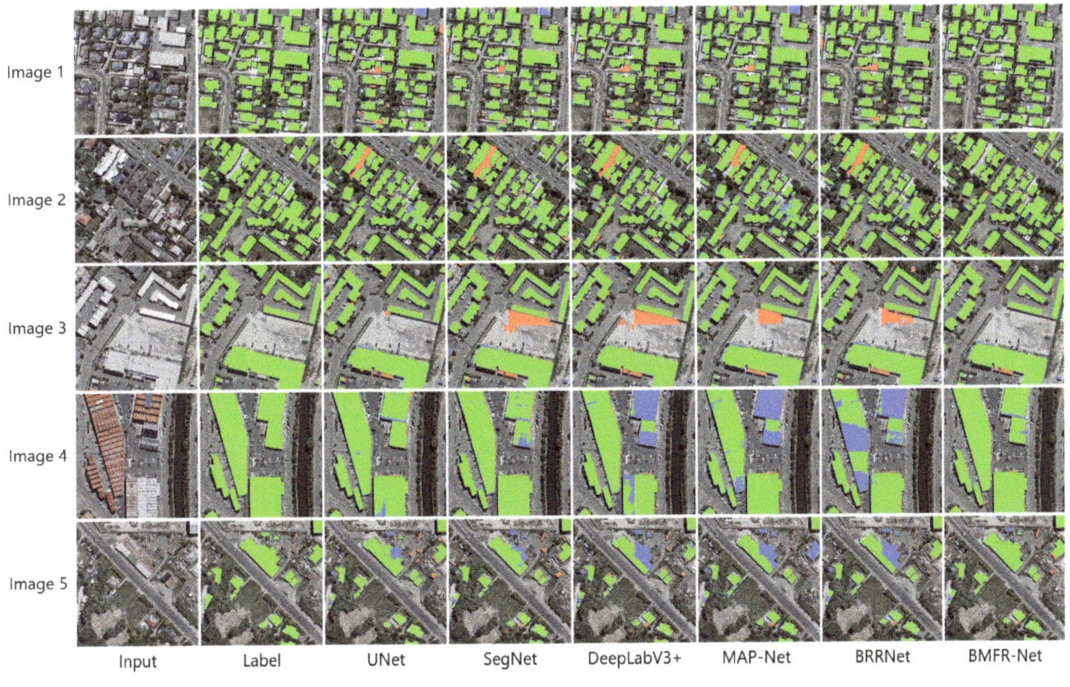

Figure 10. Typical building extraction results by different methods on WHU Buildings Dataset. In the graph, green represents TP, red represents FP, and blue represents FN.

Furthermore, as shown in Figure 10, images 4 and 5, other methods cannot recognize a building roof with complex structures and inconsistent textures and materials as one entity, resulting in several undetected holes and deficiencies in the results, whereas BMFR-Net extracted the building entirely. That is because the U-Net and SegNet are not equipped with multiscale feature aggregation modules at the end of the contracting path. Therefore, they can only extract some scattered texture and geometry information, resulting in the lack of continuity between the information. In addition, the DeepLabV3+, MAP-Net, and BRRNet all adopt a large-scale dilation rate or pooling window, which discards too much building feature information and breaks texture and geometry information continuity. In contrast, the CACP module in BMFR-Net can integrate multiscale features and enhance the continuity of local information such as texture and geometry in the feature map, making it easier to extract a complete building.

2.   The comparative experiments on the Massachusetts Building Dataset

The quantitative evaluation results on the Massachusetts Building Dataset are shown in Table 3. Since the image resolution is lower and the building scenes are more complex in the Massachusetts Building Dataset than in the WHU Building Dataset, the quantitative assessment results were lower overall. Nevertheless, BMFR-Net still had the best performance in all evaluation metrics. Compared with MAP-Net, BMFR-Net was 1.17% and 0.78% higher in the pixel-based IoU and F_1-score, respectively. In terms of the object-based evaluation, U-Net and SegNet performed better among the five SOTA methods. This is due to the fact that while U-Net can efficiently detect buildings, the integrity of the building is insufficient. In contrast, SegNet can entirely extract buildings but has a high rate of false alarms. Compared with U-Net, BMFR-Net was 0.17% and 0.34% higher in the object-based IoU and F_1-score, respectively.

Table 3. Quantitative evaluation (%) of several SOTA methods on the Massachusetts Building Dataset. The best metric value is highlighted in bold and underline.

Methods	Pixel-Based Performance Parameter					Object-Based Performance Parameter			
	OA	Precision	Recall	IoU	F_1-Score	Precision	Recall	IoU	F_1-Score
U-Net [34]	94.01	85.33	82.06	71.91	83.66	89.44	80.10	75.39	83.80
SegNet [33]	93.66	81.42	**85.60**	71.61	83.46	88.68	**80.29**	75.21	83.70
DeepLabV3+ [48]	93.39	81.62	83.39	70.21	82.50	88.30	77.64	72.42	81.93
MAP-Net [31]	94.18	84.72	84.00	72.95	84.36	88.28	78.57	73.36	82.35
BRRNet [32]	94.12	85.03	83.17	72.55	84.09	86.86	77.28	71.45	80.95
BMFR-Net (ours)	**94.46**	**85.39**	84.89	**74.12**	**85.14**	**90.49**	79.78	**75.56**	**84.14**

Extensive area building extraction examples by different methods are shown in Figures 11 and 12. Some typical detailed building extraction results are shown in Figure 13. Visually, compared with other methods, BMFR-Net had the best global extraction results. For those buildings with simple structures and single spectral characteristics, all methods can effectively extract them. However, for non-building objects with similar spectrums with buildings, such as images 2, 3, and 4 in Figure 13, these background objects are easily wrongly divided into buildings or a part of buildings is missing in the five comparison methods. BMFR-Net aggregated more semantic information from the context in the expanding path through the MOFC structure and obtained accurate building extraction results. In addition, as shown in images 1 and 5 in Figure 13, in the results of buildings with complex structures or variable spectral characteristics, the other five methods had more errors or omissions. However,, BMFR-Net uses the CACP module to fuse multiscale features and obtains rich information, effectively reducing the interference caused by shadows and inconsistent textures. As a result, its extracted building results were closer to the true results.

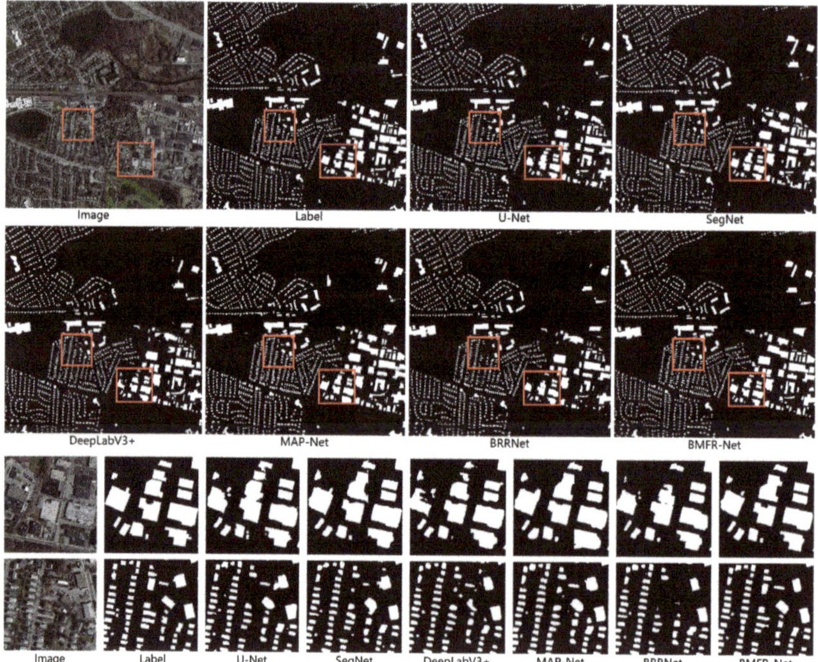

Figure 11. Extensive area building extraction results by different methods on the Massachusetts Building Dataset. The bottom column represents the corresponding partial details.

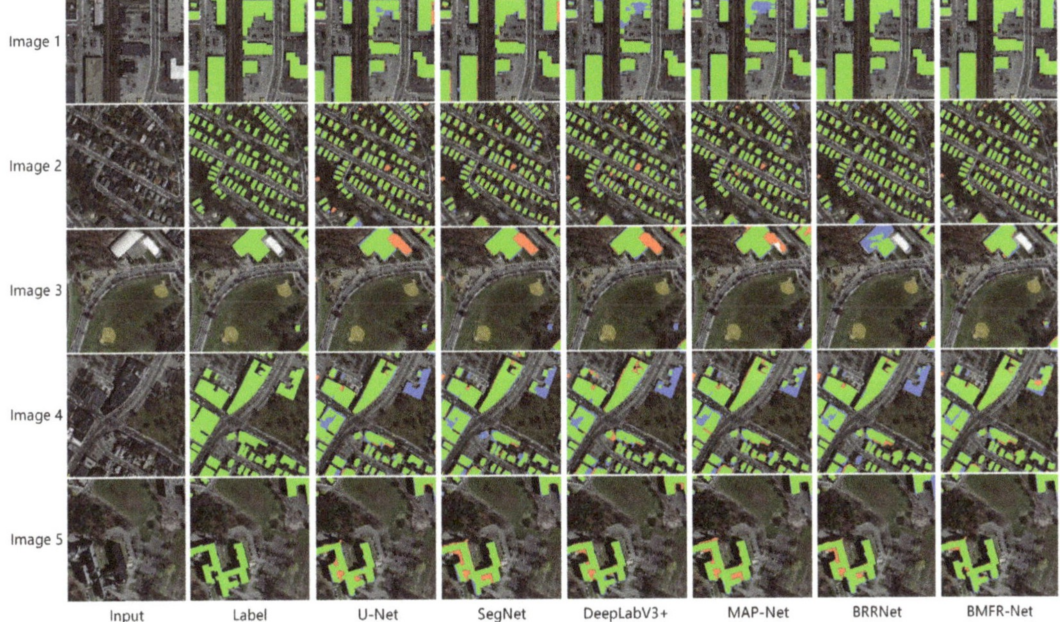

Figure 12. Extensive area building extraction results by different methods on the Massachusetts Building Dataset. The bottom column represents the corresponding partial details.

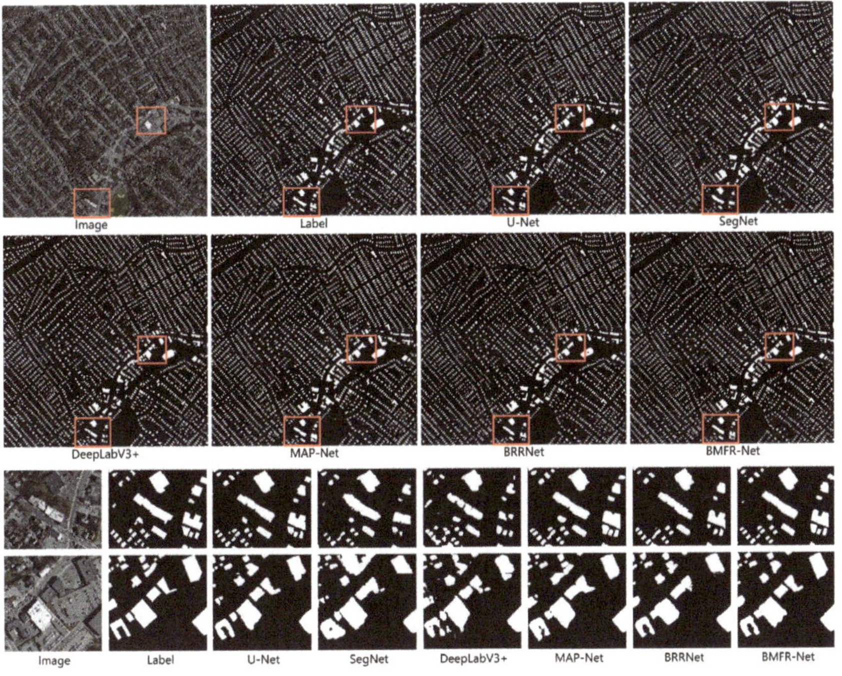

Figure 13. Typical building extraction results by different methods on Massachusetts Building Dataset. In the graph, green represents TP, red represents FP, and blue represents FN.

The results of the above experiments show that BMFR-Net outperformed the competition on two separate datasets, demonstrating that BMFR-Net is capable of extracting buildings from high-resolution remote sensing images of complex scenes. Following that, we will analyze the causes of the above results in detail. U-Net with the skip connection structure can integrate partial low-level features into the expanding path and improve its extraction accuracy. However, due to the poor ability of multiscale feature extraction and fusion, the building extraction result is not complete enough. SegNet can avoid the loss of partial effective information by using the MaxPooling indices structure. At the same time, it does not take into account multiscale feature extraction and fusion. It eliminates the skip connection structure, resulting in difficulty synthesizing the rich detail information in the low-level feature and the abstract semantic information in the high-level feature. As a consequence, the extraction results have the problems of a false alarm and missing alarm. DeepLabV3+ and MAP-Net enhance the ability of multiscale feature extraction and fusion by fusing the ASPP module and PSP module, respectively. However, they use large-scale dilation rates or pooling windows to obtain more global information, making the detection of large buildings with variable spectral characteristics incomplete. BRRNet uses atrous convolution and a residual structure to achieve multiscale feature extraction and fusion. Then the residual refinement module is used to optimize the extraction results at the end of the network. However, its ability to aggregate multiscale semantic information from the context is insufficient, making it difficult to distinguish buildings with similar spectral features from nearby objects. In addition, all these approaches only produce one output at the end of the network. The BMFR-Net realizes multiscale feature extraction and fusion by combining the CACP module at the end of the contracting path, minimizing high-level semantic information such as texture and geometry loss. Then, the MOFC structure is constructed in the expanding path of BMFR-Net. By integrating the output result of each level into the network and combining the multilevel loss functions, the MOFC structure provides more multiscale semantic information from the context for the upsampling stage and makes the parameters in the contracting path layers to be efficiently modified. Therefore, BMFR-Net can effectively distinguish feature differences between buildings with variable texture materials or non-building with similar spectrums, and it can obtain more accurate and complete building extraction results.

3.4.3. Comparison of Parameters and the Training Time of Different Methods

In general, as network parameters are increased, more memory is consumed during the training and prediction process. Besides, the training time is also one of the primary metrics in the assessment model. So, we compared the total parameters and training time of BMFR-Net and the five SOTA methods. The comparison results are as shown in Figure 14.

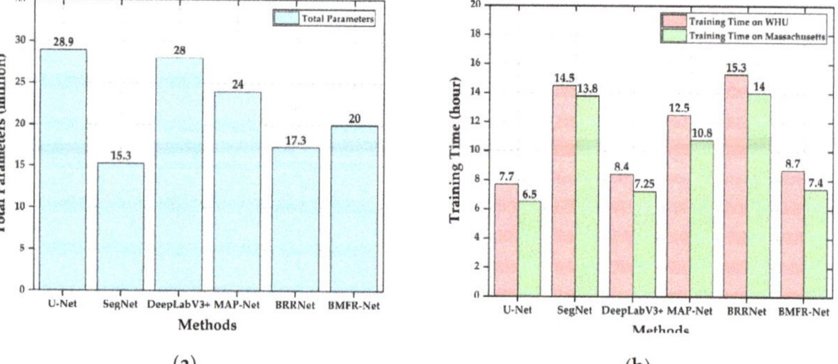

Figure 14. Comparison chart of different methods: (**a**) comparison chart of total parameters; (**b**) comparison chart of training time.

As shown in Figure 14a, the SegNet with the last encoding and first decoding stages removed had the least parameters. Although BMFR-Net had around 5 million more parameters than SegNet and the total amount of parameters reached 20 million, it still ranked in the middle of the five SOTA methods. As shown in Figure 14b, U-Net had the shortest training time for its simple network structure. Since BMFR-Net has a more powerful CACP module and a new MOFC structure, it took slightly longer to train than U-Net. Compared to SegNet with the fewest parameters, the training time on the WHU Building Dataset and the Massachusetts Building Dataset for BMFR-Net was about 6 h less on average under the same conditions, due to the sophisticated MaxPooling indices structure. Compared with DeepLabV3+, which had the second least training time, BMFR-Net had fewer parameters and better building extraction results. According to the above analysis, we could find that the BMFR-Net proposed in this paper had a more balanced efficiency performance. Even though the BMFR-Net had more parameters, it took less time to complete training under the same conditions and produced better building extraction performance.

4. Discussion

In this section, we used ablation studies to discuss the effect of the CACP module, the MOFC structure, and the multilevel weighted combination on the performance of the network. The ablation studies in this section were divided into three parts: (a) investigating the impact of the CACP module on the performance of the network; (b) verifying the correctness and effectiveness of MOFC structure; (c) exploring the influence of weight combination changes of multilevel joint weight loss function on the performance of the network. The experimental data was the WHU Building Dataset described in Section 3.1. Unless otherwise stated, all experimental conditions were consistent with Section 3.2.

4.1. Ablation Experiments of Multiscale Feature Extraction and the Fusion Module

We took U-Net as the main backbone and conducted four groups of comparative experiments to verify the effectiveness of the CACP module (as shown in Figure 4). The first group is the original U-Net. In the second group of experiments, we integrated the ASPP module into the end of the contracting path and the dilation rate of the convolution branch was set as 1, 12, and 18. In the third group of experiments, we used two groups of small-scale continuous atrous convolution with dilation rates of (1,2,3) and (1,3,5) to substitute the atrous convolution with the large-scale dilation rate in the ASPP module. The convolution layer with a kernel size of 1×1 branch in the ASPP module was replaced with the residual branch. In the last group of experiments, we first eliminated the last level of U-Net, then integrated the CACP module into the end of the U-Net contracting path. The dilation rate of the three groups of continuous atrous convolution in the CACP module was set as (1,2,3), (1,3,5), and (1,3,9) in turn. The multiscale features fusion was finally realized by adding each pixel. The experimental results and some building extraction results are shown in Table 4 and Figure 15.

Table 4. Quantitative evaluation (%) of U-Net with different multiscale feature extractions and fusion modules. The best metric value is highlighted in bold and underline.

Methods	Pixel-Based Performance Parameter					Object-Based Performance Parameter			
	OA	Precision	Recall	IoU	F_1-Score	Precision	Recall	IoU	F_1-Score
U-Net	98.20	90.25	94.00	85.34	92.09	90.01	88.60	85.46	88.85
U-Net-ASPP	98.54	93.40	93.70	87.62	93.40	90.78	89.46	85.85	89.59
U-Net-CACP	98.60	93.54	93.87	88.15	93.70	90.80	89.33	86.14	89.56
FCN-CACP	98.62	93.05	**94.65**	88.39	93.84	91.29	89.55	86.56	89.93
BMFR-Net	**98.74**	**94.31**	94.42	**89.32**	**94.36**	**91.67**	**89.61**	**86.68**	**90.12**

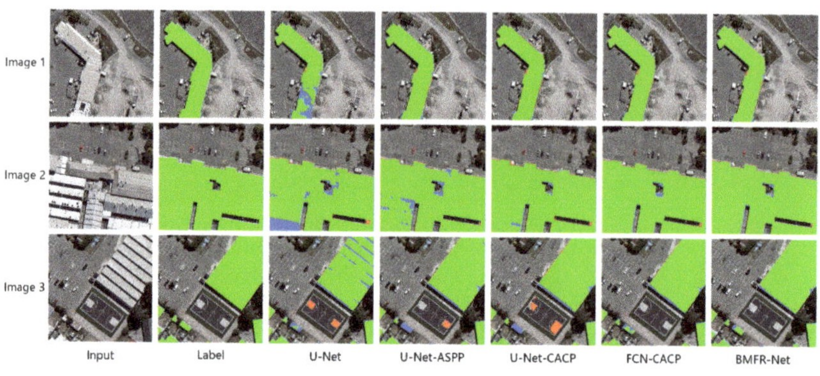

Figure 15. Typical building extraction results of U–Net with different multiscale feature extractions and fusion modules. In the graph, green represents TP, red represents FP, and blue represents FN.

According to the results listed in Table 4 and Figure 15:

- Compared with the other four networks, the evaluation metrics of the original U-Net were improved by adding the multiscale feature extraction and fusion module, demonstrating the efficacy of the multiscale feature extraction and fusion module.
- By comparing the experimental results of U-Net-CACP and U-Net-ASPP, the pixel-based IoU and F_1-Score of the network were improved by 0.53% and 0.3%, respectively, after replacing the ASPP module with the CACP module. Since the CACP module utilized the continuous small-scale atrous convolution in line with HDC constraints, it effectively slowed down the loss of high-level semantic information unique to buildings and enhanced the consistency of local information such as texture and geometry. Thus, the accuracy and recall of building extraction were improved.
- In contrast with the first three networks, the FCN-CACP had the best performance in the quantitative evaluation results, with the pixel-based and object-based F_1-score reaching the highest of 93.84% and 89.93%, respectively. As shown in Figure 15, FCN-CACP had the highest accuracy and contained the fewest holes and defects. By removing the last stage of U-Net, FCN-CACP retained the scale of the input CACP module function feature at 32×32. Consequently, it will reduce the calculation, minimize information loss of small-scale buildings and make multiscale feature extraction easier. Except for pixel-based recall, FCN-CACP had lower evaluation metrics than BMFR-Net because the addition of the MOFC structure to the BMFR-Net enhanced network performance.

4.2. Ablation Experiments of Multiscale Output Fusion Constraint

In order to validate the efficacy of the MOFC structure (as shown in Figure 5), two other kinds of multiscale output fusion constraint structures, MA-FCN [36] (as shown in Figure 16a) and MOFC_Add (as shown in Figure 16b), were introduced for comparison and analysis. MA-FCN and MOFC_Add are constructed differently in terms of how the output results are combined. In the processes of MA-FCN, we showed the production at each level of the expanding path to get the predicted results and upsampled the results to the resolution of the original image except for the last level. Then the four predicted results were fused by connecting at the end of the expanding path to obtain the final building extraction results. In the processes of MOFC_Add, we got the predicted results in the same way as MA-FCN. Then, starting from the first level of the expanding path, the first predicted result was upsampled twice and fused pixel by pixel with the second predicted outcome. The other results were upsampled in the same way until the last level. In the end, the building extraction results were generated at the end of the network. Based on U-Net, MOFC, MA-FCN, and MOFC_Add structures were constructed, respectively. In the ablation experiment, they were compared with the original U-Net. The experimental results and some building extraction results are shown in Table 5 and Figure 17.

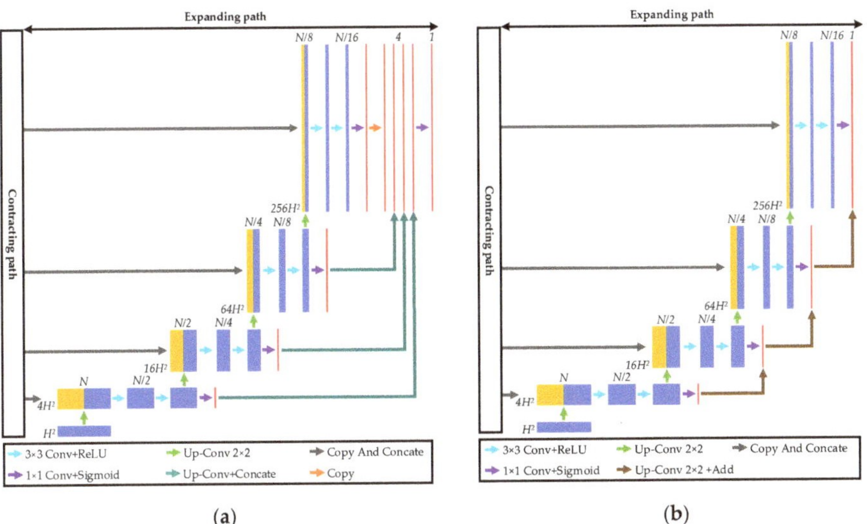

Figure 16. Two other kinds of multiscale output fusion constraint structures: (**a**) structure diagram of MA-FCN; (**b**) structure diagram of MOFC_Add.

Table 5. Quantitative evaluation (%) of U-Net with different multiscale output fusion constraint structures. The best metric value is highlighted in bold and underline.

Methods	Pixel-Based Performance Parameter					Object-Based Performance Parameter			
	OA	Precision	Recall	IoU	F_1-Score	Precision	Recall	IoU	F_1-Score
U-Net	98.20	90.25	94.00	85.34	92.09	90.01	88.60	85.46	88.85
MA-FCN	98.34	91.24	94.14	86.33	92.66	90.84	89.30	86.25	89.60
MOFC_Add	98.08	88.39	**95.31**	84.70	91.72	89.22	88.74	85.31	88.52
U-Net-MOFC	98.61	93.41	94.14	88.28	93.77	91.00	**90.05**	**86.74**	90.09
BMFR-Net	**98.74**	**94.31**	94.42	**89.32**	**94.36**	**91.67**	89.61	86.68	**90.12**

Figure 17. Typical building extraction results of U-Net with different multiscale output fusion constraint structures. In the graph, green represents TP, red represents FP, and blue represents FN.

According to the results listed in Table 5 and Figure 17:

- Compared with the original U-Net, the evaluation metrics of U-Net-MOFC and MA-FCN were significantly improved, especially the pixel-based IoU and F_1-score of U-Net-MOFC that increased by 2.94% and 1.68%, respectively. In contrast, most of the evaluation metrics of MOFC_Add were reduced. It indicates that the MOFC structure was better at aggregating multiscale meaning semantic information than the others.
- The MA-FCN performed better in pixel-based and object-based evaluation indexes than the original U-Net, but the network performance was still not as good as U-Net-MOFC. At each step of the expanding path, MA-FCN will improve the use of feature information. However, the upsampling scale was too large, resulting in a loss of effective information and a decrease in network performance. MOFC_Add had a higher recall but a lower precision, which was a significant difference. Aside from that, global performance was the worst. This is because MOFC_Add did not actively add results, making it challenging to synthesize the semantic information from different levels.
- The second-best overall performer was U-Net-MOFC. The MOFC structure enhanced the ability to aggregate multiscale semantic information frsom the context of the network by fusing the output results of each level into the network. Furthermore, multilevel joint constraints will effectively update parameters in the contracting path layer, improving the object-based IoU and F_1-score from the original U-Net by 1.28% and 1.24%, respectively. In terms of buildings with complex architectures or variable spectral characteristics in Figure 17, U-Net-MOFC can achieve more complete extraction outcomes. The highest score of F_1-score belonged to BMFR-Net. After removing the last level of U-Net-MOFC and adding the CACP module, F1 increased by 0.59%.

4.3. Ablation Experiments of the Weighted Combination of the Multilevel Joint Constraint

To check the efficacy of the multilevel joint constraint and investigate the impact of the weight combination change of loss function on the performance of BMFR-Net, we used a principal component analysis to determine five different weight combinations for comparative experiments. The weight of loss function from the end level of BMFR-Net to the beginning level of the expanding path was marked as ω_1, ω_2, ω_3, ω_4 and we ensured the sum of them was 1. The ablation experiment results of the five groups with different weight combinations are shown in Table 6 and Figure 18.

Table 6. Quantitative evaluation (%) of BMFR-Net with different weight combinations of the multilevel joint constraint. The best metric value is highlighted in bold and underline.

$(\omega_1, \omega_2, \omega_3, \omega_4)$	Pixel-Based Performance Parameter					Object-Based Performance Parameter			
	OA	Precision	Recall	IoU	F_1-Score	Precision	Recall	IoU	F_1-Score
(1,0,0,0)	98.70	94.56	93.71	88.92	94.13	90.52	88.38	85.34	88.92
(0.25,0.25,0.25,0.25)	98.73	**94.62**	93.91	89.15	94.26	90.99	**89.74**	86.24	89.84
(0.4,0.3,0.2,0.1)	**98.74**	94.31	**94.42**	**89.32**	**94.36**	**91.67**	89.61	**86.68**	**90.12**
(0.1,0.2,0.3,0.4)	98.71	93.99	**94.42**	89.04	94.20	91.27	89.56	86.57	89.92
(0.7,0.1,0.1,0.1)	98.72	94.22	94.25	89.10	94.23	90.45	89.14	85.93	89.30

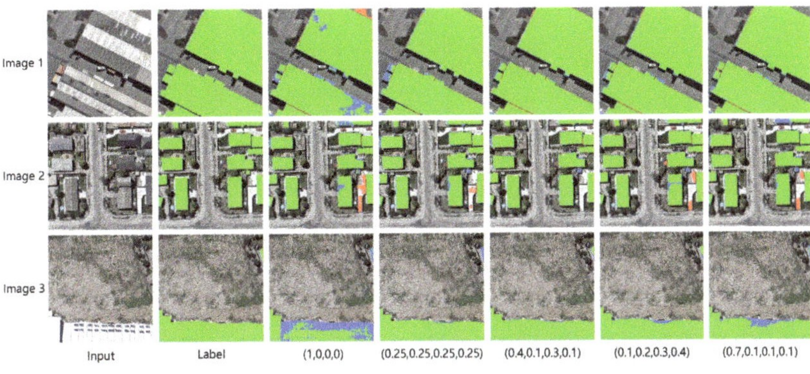

Figure 18. Typical building extraction results of BMFR-Net with different weight combinations of the multilevel joint constraint. In the graph, green represents TP, red represents FP, and blue represents FN.

According to the results listed in Table 6 and Figure 18:

- The global extraction effect of (1,0,0,0) was the worst. It had a poor pixel-based recall of 93.71% but a high pixel-based precision of 94.56%. The explanation for this is that BMFR-Net has deep layers and it is difficult to effectively update the parameters in the contracting path in BMFR-Net, which are located far away from the output layer due to the single level loss constraint. As a result, the ability of the network to learn local information such as the color and edge from low-level features is harmed and the recall of building extraction results is reduced. As shown in image 1 in Figure 18, the BMFR-Net buildings with the weight combination of (1,0,0,0) were missing, while the buildings extracted by the BMFR-Net with multilevel joint constraints were more complete.
- By contrast, the pixel-based and object-based F_1-score of (0.4,0.3,0.2,0.1) was the highest, reaching 94.23% and 89.30%, respectively. From the bottom to the top of the BMFR-Net expanding path, the resolution and global meaning semantic information of the feature maps gradually increased and were enriched. The loss function became increasingly influential in updating the parameters as it progressed from the low-level to high-level. Therefore, the weight combination of (0.4,0.3,0.2,0.1) was best for balancing the requirement of primary and secondary constraints in the network, and the building extraction effect was better. As shown in images 2 and 3 in Figure 18, the accuracy and integrity of building extraction results in (0.4,0.3,0.2,0.1) were higher than others.
- Comparing the results of (0.4,0.3,0.2,0.1) with (0.7,0.1,0.1,0.1), it is clear that when ω_1 is enlarged, the overall performance of the network will decrease. Although the last level loss function is the primary constraint of the network, an unrestricted increase in its weight and a decrease in the weight of other levels would cause the network parameters to overfit the key constraints. Therefore, the parameters in the contracting path layers can not be effectively updated, limiting the accuracy of building extraction.

5. Conclusions

In this paper, we designed an improved full convolutional network named BMFR-Net to address the issue of incomplete and incorrect identification in extraction results caused by buildings with variable texture materials and foreign objects with the same spectrum. The main backbone of BMFR-Net is U-Net, where the last level has been removed. BMFR-Net mainly includes the CACP module and the MOFC structure. By performing parallel small-scale atrous convolution operations in accordance with HDC constraints, the CACP module effectively slowed down the loss of adequate information in the process of multiscale function extraction. The MOFC structure integrated the multiscale output results into the network to strengthen the ability to aggregate the semantic information from the context and it employed the multilevel joint weighted loss function to update the parameters in the contracting path in BMFR-Net, which were located far away from the output layer effectively. Both of them collaborated to increase building extraction precision. The pixel-based and object-based F_1-score of BMFR-Net on the WHU Building Dataset and Massachusetts Building Dataset reached 94.36% and 90.12% and 85.14% and 84.14%, respectively. Compared with the other five SOTA approaches, BMFR-Net outperformed them all in both visual interpretation and quantitative evaluation. The extracted buildings were more accurate and complete. In addition, we experimentally validated the effectiveness of the multilevel joint weighted dice loss function, which could on average improve the pixel-based F_1-score and IoU by about 0.4% and 0.67% of the model, respectively. Additionally, the precision and recall were better balanced. Furthermore, the ablation studies confirmed the effectiveness of the CACP module and the MOFC structure efficacy and clarified the relationship between different weight coefficients and network performance.

Although the proposed network performed well on two public datasets, there were still some shortcomings. First of all, the number of network parameters was still rather high, at 20.0 million, which necessitates additional memory and training time, reducing deployment efficiency. Furthermore, the BMFR-Net and other existing models rely too much on the training and learning of a massive amount of manual label data, resulting in a significant rise in network training costs. In the future, we will improve BMFR-Net and create a lightweight semi-supervised building extraction neural network to improve computational efficiency and reduce the dependence on manual label data.

Supplementary Materials: Codes and models that support this study are available at the GitHub link: https://github.com/RanKoala/BMFR-Net.

Author Contributions: S.R., Y.Y. and X.G. conceived and conducted the experiments and performed the data analysis. G.Z. assisted in collating experiment data. S.L. and P.W. revised the manuscript. S.R. wrote the article. All authors have read and agreed to the published version of the manuscript.

Funding: This research is supported by Open Fund of Key Laboratory of National Geographic Census and Monitoring, Ministry of Natural Resources Open Fund (No. 2020NGCM07); Open Fund of National Engineering Laboratory for Digital Construction and Evaluation Technology of Urban Rail Transit(No.2021ZH02); Open Fund of Hunan Provincial Key Laboratory of Geo-Information Engineering in Surveying, Mapping and Remote Sensing, Hunan University of Science and Technology (No.E22133); Open Fund of Beijing Key Laboratory of Urban Spatial Information Engineering (No.20210205); the Open Research Fund of Key Laboratory of Earth Observation of Hainan Province (No.2020LDE001); the National Natural Science Foundation of China (No. 41872129).

Institutional Review Board Statement: Not applicable.

Informed Consent Statement: Not applicable.

Data Availability Statement: The WHU Building Dataset from http://gpcv.whu.edu.cn/data/building_dataset.html, (accessed on 15 July 2021). The Massachusetts Building Dataset from https://www.cs.toronto.edu/~vmnih/data/, (accessed on 15 July 2021).

Acknowledgments: We would like to thank the anonymous reviewers for their constructive and valuable suggestions on the earlier drafts of this manuscript.

Conflicts of Interest: The authors declare no conflict of interest.

References

1. Shrestha, S.; Vanneschi, L. Improved Fully Convolutional Network with Conditional Random Fields for Building Extraction. *Remote Sens.* **2018**, *10*, 1135. [CrossRef]
2. Huang, X.; Zhang, L. Morphological Building/Shadow Index for Building Extraction From High-Resolution Imagery Over Urban Areas. *IEEE J. Sel. Top. Appl. Earth Obs. Remote Sens.* **2012**, *5*, 161–172. [CrossRef]
3. Huang, X.; Zhang, L. A Multidirectional and Multiscale Morphological Index for Automatic Building Extraction from Multispectral GeoEye-1 Imagery. *Photogramm. Eng. Remote Sens.* **2011**, *77*, 721–732. [CrossRef]
4. Li, W.; He, C.; Fang, J.; Zheng, J.; Fu, H.; Yu, L. Semantic Segmentation-Based Building Footprint Extraction Using Very High-Resolution Satellite Images and Multi-Source GIS Data. *Remote Sens.* **2019**, *11*, 403. [CrossRef]

5.   Jung, C.R.; Schramm, R. Rectangle Detection based on a Windowed Hough Transform. In Proceedings of the 17th Brazilian Symposium on Computer Graphics & Image Processing, Curitiba, Brazil, 17–20 October 2004; pp. 113–120.
6.   Sirmacek, B.; Unsalan, C. Urban-Area and Building Detection Using SIFT Keypoints and Graph Theory. *IEEE Trans. Geosci. Remote Sens.* **2009**, *47*, 1156–1167.
7.   Gao, X.; Wang, M.; Yang, Y.; Li, G. Building Extraction From RGB VHR Images Using Shifted Shadow Algorithm. *IEEE Access.* **2018**, *6*, 22034–22045. [CrossRef]
8.   Inglada, J. Automatic recognition of man-made objects in high resolution optical remote sensing images by SVM classification of geometric image features. *ISPRS J. Photogramm. Remote Sens.* **2007**, *62*, 236–248. [CrossRef]
9.   Boulila, W.; Sellami, M.; Driss, M.; Al-Sarem, M.; Ghaleb, F.A. RS-DCNN: A novel distributed convolutional-neural-networks based-approach for big remote-sensing image classification. *Comput. Electron. Agric.* **2021**, *182*, 106014. [CrossRef]
10.   Han, W.; Feng, R.; Wang, L.; Cheng, Y. A semi-supervised generative framework with deep learning features for high-resolution remote sensing image scene classification. *ISPRS J. Photogramm. Remote Sens.* **2018**, *145*, 23–43. [CrossRef]
11.   Ma, A.; Wan, Y.; Zhong, Y.; Wang, J.; Zhang, L. SceneNet: Remote sensing scene classification deep learning network using multi-objective neural evolution architecture search. *ISPRS J. Photogramm. Remote Sens.* **2021**, *172*, 171–188. [CrossRef]
12.   Mnih, V. Machine Learning for Aerial Image Labeling. Ph.D. Thesis, University of Toronto, Toronto, ON, Canada, 2013.
13.   Saito, S.; Yamashita, T.; Aoki, Y. Multiple Object Extraction from Aerial Imagery with Convolutional Neural Networks. *J. Imaging Sci. Technol.* **2016**, *60*, 10402.10401–10402.10409. [CrossRef]
14.   Long, J.; Shelhamer, E.; Darrell, T. Fully convolutional networks for semantic segmentation. In Proceedings of the 2015 IEEE Conference on Computer Vision and Pattern Recognition (CVPR), Boston, MA, USA, 7–12 June 2015; pp. 3431–3440.
15.   Liu, W.; Yang, M.; Xie, M.; Guo, Z.; Li, E.; Zhang, L.; Pei, T.; Wang, D. Accurate Building Extraction from Fused DSM and UAV Images Using a Chain Fully Convolutional Neural Network. *Remote Sens.* **2019**, *11*, 2912. [CrossRef]
16.   Maggiori, E.; Tarabalka, Y.; Charpiat, G.; Alliez, P. Convolutional Neural Networks for Large-Scale Remote-Sensing Image Classification. *IEEE Trans. Geosci. Remote Sens.* **2017**, *55*, 645–657. [CrossRef]
17.   Marmanis, D.; Wegner, J.D.; Galliani, S.; Schindler, K.; Datcu, M.; Stilla, U. Semantic Segmentation of Aerial Images with an Ensemble of CNSS. In Proceedings of the ISPRS Annals of the Photogrammetry, Remote Sensing and Spatial Information Sciences, Prague, Czech Republic, 12–19 July 2016; pp. 473–480.
18.   Alshehhi, R.; Marpu, P.R.; Woon, W.L.; Mura, M.D. Simultaneous extraction of roads and buildings in remote sensing imagery with convolutional neural networks. *ISPRS J. Photogramm. Remote Sens.* **2017**, *130*, 139–149. [CrossRef]
19.   Yu, Y.; Ren, Y.; Guan, H.; Li, D.; Yu, C.; Jin, S.; Wang, L. Capsule Feature Pyramid Network for Building Footprint Extraction From High-Resolution Aerial Imagery. *IEEE Geosci. Remote Sens. Lett.* **2021**, *18*, 895–899. [CrossRef]
20.   Hui, J.; Du, M.; Ye, X.; Qin, Q.; Sui, J. Effective Building Extraction From High-Resolution Remote Sensing Images With Multitask Driven Deep Neural Network. *IEEE Geosci. Remote Sens. Lett.* **2019**, *16*, 786–790. [CrossRef]
21.   Bittner, K.; Adam, F.; Cui, S.; Körner, M.; Reinartz, P. Building Footprint Extraction From VHR Remote Sensing Images Combined With Normalized DSMs Using Fused Fully Convolutional Networks. *IEEE J. Sel. Top. Appl. Earth Obs. Remote Sens.* **2018**, *11*, 2615–2629. [CrossRef]
22.   Huang, J.; Zhang, X.; Xin, Q.; Sun, Y.; Zhang, P. Automatic building extraction from high-resolution aerial images and LiDAR data using gated residual refinement network. *ISPRS J. Photogramm. Remote Sens.* **2019**, *151*, 91–105. [CrossRef]
23.   Hu, Q.; Zhen, L.; Mao, Y.; Zhou, X.; Zhou, G. Automated building extraction using satellite remote sensing imagery. *Autom. Constr.* **2021**, *123*, 103509. [CrossRef]
24.   Zhu, Q.; Li, Z.; Zhang, Y.; Guan, Q. Building Extraction from High Spatial Resolution Remote Sensing Images via Multiscale-Aware and Segmentation-Prior Conditional Random Fields. *Remote Sens.* **2020**, *12*, 3983. [CrossRef]
25.   Szegedy, C.; Liu, W.; Jia, Y.; Sermanet, P.; Reed, S.; Anguelov, D.; Erhan, D.; Vanhoucke, V.; Rabinovich, A. Going deeper with convolutions. In Proceedings of the 2015 IEEE Conference on Computer Vision and Pattern Recognition (CVPR), Boston, MA, USA, 7–12 June 2015; pp. 1–9.
26.   Zhao, H.; Shi, J.; Qi, X.; Wang, X.; Jia, J. Pyramid Scene Parsing Network. In Proceedings of the 2017 IEEE Conference on Computer Vision and Pattern Recognition (CVPR), Honolulu, HI, USA, 21–26 July 2017; pp. 6230–6239.
27.   Yu, F.; Koltun, V. Multi-Scale Context Aggregation by Dilated Convolutions. *arXiv* **2015**, arXiv:1511.07122.
28.   Chen, L.C.; Papandreou, G.; Kokkinos, I.; Murphy, K.; Yuille, A.L. DeepLab: Semantic Image Segmentation with Deep Convolutional Nets, Atrous Convolution, and Fully Connected CRFs. *IEEE Trans. Pattern Anal. Mach. Intell.* **2018**, *40*, 834–848. [CrossRef]
29.   Deng, W.; Shi, Q.; Li, J. Attention-Gate-Based Encoder–Decoder Network for Automatical Building Extraction. *IEEE J. Sel. Top. Appl. Earth Obs. Remote Sens.* **2021**, *14*, 2611–2620. [CrossRef]
30.   Liu, W.; Xu, J.; Guo, Z.; Li, E.; Liu, W. Building Footprint Extraction From Unmanned Aerial Vehicle Images Via PRU-Net: Application to Change Detection. *IEEE J. Sel. Top. Appl. Earth Obs. Remote Sens.* **2021**, *14*, 2236–2248. [CrossRef]
31.   Zhu, Q.; Liao, C.; Hu, H.; Mei, X.; Li, H. MAP-Net: Multiple Attending Path Neural Network for Building Footprint Extraction From Remote Sensed Imagery. *IEEE Trans. Geosci. Remote Sens.* **2021**, *59*, 6169–6181. [CrossRef]
32.   Shao, Z.; Tang, P.; Wang, Z.; Saleem, N.; Sommai, C. BRRNet: A Fully Convolutional Neural Network for Automatic Building Extraction From High-Resolution Remote Sensing Images. *Remote Sens.* **2020**, *12*, 1050. [CrossRef]

33. Badrinarayanan, V.; Kendall, A.; Cipolla, R. SegNet: A Deep Convolutional Encoder-Decoder Architecture for Image Segmentation. *IEEE Trans. Pattern Anal. Mach. Intell.* **2017**, *39*, 2481–2495. [CrossRef]
34. Ronneberger, O.; Fischer, P.; Brox, T. U-Net: Convolutional Networks for Biomedical Image Segmentation. In Proceedings of the International Conference on Medical Image Computing and Computer-Assisted Intervention, Munich, Germany, 5–9 October 2015; pp. 234–241.
35. Diakogiannis, F.I.; Waldner, F.; Caccetta, P.; Wu, C. ResUNet-a: A deep learning framework for semantic segmentation of remotely sensed data. *ISPRS J. Photogramm. Remote Sens.* **2020**, *162*, 94–114. [CrossRef]
36. Wei, S.; Ji, S.; Lu, M. Toward Automatic Building Footprint Delineation From Aerial Images Using CNN and Regularization. *IEEE Trans. Geosci. Remote Sens.* **2020**, *58*, 2178–2189. [CrossRef]
37. Hosseinpoor, H.; Samadzadegan, F. Convolutional Neural Network for Building Extraction from High-Resolution Remote Sensing Images. In Proceedings of the 2020 International Conference on Machine Vision and Image Processing (MVIP), Tehran, Iran, 18–20 February 2020; pp. 1–5.
38. Ji, S.; Wei, S.; Lu, M. Fully Convolutional Networks for Multisource Building Extraction From an Open Aerial and Satellite Imagery Data Set. *IEEE Trans. Geosci. Remote Sens.* **2018**, *57*, 574–586. [CrossRef]
39. Wang, P.; Chen, P.; Yuan, Y.; Liu, D.; Huang, Z.; Hou, X.; Cottrell, G. Understanding Convolution for Semantic Segmentation. In Proceedings of the 2018 IEEE Winter Conference on Applications of Computer Vision (WACV), Lake Tahoe, NV, USA, 12–15 March 2018; pp. 1451–1460.
40. He, K.; Zhang, X.; Ren, S.; Jian, S. Deep Residual Learning for Image Recognition. In Proceedings of the 2016 IEEE Conference on Computer Vision and Pattern Recognition (CVPR), Las Vegas, NV, USA, 27–30 June 2016; pp. 770–778.
41. Lin, T.Y.; Dollar, P.; Girshick, R.; He, K.; Hariharan, B.; Belongie, S. Feature Pyramid Networks for Object Detection. In Proceedings of the 2017 IEEE Conference on Computer Vision and Pattern Recognition (CVPR), Honolulu, HI, USA, 21–26 July 2017; pp. 936–944.
42. Boer, P.T.D.; Kroese, D.P.; Mannor, S.; Rubinstein, R.Y. A Tutorial on the Cross-Entropy Method. *Ann. Oper. Res.* **2005**, *134*, 19–67. [CrossRef]
43. Milletari, F.; Navab, N.; Ahmadi, S.A. V-Net: Fully Convolutional Neural Networks for Volumetric Medical Image Segmentation. In Proceedings of the 2016 Fourth International Conference on 3D Vision (3DV), Stanford, CA, USA, 25–28 October 2016; pp. 565–571.
44. Google. TensorFlow 1.14. Available online: https://tensorflow.google.cn/ (accessed on 15 July 2021).
45. Chollet, F. Keras 2.2.4. Available online: https://keras.io/ (accessed on 15 July 2021).
46. Kingma, D.; Ba, J. Adam: A Method for Stochastic Optimization. *arXiv* **2014**, arXiv:1412.6980.
47. Ok, A.O.; Senaras, C.; Yuksel, B. Automated Detection of Arbitrarily Shaped Buildings in Complex Environments From Monocular VHR Optical Satellite Imagery. *IEEE Trans. Geosci. Remote Sens.* **2013**, *51*, 1701–1717. [CrossRef]
48. Chen, L.-C.; Zhu, Y.; Papandreou, G.; Schroff, F.; Adam, H. Encoder-Decoder with Atrous Separable Convolution for Semantic Image Segmentation. In Proceedings of the European Conference on Computer Vision (ECCV), Munich, Germany, 8–14 September 2018; pp. 833–851.

Article

Reconstruction of Complex Roof Semantic Structures from 3D Point Clouds Using Local Convexity and Consistency

Pingbo Hu [1,2,*], **Yiming Miao** [1,2] and **Miaole Hou** [1,2]

1   School of Geomatics and Urban Spatial Informatics, Beijing University of Civil Engineering and Architecture, Beijing 100044, China; 2108160119008@stu.bucea.edu.cn (Y.M.); houmiaole@bucea.edu.cn (M.H.)
2   Beijing Key Laboratory for Architectural Heritage Fine Reconstruction & Health Monitoring, Beijing University of Civil Engineering and Architecture, Beijing 102616, China
*   Correspondence: hupingbo@bucea.edu.cn

Abstract: Three-dimensional (3D) building models are closely related to human activities in urban environments. Due to the variations in building styles and complexity in roof structures, automatically reconstructing 3D buildings with semantics and topology information still faces big challenges. In this paper, we present an automated modeling approach that can semantically decompose and reconstruct the complex building light detection and ranging (LiDAR) point clouds into simple parametric structures, and each generated structure is an unambiguous roof semantic unit without overlapping planar primitive. The proposed method starts by extracting roof planes using a multi-label energy minimization solution, followed by constructing a roof connection graph associated with proximity, similarity, and consistency attributes. Furthermore, a progressive decomposition and reconstruction algorithm is introduced to generate explicit semantic subparts and hierarchical representation of an isolated building. The proposed approach is performed on two various datasets and compared with the state-of-the-art reconstruction techniques. The experimental modeling results, including the assessment using the International Society for Photogrammetry and Remote Sensing (ISPRS) benchmark LiDAR datasets, demonstrate that the proposed modeling method can efficiently decompose complex building models into interpretable semantic structures.

Keywords: compound building reconstruction; LiDAR; point clouds; semantic decomposition

Citation: Hu, P.; Miao, Y.; Hou, M. Reconstruction of Complex Roof Semantic Structures from 3D Point Clouds Using Local Convexity and Consistency. *Remote Sens.* **2021**, *13*, 1946. https://doi.org/10.3390/rs13101946

Academic Editors: Wanshou Jiang, San Jiang and Xiongwu Xiao

Received: 20 March 2021
Accepted: 10 May 2021
Published: 17 May 2021

Publisher's Note: MDPI stays neutral with regard to jurisdictional claims in published maps and institutional affiliations.

1. Introduction

Buildings are the most prominent features in an urban environment. Due to its vast application demands, such as solar radiation estimation [1], visibility analysis [2], and disaster management [3]. The three-dimensional (3D) reconstruction and modeling have received intensive attention in city planning, geomatics, architectonics, computer vision, photogrammetry, and remote sensing. The rapid data acquisition technology from the optical image and light detection and ranging (LiDAR) can produce increasingly dense and reliable point clouds, making it possible to automatically reconstruct 3D building models in a large area. During past decades, various 3D modeling approaches in interactive [4] or automatic [5,6] have been proposed to reconstruct 3D building models using the satellite and aerial optical images [7–9], LiDAR [5,10], and combined images and LiDAR [11,12], resulting in full 3D or 2.5D building models at the scale of a city [13,14] and an individual building [4,15]. Even though much progress has been made to produce building models better and faster, the reliable and automatic reconstruction of detailed building models remains a challenging issue [16,17]. In particular, the reconstructed purely geometric model is usually a combination of planar patches or a set of polygons, which is difficult for many disciplines such as urban planning and land management to further semantically interpret and edit the types and structures of buildings.

In this work, we present a novel unsupervised approach to reconstruct the point clouds into semantic structures (e.g., dormer, hipped roof), purely using the geometric

constraints. Departing from previous studies, our approach can directly recognize and interpret meaningful roof semantic subparts and their hierarchical topology for a compound 3D building, which can be further used to enrich the building model library or construct public training data for supervised learning. The main contributions of this work are twofold:

(1) a progressive grouping algorithm is applied to automatically decompose compound buildings into subparts, thereby generating a structured unit block without any independent overlapping elements,

(2) a hierarchical topology tree model is introduced, as well as a roof connection graph and its decomposed subgraphs, to simplify the complexity of the building reconstruction.

The remainder of this paper is organized as follows. An overview of the related unsupervised and supervised approaches for building reconstruction is introduced in Section 2. The detailed reconstruction steps are described in the next Section 3. Experimental results and discussions are presented in Sections 4 and 5, respectively. Conclusions including future work are summarized in Section 6.

2. State-of-the-Art Methods

Over the past few decades, the issue of 3D building reconstruction has received considerable attention probably due to the advancement of photogrammetry and active sensors, producing a wealth of research work on this broad topic. Among these huge varieties of reconstruction methods, we can distinguish two categories for building modeling, unsupervised and supervised methods, which is the first time for such a summary as we know. In this section, the most related literature on 3D building reconstruction using ALS point clouds is discussed.

2.1. Unsupervised Methods

Unsupervised building reconstruction has been studied extensively in the fields of city planning, geomatics, architectonics, computer vision, photogrammetry, and remote sensing. These 3D building modeling approaches can be divided into data-driven and model-driven methods. A comparison on the data-driven and model-driven approaches can be found in Haala and Kada [18]. Interested readers are referred to some review literature [17,19]. Due to insufficient input data and complex building types, 3D building reconstruction remains an open problem even if only simple flat roof surfaces are considered [5]. Thus, hybrid-driven methods are gradually being concerned, which integrate additional information from both data- and model-driven methods.

For data-driven approaches, also called non-parametric or bottom-up approaches, it assumes that a building is a polyhedral model, which can be directly modeled by geometric information such as the intersection and regularization. It usually starts with the extraction of roof planar patches by region growing [20,21], feature clustering [22,23], model fitting [24,25], and global energy optimization [26–29], and then assembling these extracted roof planes to form a polygon building model. To improve the shape of the reconstructed 3D models, some regularization rules, such as parallel and perpendicular are often applied, resulting in a compact 3D building polygonal model with roof ridges and boundaries. These data-driven methods [13,30–32] have succeeded in the reconstruction of simple Manhattan-like objects but are unstable in the presence of noisy or incomplete point clouds. In order to sufficiently utilize the prior knowledge for building reconstruction, Zhou and Neumann [32] improved the quality of roof models by discovering the global regularities of the similarities between roof planar patches and roof boundaries, which can significantly reduce the complexity of 3D reconstructions. Poullis [33] developed a complete framework to automatically reconstruct urban building models from point clouds by combining a hierarchical statistical analysis of the data geometric properties and a fast energy minimization process for the boundary extraction and refinement. To generate more detailed roof models, Dehbi et al. [34] propose a novel method for roof reconstruction using active sampling, and it is limited to only dormer types. The main advantage is that it can reconstruct a polyhedral building with complex shapes, while the main drawback is the sensitivity to the incompleteness of the point cloud caused by occlusions, shadows, etc. Besides, the generated models from line segments and planar patches are usually purely geometric models, and the semantics of the roof structures are always missed.

The model-driven methods [35–39], known as top-down or parametric methods, start from predefined parametric 3D roof structures and then fit a building model that is best-fitted the input point cloud, resulting in some simple roof blocks in the early stage. Kada and McKinley [40] decomposed the LiDAR point cloud data into multiple objects and then combined them into a whole model. It performs well for automatic reconstruction on a large scale. In addition, 3D building roof structures can be fitted and recognized from the Reversible Jump Markov Chain Monte Carlo algorithm [41] and EM-based Gaussian mixture technique [42]. Moreover, constructive solid geometry (CSG) [43–46] primitives are always introduced in the process of model-driven building reconstruction, which can produce complete 3D building models through Boolean operators (union, intersection, and subtraction). The predefined CSG components organized by semantic information and shape parameters are suitable for 3D roof reconstruction for buildings with fixed styles. However, it is difficult for us to automatically decompose a complex building into predefined CSG primitives, thus, a semi-automatic process is usually adopted using the external ground plans or building footprints [44,46,47]. To deal with more complex buildings, Wang [46] proposed a novel method to reconstruct a compound building with semantic structures using roof local symmetries. The model-driven approaches can robustly reconstruct building models with simple roof styles by utilizing prior knowledge like parallel and symmetry, generating watertight building models. However, roof primitives or structures in the true world reveal a huge diversity, thus, it will fail when a searched roof cannot be described by any of the predefined primitive [34,48,49].

The hybrid-driven 3D reconstruction approaches [10,34,48,50], combining the conventional data- and model-driven strategies, aim to recognize building roof structures or search the best-fitted roof primitives from a predefined roof library. These hybrid-based algorithms can benefit from the roof topology graphs (RTG), which is a graph reflecting neighboring relations between extracted roof planes. Once the RTG is obtained, the searching and fitting process is performed in the topological space. The first RTG-based reconstruction can be found in Verma et al. [51], where the normal vector is added as an attribute of RTG. However, the scope of its application has been significantly reduced because the predefined roof primitive types are simple and fixed. Oude Elberink and Vosselman [52] expanded the parametric roof primitives' library and added more connected attributes such as convexity and concavity for building reconstruction. Similar works by Perera and Maas [50] and Xiong et al. [53] were proposed to distinguish the roof elements and interpret building structures by the improvements in reliability and availability of RTG. The circle graph analysis [5] by minimal-close circles and outer-most circles are more adaptive than previous modeling methods. This sub-graph matched approach is easily prone to errors if mismatches of a sub-graph. In order to avoid multiple searching and matching of the same roof element, Rychard and Borkowski [10] propose a novel building reconstruction method to generate interpretable roof semantic blocks using a new roof structure library. Instead of using the roof topology graph, Xu et al. [54] developed a hierarchical topology tree (HRTT) model expressed by roof planar topological connections on plane–plane, plane–model, and model–model to reconstruct 3D building models. It can produce more accurate building models and can obviously improve the topological quality. However, an inherent problem of hybrid reconstruction is that it can be easily suffered error-prone in itself (e.g., incomplete roof planes extraction) or mismatches of roof sub-graphs. It is difficult to describe various building styles in the human world using limited roof primitives from a predefined RTG library. Moreover, they cannot interpret the semantic structures and maintain a valid topology of the building.

2.2. Supervised Methods

Supervised 3D building reconstruction approaches have gradually attracted widespread attention, especially the emergence of convolutional neural networks (CNN). Similar to the two-dimensional image semantic labeling methods, it assigns the most probable label to each 3D element (e.g., point, planes, roof subparts) using a labeling model learned from a huge number of training data. These labeled semantic features can be classified by a machine learning-based method [55,56]. The random forest [13] and support vector machines [55] are often used to identify the main building components, such as floors, ceilings, and roofs, which can be further assembled into a semantically enriched 3D model. However, the inputs for these methods are the encoded training features derived from local (e.g., surface area, orientation) and contextual (e.g., coplanarity, parallelism) information, which needs to be designed by hand. Recently, the emerging deep learning techniques have reached human-level performance in the domain of computer vision and natural language processing and have gradually been introduced to building reconstruction. Wichmann et al. [57] developed and released an available training dataset named RoofN3D, which can be used to train CNN for different 3D building reconstruction tasks. Axelsson et al. [58] have presented a deep convolutional neural network to automatically classify roof types into ridges and flat roofs, which can further support the generation of a nationwide 3D landscape model. It cannot interpret large buildings with small meaningful subparts automatically. Zhang and Zhang [59] introduced a deep-learning-based approach to successfully reconstruct urban building mesh models at level of detail 2 (LOD2) according to the CityGML specification but cannot interpret the building structures and roof topology. In addition, Yu et al. [60] developed a new fully automatic 3D building reconstruction approach that can generate the LOD1 building models in a large area but cannot generate complicated building structures. Although various supervised solutions have been proposed in recent years to reconstruct buildings with simple roof types, they are still hindered by the lack of public training data, especially the semantic subparts of complex buildings.

3. Methodology

The framework of the proposed complex building reconstruction approach from 3D point clouds, as shown in Figure 1, encompasses four key components, data preprocessing (Section 3.1), roof plane extraction and semantic labeling (Section 3.2), building semantic decomposition (Section 3.3), and generation of building models (Section 3.4).

3.1. Data Preprocessing

During data preprocessing, regions of building blocks are first detected from 3D point clouds. Taking the original point cloud as input, terrain points are firstly separated from non-terrain points using filter approaches like the adaptive TIN [61], cloth simulation filter (CSF) [62], and two-step adaptive extraction [63]. To obtain building and vegetation point clouds, a height threshold (1.5 m–2.5 m) processing was used to identify the high-rise points from the non-terrain points, then the obtained high-rise points can be further used to extract building point cloud using a top-down extraction approach [64]. Moreover, the extraction of building point cloud will benefit from the corresponding image data, as points can be projected back on imagery and cleaned using the normalized difference vegetation index (NDVI) threshold (0.1–0.15). With the extracted building point cloud, a Euclidean clustering method is applied to group the into individual clusters, and the clusters with small area (3–5 m^2) are removed as tree clusters. The threshold of a small area is determined by the point density and the minimum number of points per cluster; e.g., if the point density of the point cloud is 4 points/m^2, and a cluster with 12 points indicate an area of approximate 3 m^2. After the aforementioned process, these segmented individual buildings can be subsequently reconstructed. It should be emphasized that these parameters (height, NDVI, area threshold) are selected empirically, and the details for these operations and parameters are beyond the scope of this paper.

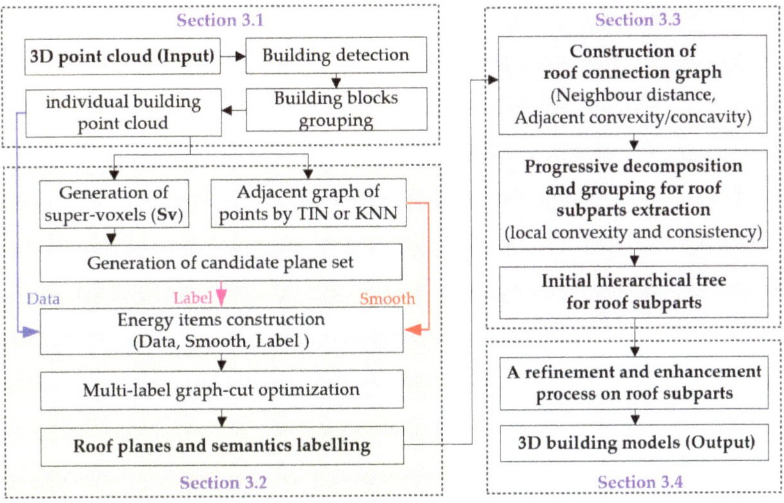

Figure 1. Methodological framework of the complete building reconstruction.

3.2. Roof Plane Extraction and Semantic Labeling

This is a preliminary step for building reconstruction; we first segment the roof point clouds into individual planes by minimizing a global energy function, and then tag its semantics as roof or attachment (e.g., dormer, gable, hipper, chimney). Roof plane segmentation from point clouds is crucial to 3D building reconstruction and is still challenging due to noisy, incomplete, and outlier-ridden data. To achieve accurate and reliable roof planes, a multi-label optimization model [65] was applied, as presented in Equation (1).

$$E(L) = \overbrace{\sum_{p} D(p, L_p)}^{data\ \cos t} + \overbrace{\sum_{(p,q) \in TinEdge} w_{pq} \cdot \delta(p, q)}^{smooth\ \cos t} + \overbrace{\sum_{L_i \in L} |\xi_L|}^{label\ \cos t} \quad (1)$$

The introduced model can transform the plane extraction problem into the best matching issue by balancing different energy costs on geometric data errors (*data* cos *t*), spatial smooth coherence (*smooth* cos *t*), and the number of planes (*label* cos *t*). The objective of this optimization framework is to assign every point (Data) to the most suitable plane (Label). To get the initial candidate labels for the energy model, the input point cloud will be firstly over-segmented into a set of patches by the Voxel Cloud Connectivity Segmentation (VCCS) algorithm [66] or our previous work [27], where each patch represents a local surface with centroid c, curvature f, and normal vector n. The initial candidate labels can be generated using centroids and normal vectors of surface patches, which are selected from all segmented patches or some patches with small curvature f. In addition, an operation by randomly sampling a subset of patches centroids c is done to enrich the potential candidate labels.

The first data cost term in Equation (1), a geometric error measurement is calculated as the quadratic perpendicular distance between a point to a potential label L_p. The second term in (1) is the smoothness between the neighbored point pairs, and the neighborhood can be achieved from triangulated irregular networks (TIN) or k-nearest neighbors (KNN). The indicator function $\delta(\cdot)$ for the adjacent points is selected as Potts model [67], and is set to 1 if a pair of points (p, q) belong to the same label, otherwise, it is 0. Intuitively, a pair of points that are closer together are more likely to be on the same plane, thereby the weight function ω_{pq} can be set as an inverse function of the distance between adjacent points (p, q) located on a TIN edge.

$$\omega_{pq} = \exp(-\|p - q\|) \quad (2)$$

The label cost item is a penalty for the number of input potential labels. In order to compactly represent the input scene, it is encouraged to use fewer labels. The proposed label model can be written by

$$|\zeta_L| = \exp(-|L_i|) \tag{3}$$

where $|L_i|$ is the numbers of inlier points on the plane with index i. The proposed global energy optimization is an iterative process along with the framework of Propose Expand and Re-estimate Labels ("PEaRL") [68] and terminates only if the energy is no more decreased, resulting in a set of labels for the input point cloud. The final planes can be achieved by fitting the points with the same label index.

Similar to the work of Pu and Vosselman [69], the plane semantic features can be further inferred from the knowledge rules (Area, Orientation, Position) into roof or attachment (e.g., dormer, gable, hipper, chimney). It should be mentioned that the plane semantics are optional for the next decomposition process, and the semantics of roof planes can be also classified by the supervised method [70].

3.3. Semantic Decomposition of Compound Building

The semantic decomposition is to generate building subparts using a progressive decomposition and grouping algorithm. For each complex building, a roof connection graph is firstly constructed using the extracted planar primitives, and then a decomposing and grouping operation on the roof connection graph will be performed to generate subgraphs, which are potential roof semantic structures (e.g., dormer, gable, hipper, chimney).

3.3.1. Hierarchy Tree Representation of Complex Buildings

A general assumption of the proposed modeling algorithm is that a compound building roof can be represented by various simple and meaningful subparts. Although the styles of buildings are diverse, the basic units (*subparts*) are similar: a subpart (structure) is a visual-pleased box composed of two or more parametric plane primitives with semantic features.

$$Building = \bigcup_i Subpart_i$$
$$Subpart = \bigcup_j \{Plane_j\}$$
$$Plane_j = \{GeoData; Semantics\} or \{GeoData; \varnothing\}$$

It should be noted that the plane semantics can be allowed to be empty $\varnothing$. Moreover, a general hierarchical-tree-based representation for a complex building was introduced, shown in Figure 2.

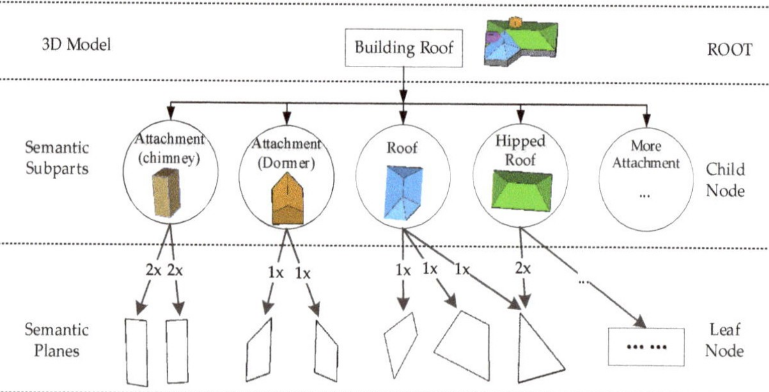

Figure 2. The hierarchical tree representation of a building.

The root of the hierarchical tree, illustrated in Figure 2, is a 3D building model organized by roof semantic subparts, and the planar primitives and roof subparts are treated as leaf nodes and child nodes, respectively. For example, the roof attachment named vertical chimney in the second row is two pairs of parallel planes, while a dormer is a combination of two adjacent planes.

3.3.2. Construction of the Roof Connection Graph

To generate a reasonable decomposition and grouping for roof subparts extraction, the roof connection graph C, a weighted undirected connected graph, is obtained by the extracted roof planes in the Section 3.1. The vectices (V) in C are roof planar primitives, and an edge E between the two vertices represents spatial connectivity. In addition, local geometric convexity and consistency will be calculated for each edge.

The Euclidian distance F_{Dist} between adjacent planar primitives is firstly set as an attribute of the edge E. The psychophysical studies [71–73] suggest that the transition between convex and concave parts might be indicative of the separation between objects and/or their parts. In other words, concave-convex features are the cues for the decomposition objects into semantic subparts. Thus, edges in C are equipped with 3D concave or convex attributes to ensure the reliability and efficiency of building roof decomposition, and the local convexity F_{con} for adjacent planes is calculated by:

$$F_{con} = \begin{cases} True & F_{Convex} \Rightarrow \theta_1 < \theta_2 \Leftrightarrow \vec{n}_1 \cdot \vec{d} > \vec{n}_2 \cdot \vec{d} \\ False & F_{Concave} \Rightarrow \theta_1 > \theta_2 \Leftrightarrow \vec{n}_1 \cdot \vec{d} < \vec{n}_2 \cdot \vec{d} \end{cases} \tag{4}$$

where $\vec{n}_1, \vec{n}_2$ and $\vec{x}_1, \vec{x}_2$ are normals and centroids of adjacent planes, respectively. The angle (θ) of the normals to the vector $d = \vec{x}_1 - \vec{x}_2$ joining the centroids can be calculated using the dot product. For a convex connection, the angle θ_1 is smaller than θ_2, while for concave, the opposite is true. The convexity/concavity are shown in Figure 3.

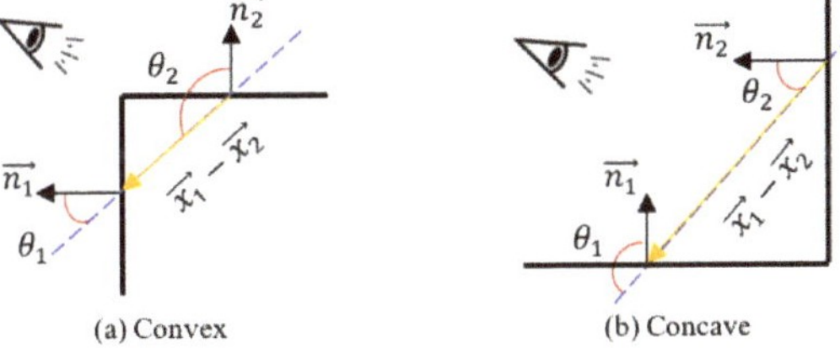

Figure 3. Illustration of the convexity/concave criterion.

The F_{Dist} and F_{Con} can be directly marked as attributes for each edge E and stored in roof connection graph C, that is $C = \{V, E\}$, $E = [F_{Dist}, F_{Con}]$. Finally, the generated graph C will be the foundation for the next progressive decomposition and grouping operation.

3.3.3. Progressive Decomposition for Subparts Extraction

The commonly used methods for roof subparts extraction are usually accomplished by searching and matching the sub-graph element from a predefined library, and it is usually hampered by some significant problems, such as the completeness of the library, the ambiguous definition, and errored sub-graph recognition. When observing objects, we will attempt to group similar elements, recognize patterns, and simplify complex images as we look at objects. To achieve the sub-convex building subparts, visual perception constraints (e.g., proximity, similarity, consistency) derived from the Gestalt Principles are applied [71–73]. The predefined Euclidian distance F_{Dist} and local convexity F_{Con} are the proximity and similarity constraints, respectively. The proximity constraint is straightforward, that is, the planes that are close to each other are more likely to be treated as one group. While the local convexity represents adjacent primitives sharing visual features (such as shape, convexity, and concavity) can group into a perceptive group. Therefore, the consistency constraint F_{CC} (shown in Figure 4), preferences to establish a sub-convex box, is introduced to represent continuous concavity/convexity during the progressive decomposition processing.

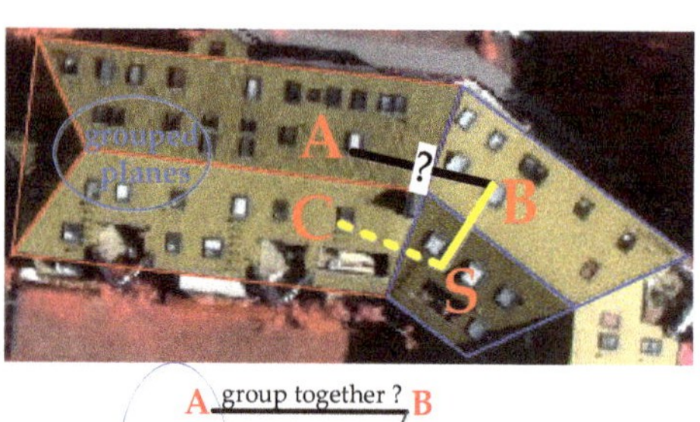
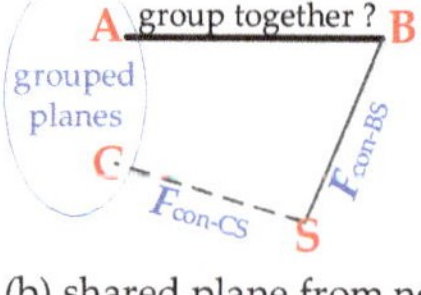

(a) shared plane

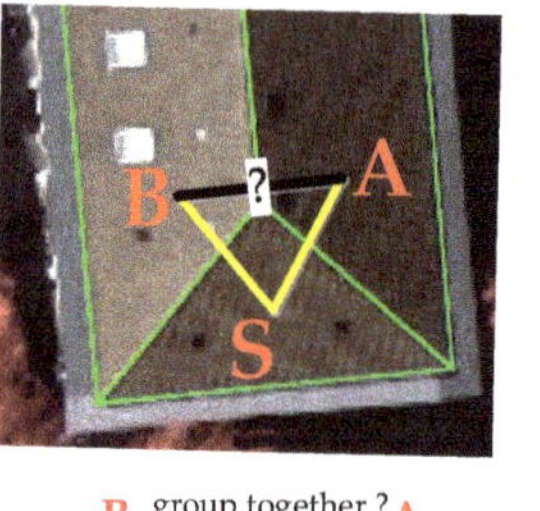

(b) shared plane from neighboring

Figure 4. Illustration of consistency constraint to group plane *A* and *B*. The $F_{Con\text{-}AS}$ and $F_{Con\text{-}BS}$ are 3D concave/convex relationships (similarity constraint) for adjacent planes.

As presented in Figure 4, it indicated that adjacent plane A and B can be grouped only if it satisfies: (1) the edge between plane A and B is labeled as convex, (2) the similarity F_{Con} between plane pairs (A-S, B-S) or (C-S, B-S) should be exactly the same, where S is a shared plane and C is the neighbor of planes A and B that need to be grouped. The consistency constraint criterion F_{CC} is then defined as:

$$F_{CC} = \begin{cases} \text{true} & F_{Con-AS} = F_{Con-BS} & \text{(a)} \\ \text{true} & F_{Con-AB} = convex\&(F_{Con-CS} = F_{Con-BS}) & \text{(b)} \\ \text{false} & \text{otherwise} \end{cases} \tag{5}$$

To achieve the roof subparts, a progressive iterative decomposition of the roof connection graph is introduced, as illustrated in Algorithm 1.

Algorithm 1. Progressive Decomposition

Input: a roof connection graph G and roof planar primitives $PS = [L_p]$
Output: roof subparts $[G_{Sub}]$ and an initial building hierarchical tree T

1: **While** $PS \neq \varnothing$ **do**
2:     Find a planar primitive L_{p0} with the largest area from PS
3:     Create an empty roof plane set G_{Sub} and initial it with plane L_{p0}
4:     Generate G_{Sub} by the iteratively decomposing G using (F_{Dist}, F_{Con}, F_{CC})
5:     Update the nodes of building hierarchical tree T from G_{Sub}
6:     **For** $L_{pi} \in G_{Sub}$ **do**
7:       remove plane L_{pi} from PS
8:       update the nodes and edges of G
9:     **End For**
10: **End While**

The extraction of building subparts as well as the hierarchical tree is carried out in a progressive manner, which aims to search and find the best set of roof planar primitives that are potential to the same sub-convex box. The detailed iteration of the decomposition and grouping to generate a building roof structure is elaborated in the following:

(1) Start from a planar primitive L_{p0}, which has the largest geometric area, and initial the current group $G_{Sub} = [L_{p0}]$;

(2) Create a candidate plane set $G_{candidate}$ that all primitives are connected to the last added element of G_{Sub}, which means that there exists an edge in G. If the candidate set $G_{candidate}$ is empty or all candidate elements are grouped, the current grouping loop ends;

(3) Calculate the convexity F_{Con} and consistency F_{CC} for each element in $G_{candidate}$, and remove the ones that cannot meet these constraints;

(4) Sort the remaining candidate primitives in $G_{candidate}$ according to principles of the closest connected distance F_{Dist} and the same semantics, and the candidate element with the minimum F_{Dist} will be grouped into G_{Sub}. If the set $G_{candidate}$ is empty, the decomposition will terminate;

(5) Go to step (2).

When a building roof subpart is grouped according to the aforementioned decomposition steps, the nodes of the building hierarchical tree will be generated, and the information of the grouped primitives will be simultaneously updated from G. Moreover, this iterative decomposition will be terminated when all input roof planes are grouped. Once the decomposition from G is finished, sub-graphs of different building parts can be obtained, as shown in Figure 5.

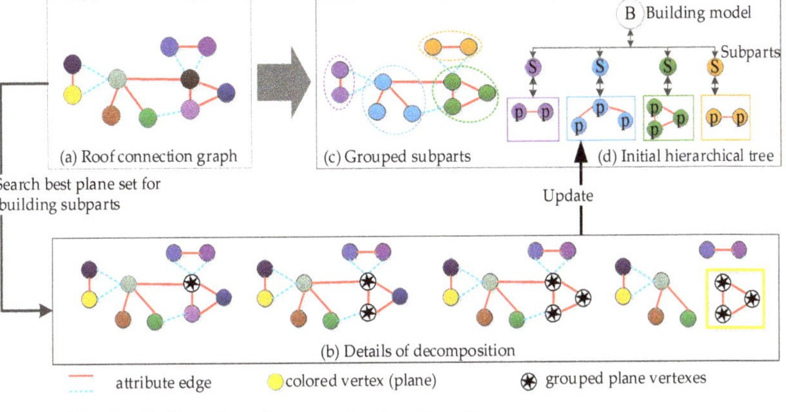

Figure 5. The detailed iteration of progressive decomposition.

3.4. Generation of 3D Building Models

Due to the limit of acquisition devices and scene occlusions, these subparts cannot be correctly interpreted or identified as unambiguous building structures. Thus, a refinement step will be introduced for each grouped subpart using the constraints of symmetry and closure to produce a visually pleasing 3D model in the final. As each primitive connected to its adjacent planes ought to be convex, thus, the closure is that the projected primitives should be connected in sequence and made a closed loop. The local symmetry is used to fulfill the missing or extend ghost primitive based on the architecture aesthetics. As shown in Figure 6, it is performed on whether the normal vector projections of adjacent planes are parallel to each other.

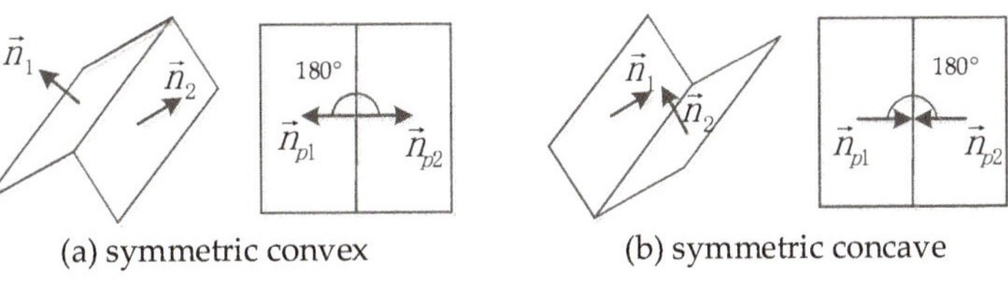

Figure 6. The local symmetry of adjacent planes.

For any pair of adjacent primitives in Figure 6, we firstly project its normal vectors $(\vec{n}_1, \vec{n}_2)$ onto the ground plane, and then an analysis is performed on whether the projected normal vectors $(\vec{n}_{p1}, \vec{n}_{p2})$ are mutually parallel with respect to its intersection: if mutually parallel the adjacent planes are symmetric. Moreover, the details of enhancing these decomposed building subparts are elaborated as follows:

(1) For any extracted subpart, we firstly extract its corresponding sub-node and inlier leaf nodes in Figure 5d.

(2) Calculate the local symmetry indicators of adjacent primitives and perform it.

(3) A closed hull loop detection, stitching together the projected primitives in sequence, will be performed based on closure perception laws. Moreover, an add and union primitive operation will be carried out.

Subpart labeled as the roof: Check the outer border ring of the primitives projected to the ground plane, and if the loop is a concave hull, which means that there exists an incomplete closed loop, a "extend ghost" primitive will be accomplished by searching a connected plane from the roof connection graph or stitching the vertexes of the nearest primitives along the loop. Especially, if the newly added "extend ghost" primitive is parallel to its adjacent, a plane union operation will be performed.

Subpart labeled as dormer: We usually handle the missing vertical primitive along with the boundary loop, where the projected plane is perpendicular to the normal vector.

Subpart labeled as chimney: There exist two types of chimney: column and cone. The missed primitives will be fulfilled along the boundary, and the projection plane is the ground plane. The difference between them is that the fixed source vertex is the same for a cone part.

(4) A similar regularization process [74] is applied to the refined building subparts to produce a 3D geometric vector boundary, and the changed information in the hierarchical tree will be synchronously updated.

4. Experimental Results

4.1. Description of the Datasets

The proposed approach has been implemented with the computational geometry algorithms library (CGAL) [75] and the point cloud library (PCL) [66], and mainly tested on datasets with different point densities and urban characteristics. An overview of the tested datasets is shown in Figure 7. The first dataset is the Guangdong data in China, which has a high point density and buildings of various shapes and sizes, and the next one is the NYU ALS dataset released by the center for urban science and progress of the New York University [76]. The last widely adopted benchmark dataset, obtained from ISPRS Test Project on Urban Classification and 3D Building Reconstruction [16], is located in the city of Vaihingen, German.

The Guangdong dataset was obtained in 2016 using the Trimble Harrier 68i with an average height of 800 m. It is located in a rural region covering an area of approximately 340×360 m^2 and includes 83 buildings with 257 planes in various shapes and sizes. The point density is approximately 13 points/m^2, but has some missing areas as the occlusion, which can easily be prone to failures using the current 3D reconstruction methods. Moreover, the NYU dataset is a high-density ALS data for urban areas and contains a complex set of roof types such as multi-layered and flat. The point density is approximately 123 points/m^2, while the ISPRS benchmark datasets in Area 1–3 were obtained by a Leica ALS50 system in 2008 with a point density of 4–7 points/m^2. There are 37 historic buildings with complex structures and irregular boundaries in Area 1, while Area 2 is characterized by high-rise residential buildings. The roof boundaries are very complex, and the gaps between adjacent roofs have large height differences. Area 3 is a purely residential area, including 56 buildings with many small roof structures. Moreover, the modeling results derived from the Vaihingen benchmark dataset can be evaluated by ISPRS and can be compared with other methods using a unified standard [16]. As the assessment by ISPRS have terminated, we will illustrate the evaluation on Area 1 and 3, and the other three datasets will be assessed using the same geometric errors.

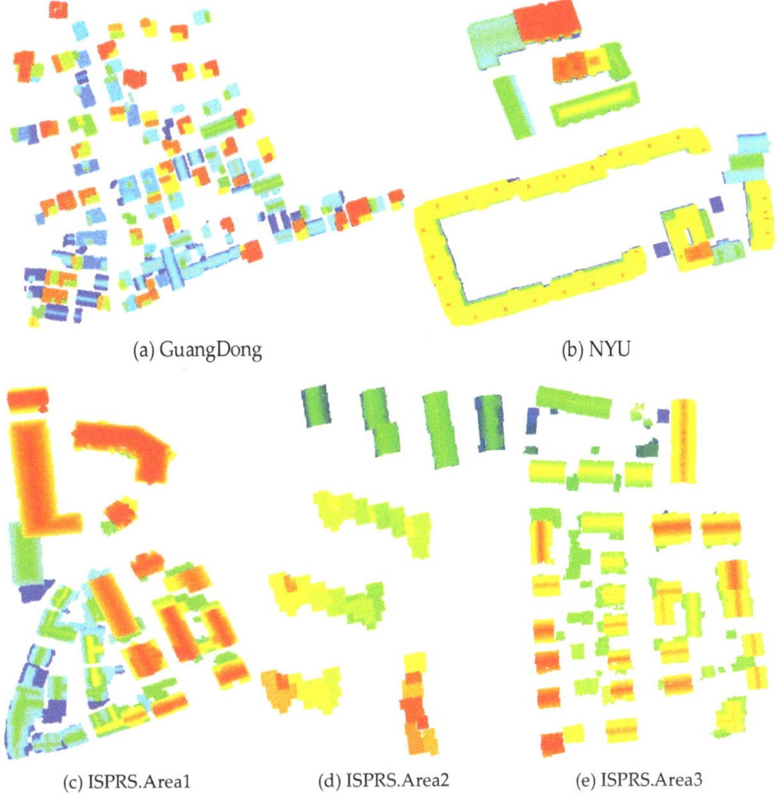

Figure 7. An overview of the tested datasets.

4.2. Results of Model Reconstruction

In the aforementioned datasets, a series of representative buildings are selected to validate the proposed approach. These compound buildings illustrated in Figure 8 are reconstructed by a set of basic building subparts, which are a combination of planar primitives.

It can be seen from Figure 8 that the generated compound buildings in part (a) are assembled by semantic building units in part (c), including hipped roof, dormer, etc. In part (b), the 3D wireframe of each reconstructed building generated is a hierarchical topology tree, which is organized by reconstructed building subparts in part (c). These generated building subparts with an explicit topology can be further used to enrich the building model library or construct public training data for supervised learning. The proposed approach aims to correctly and automatically reconstruct building subparts, and the additional semantics are to be inferred by the dominant semantics of the grouped planes. It is beyond the scope of this paper to accurately interpret the semantics data for various styles. In addition, the final 3D models are illustrated in Figure 9.

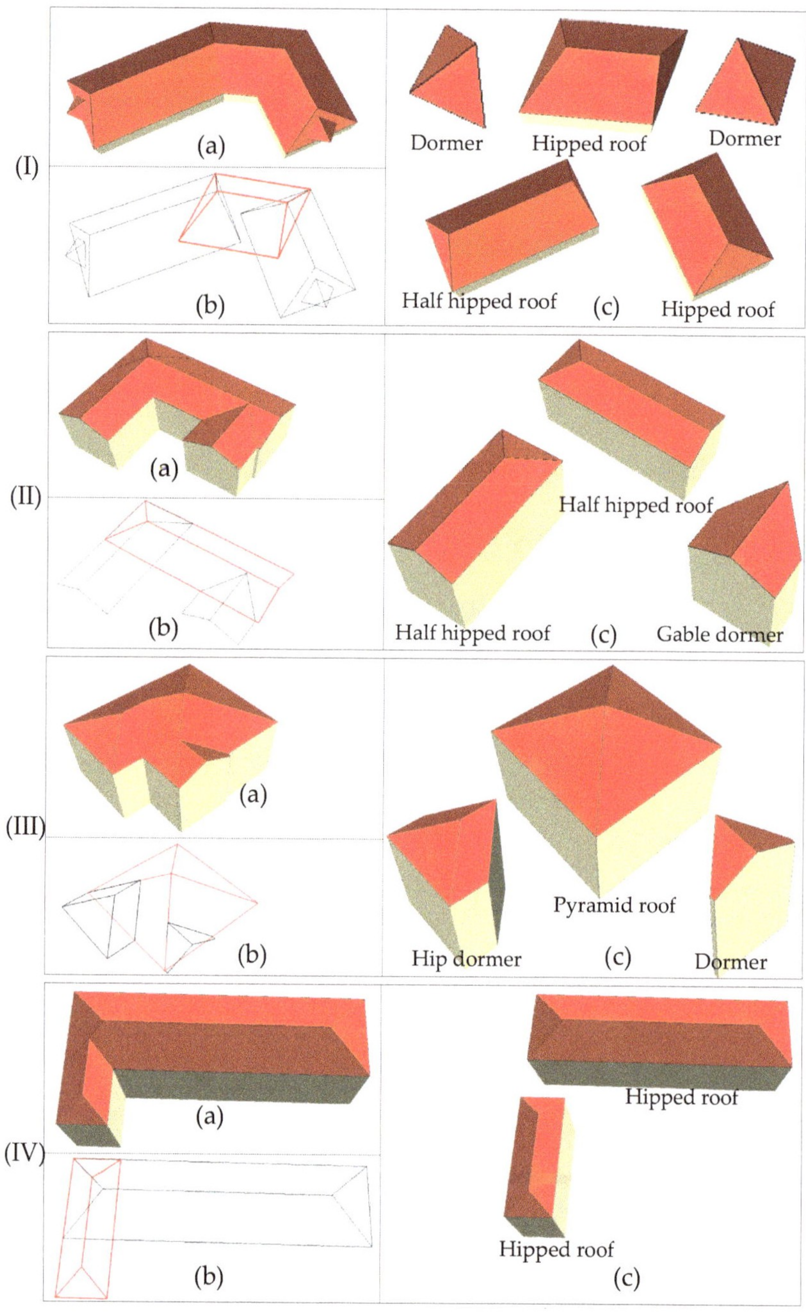

Figure 8. *Cont.*

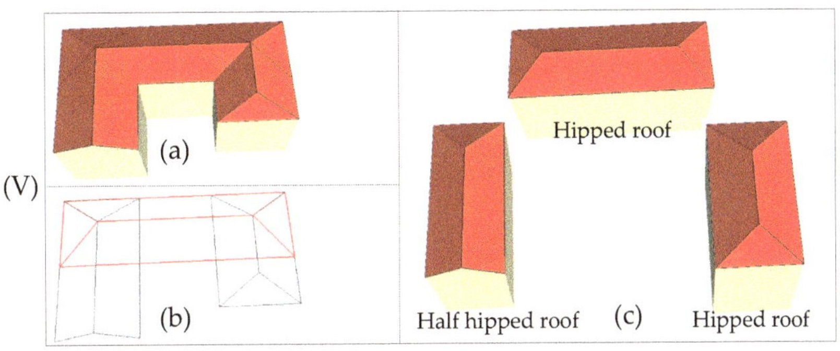

Figure 8. Experimental results of some representative complex buildings. (**a**) The decomposition and reconstruction 3D models, (**b**) the 3D geometric vector boundary, and (**c**) the generated building subparts.

(a) Guangdong

Figure 9. *Cont.*

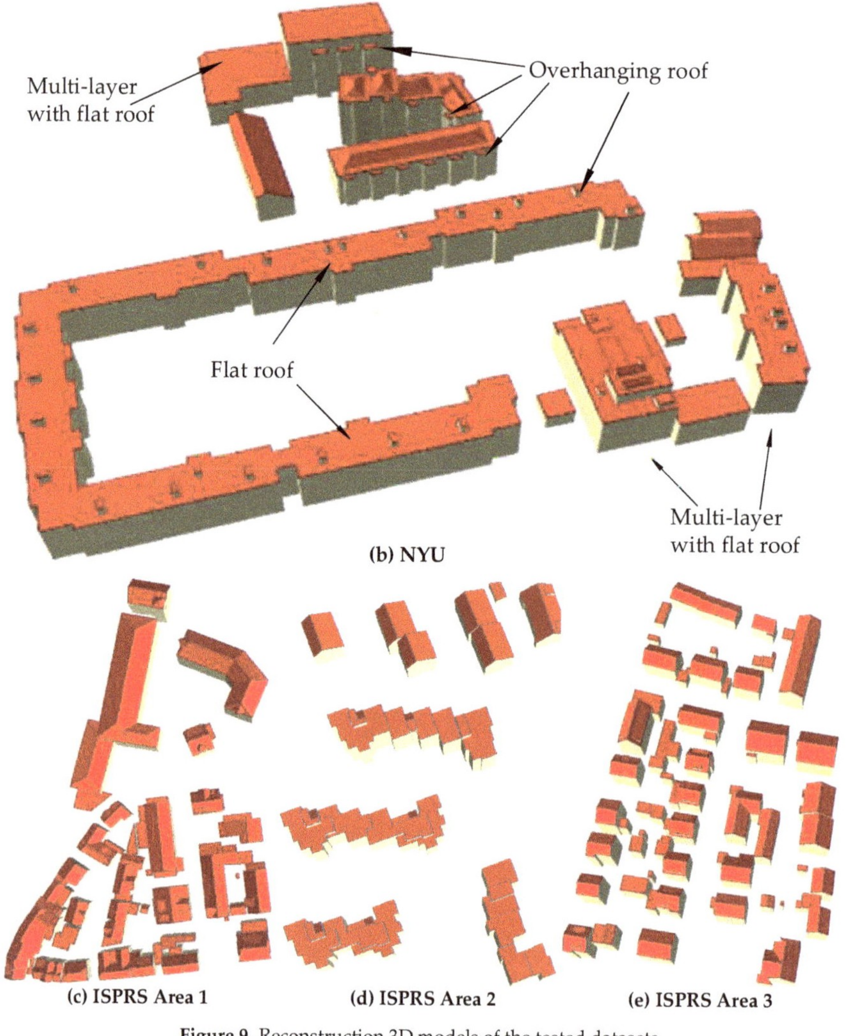

Figure 9. Reconstruction 3D models of the tested datasets.

Different from the ISPRS benchmark data (Area 1 and Area 3), the Guangdong data is a private testing data and public NYU is a non-standard dataset, thus, the various standard internal consistency metrics cannot be assessed by the ISPRS. In addition, the assessment on ISPRS Area 2 is missed as the ISPRS evaluation is stopped. Therefore, the evaluations of the three datasets are performed by a simple internal quality and a visual judgment. Results of visual judgment are shown in Figure 9a–c, while the internal quality are the reconstructed geometric reconstructed error and the rate of fully reconstructed buildings. The geometric reconstructed errors, a distance from a point to a reconstructed plane (average), are approximately 0.033 m (Guangdong), 0.021 m (NYU), and 0.2 m (ISPRS Area 2). In addition, a total of 77 buildings (252 roof planes) were successfully reconstructed, achieving a fully reconstructed 92.7% of the original 83 buildings for Guangdong data. These buildings are fully reconstructed in the public NYU and ISPRS Area 2 datasets, and the complex roof types in urban areas, like overhanging roof, multi-layer roofs, and flat roofs, are successfully modelled. The overhanging roof is usually a single plane in the constructed roof connection graph and can be easily grouped. While the complex multi-layer roofs can be reconstructed as a variety of different roof structures, and flat roofs with different height are fully modelled as different parts in the testing Areas. The facades in the NYU dataset are ignored in the current scheme.

Furthermore, for the ISPRS benchmark datasets of Vaihingen (Area 1 and 3), it allows us to use external reference data and assess the result according to unified criteria against other modeling methods [16]. The results are listed in Table 1.

Table 1. Quantitative assessment of plane extraction.

Items	Area 1	Area 3
Reconstructed planes	202	133
True Positive (TP)	201	130
False Positive (FP)	1	3
False Negative (FN)	22	34

It can be seen from Table 1 that the proposed method for building roof reconstruction has achieved 201 correctly reconstructed out of 202 in Area 1 (99.5% correctly reconstructed). While for Area 3, the number of correctly reconstructed planes is 130, reaching 97.7%. The most common reason for the false reconstructions (FP) is the lack of insufficient points.

5. Discussion

5.1. Visual Analysis of the Decomposition Results

In this section, the decomposed building subparts by the proposed approach, as shown in Figure 10, are compared with the commonly used building reconstruction methods. These different 3D models generated from the same compound building proves our novelty.

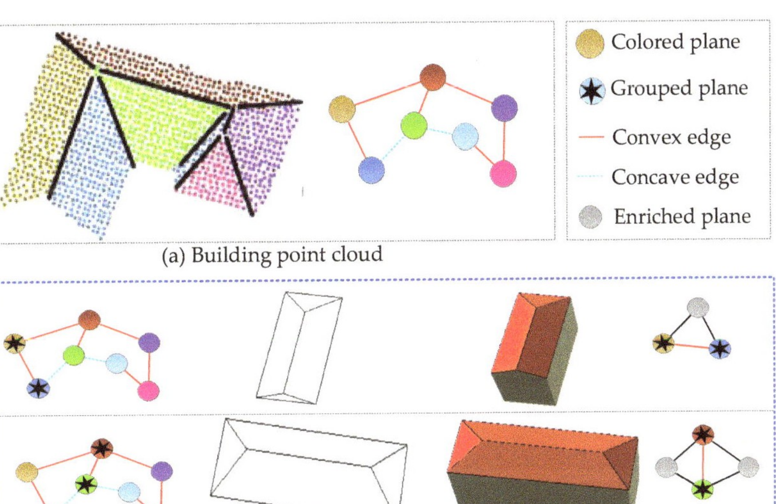

(a) Building point cloud

(b) The proposed modeling method

Figure 10. *Cont.*

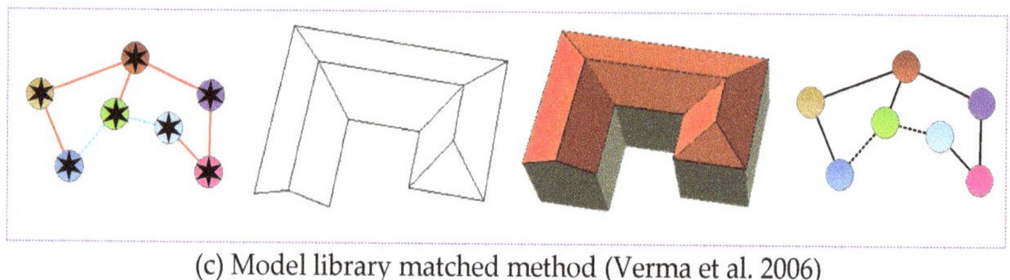

(c) Model library matched method (Verma et al. 2006)

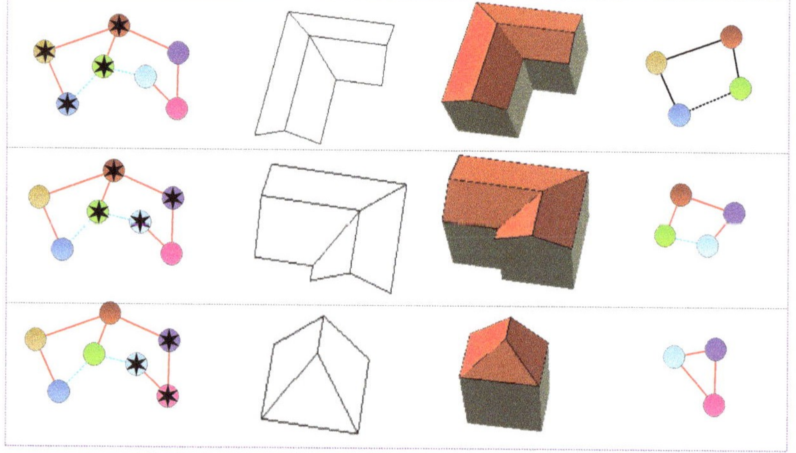

(d) Minimum cycles matching method (Xiong et at. 2014)

Figure 10. Decomposition results by different 3D building methods. (**a**) The original input point cloud and roof connection graph. (**b**) The results from the proposed approach. (**c**) A whole model by Verma et al. [51]. (**d**) Decomposition results by Xiong et al. [5]. The first to fourth columns of each row are grouped planes in the original roof connection graph, 3D wireframe, 3D model, and the final enriched subgraph of a grouped unit.

It can be seen from part (b) of Figure 10 that the proposed approach can achieve more unambiguous and meaningful building subparts. Verma et al. [51] used an exhaustive search to fit the point clouds to the best matched predefined simple GU, GL, and GI models, as shown in part (c). The final matched model is limited to a simple polygon model and cannot be used flexibly because it requires a more complex building library to be defined in advance. Xiong et al. [5] defined an improved roof topology graph to reconstruct 3D building models and, achieved the inner and outer corners from the concurrent planes or boundary points, thereby forming a combined building model linked to all inner and outer corners, as shown in part (d). It turns out to be more adaptive than similar work [77]. However, it is difficult to obtain the topological relationship of different corners and connected lines. The generated geometric models by minimum cycle analysis need to be checked because a corner may not be expressed or matched by the predefined minimum cycles. In addition, the semantics of the matched roof components are always omitted. Differing from the aforementioned approaches, the proposed automatic 3D modeling approach can reconstruct building semantic subparts, which can be easily interpreted as building structures. The decomposition results in Figure 10b are different hipped roofs, which can be easily inferred by the human being. Each roof subpart is a combination of parametric planar primitives and can be further used to assemble a hierarchy-tree of a building.

Different from the previous approach [48], we have made improvements in the current status to generate structural building models with the introduced semantics. The plane semantics are added for the roof connection graph, semantic decomposition, and roof subparts refinement to generate 3D building models; it is more helpful to reconstruct structural building subparts. The decomposed results can be interpreted as different semantic structures, where the semantics can be inferred from the largest number of semantic planes. By introducing the semantics into the iterative decomposition and grouping algorithm, it can be easily extended to house modeling from a multi-source point cloud, e.g., the ground-based point cloud can provide more details of building façades, thus, we can reconstruct more refined house models in LOD3, LOD4 by combining these different points cloud. A visual comparison on the proposed and the previous approach is shown in Figure 11, and the difference and improvement are that whether they can interpret the grouped roof subparts as different semantic structures.

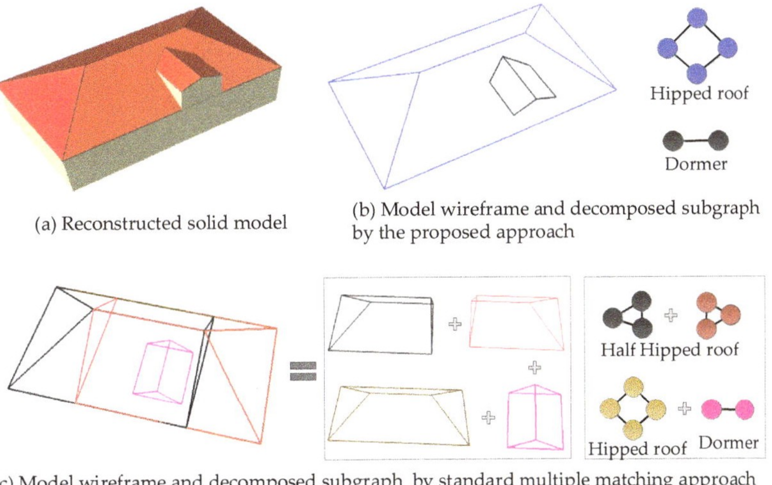

(a) 3D building model

(b) Grouped roof subparts **with** semantics by the proposed approach

(c) Grouped roof subparts **without** semantics by the previous approach

Figure 11. Comparison of experimental results (with vs. without semantic structures).

Furthermore, the decomposed roof subparts we proposed are basic building units without overlapping elements in the reconstruction process and can produce more unambiguous 3D models, as presented in Figure 12.

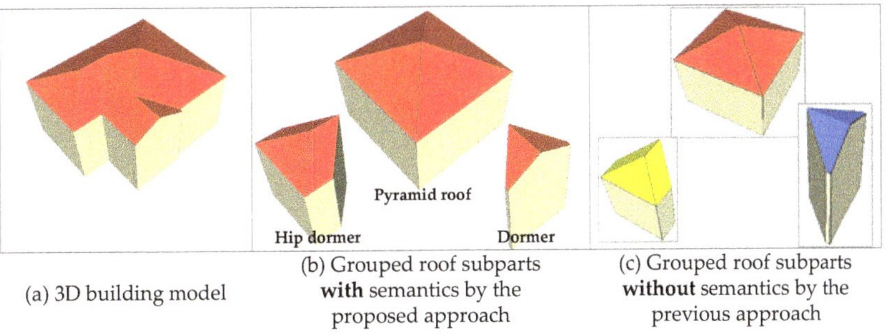

(a) Reconstructed solid model

(b) Model wireframe and decomposed subgraph by the proposed approach

(c) Model wireframe and decomposed subgraph by standard multiple matching approach

Figure 12. Comparison between standard matching approach (**b**) and the proposed unambiguous decomposition (**c**) for the same building (**a**).

It can be seen from Figure 12b that the generated 3D building subparts by the proposed approach can avoid multiple matching of the same building element. For these standard matched methods [5,77], the roof topology graph defined by different forms is searched from the predefined library and decomposed into elementary graphs. These automatically recognized subgraphs enable us to assign semantics to all extracted building planar primitives and assemble them into an appropriate 3D model. However, there will inevitably be a problem, that is, the standard-matching reconstruction methods rebuild the same roof elements repeatedly, resulting in overlapping roof primitives, as shown in the mid of Figure 12c.

5.2. Performance Analysis of Multi-Label Energy Optimization

To further investigate the effects of the global energy-based optimization procedure, we have calculated the iterations and runtimes, as shown in Table 2.

Table 2. Statistics of experiments on energy optimization.

Item	Building Points	Iterations	Run Times (min) from the Proposed	Run Times (min) from the RANSAC
ISPRS Area 1	24,971	4	1.7	0.9
ISPRS Area 2	34,364	3	2.2	1.1
ISPRS Area 3	39,938	4	2.6	1.2
Guangdong	76,824	5	4.3	2.1
NYU	111,289	3	5.7	3.9

It can be found from Table 2 that the number of iterations is located in a lower range, which means that the designed cost functions are stable and balanced. Along with the iteration, the energy will be sharply dropped, leading to a quick convergence. Compared with the RANSAC plane fitting, the running time is relatively long as the optimization is performed on each point. One possible improvement is to change the assignment issue from "point-to-plane" to "supervoxel-to-plane". Moreover, the results of roof plane extraction in Area 1 and Area 3 are compared to a traditional multi-model fitting method like RANSAC, as shown in Figure 13.

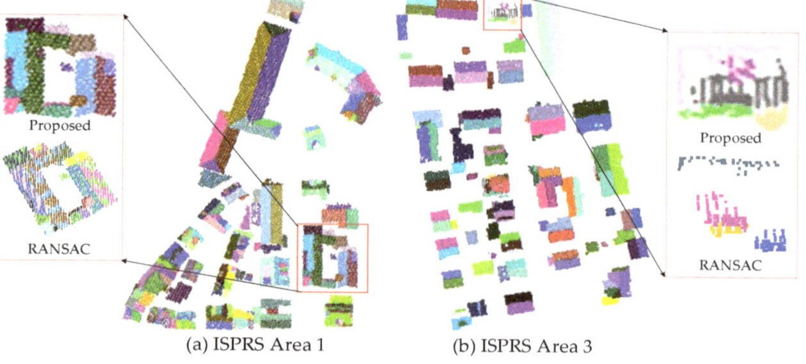

Figure 13. Results of roof plane extraction in comparison.

It can be seen from the Figure 13 that the global energy-optimized approach can be effective to extract roof planes. Compared with a traditional multi-model fitting method like RANSAC, the proposed approach can overcome inconsistencies such as noise and missing data in plane transitions and is more beneficial to construct the adjacent relationship between roof planes.

5.3. Accuracy Assessments on ISPRS Benchmark Dataset

The geometric accuracy of the reconstructed 3D models derived from the benchmark data of Vaihingen in Areas 1 and 3 are evaluated by ISPRS using the standardized validation methods. The metrics of completeness, correctness, and quality, defined by

Rutzinger et al. [78] are evaluated based on the mutual overlapping with reference data. The quality results are shown in Figure 14.

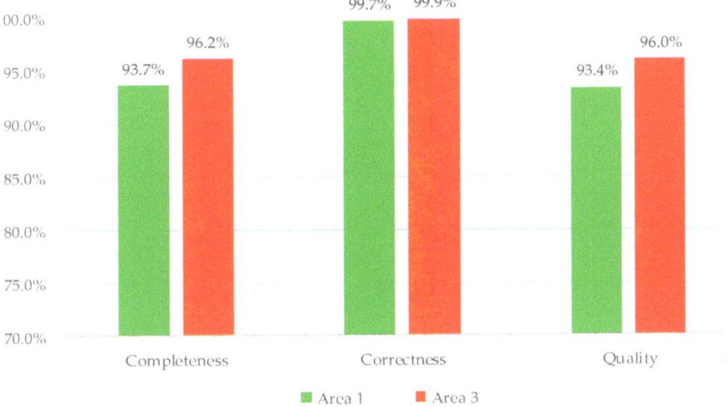

Figure 14. The quality metrics of the ISPRS benchmark dataset.

The quality of per-area level for Area 3 reaches 96.0%, while Area 1 large than 93.4%. These high precision results can be benefited from the proposed global optimization, which can preserve the correct segmentation at plane transition regions with sparse points. In addition, the comparison of the reconstructed planes with the reference information is illustrated in Figure 15, where the 3D information has been converted to a label image by ISPRS.

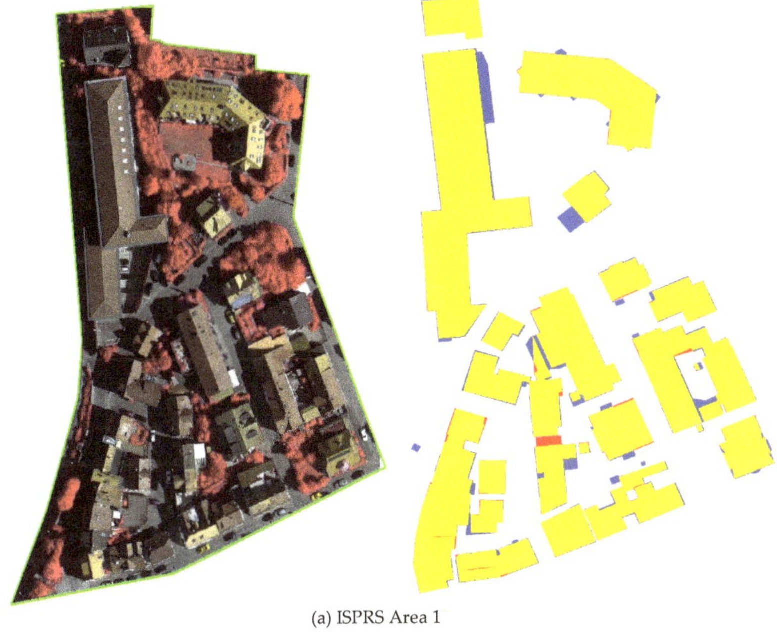

(a) ISPRS Area 1

Figure 15. *Cont.*

(b) ISPRS Area 3

Figure 15. Evaluation results of the reconstructed roof plane. Blue: false-negative pixels (FN), yellow: true-positive pixels (TP), red: false-positive pixels (FP).

The comparative verification (Figure 15) indicates that 3D building components are successfully achieved during the reconstruction process. Buildings that are not correctly reconstructed are one and three for Area 1 and 3, respectively. These failures, assessed as False Positive (FP), are filled within the buildings that undetected in the preprocessing process of point cloud classification, because houses are surrounded by trees, which makes it difficult to extract building point cloud. Moreover, the geometrical accuracy of state-of-the-art methods described on the ISPRS website is selected for comparison, as presented in Figure 16.

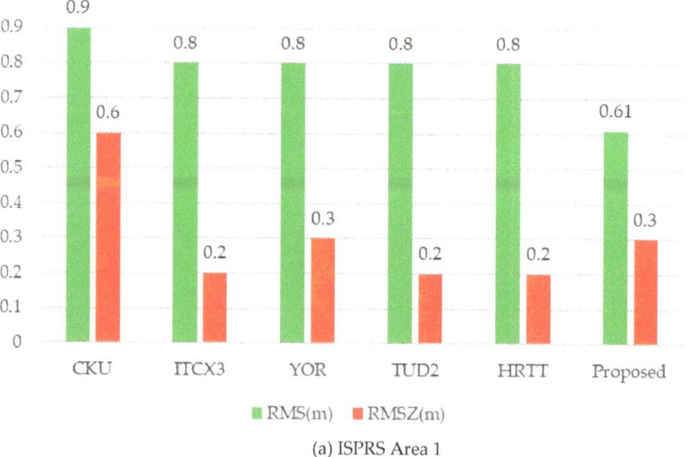

(a) ISPRS Area 1

Figure 16. *Cont.*

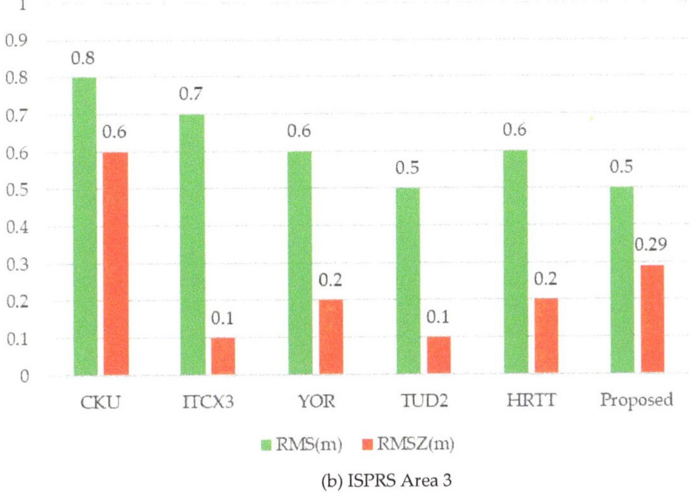

Figure 16. Geometrical accuracy comparison of the reconstructed models. The CKU [79], ITCX3 [5], YOR [80], TUD2 [50], and HRTT [54] are shorts for participant.

The accuracy of the final generated models may be affected by a variety of factors, such as building detection and segmentation, the strategy of reconstruction. Additionally, an excellent reconstruction algorithm is to find a balance to generate building models. From most of the evaluation methods presented in Figure 16, one of the two indicators (RMS and RMSZ) exceeds the median value, while the other is obviously reduced. In terms of quantitative results, none of the methods is significantly better than others. For the proposed reconstructed approach, the average horizontal error is 0.8 m (Area 1) and 0.6 m (Area 3), while the vertical error is 0.3 m (Area 1) and 0.29 m (Area 3). Even though the two metrics of the proposed method are not optimal, they are similar to the median value, which means that we have achieved a balance between the reconstructed RMS and RMSZ. The main reason for achieving the balance depends on the global optimization of roof plane extraction, and more importantly, it is easy to obtain an unambiguous principal direction of regularization from each decomposed building subpart.

6. Conclusions

In this paper, we present a novel method for complex building reconstruction from 3D point clouds using the local geometric constraints. The output of the reconstruction is a combination of unambiguous unit blocks with no overlapping elements, which are assembled in a hierarchical topology tree. By first constructing a roof connection graph using the extracted roof planar primitives, we developed semantic-specific reconstruction strategies with local geometric constraints to obtain visually attractive building models. The key aim is to decompose a compound building model into semantic subparts with fixed planar parameters and topological relationships, through a progressive hierarchical grouping operation.

The performed reconstruction experiments indicate that the proposed approach can simplify the reconstruction process and generate a combination of gabled or hipped roofs with precisely reconstructed geometric features. Moreover, these generated building subparts can be further used to enrich the building of a model library or construct public training data for supervised reconstruction. However, the proposed modeling scheme for building reconstruction has some limitations, leading to the failure of the generated 3D models. These limitations include the lack of adjacent roof segments, sparse points for the local symmetry processing, and the reconstruction of free-from objects. For future work, there are some possible improvements. For example, higher density and quality points obtained from the ubiquitous digital cameras and active 3D sensing devices can be used to avoid the former two issues, while new reconstruction strategies for free-from objects and building structural elements identification need to be developed.

Author Contributions: Conceptualization, P.H. Methodology, P.H.; software, P.H. and Y.M.; formal analysis, P.H.; writing—original draft preparation, P.H. and Y.M.; writing—review and editing, P.H.; visualization, P.H., M.H., and Y.M.; project administration, P.H.; funding acquisition, P.H. All authors have read and agreed to the published version of the manuscript.

Funding: This work is jointly supported by the National Natural Science Foundation of China (No. 41901405), China Postdoctoral Science Foundation (No. 2020M680323), Beijing Advanced Innovation Center for Future Urban Design, Beijing University of Civil Engineering and Architecture (No. UDC2020020124), and The Fundamental Research Funds for Beijing University of Civil Engineering and Architecture (No. X20045).

Acknowledgments: We would first like to thank the anonymous reviewers for their valuable feedback. The Vaihingen data set was provided by the German Society for Photogrammetry, Remote Sensing, and Geoinformation (DGPF), and The NYU ALS point cloud is released by released by New York University, the center for urban science and progress. The authors wish to thank Markus Gerke, Uwe Breitkopf, and the ISPRS Commission III/4 for the evaluation of the results.

Conflicts of Interest: The authors declare no conflict of interest.

References

1. Fuxun, L.; Bisheng, Y.; Ronggang, H.; Zhen, D.; Jianping, L. Facade Solar Potential Analysis Using Multisource Point Cloud. *Acta Geod. Cartogr. Sin.* **2018**, *47*, 225–233. (In Chinese)
2. Döllner, J.; Baumann, K.; Buchholz, H. Virtual 3D City Models as Foundation of Complex Urban Information Spaces. In Proceedings of the 11th International Conference on Urban Planning and Spatial Development in the Information Society, Vienna, Austria, 13–16 February 2006.
3. Qing, Z.; Haowei, Z.; Yulin, D.; Xiao, X.; Fei, L.; Liguo, Z.; Haifeng, L.; Han, H.; Junxiao, Z.; Li, C.; et al. A review of major potential landslide hazards analysis. *Acta Geod. Cartogr. Sin.* **2019**, *48*, 1551–1561. (In Chinese)
4. Nan, L.; Sharf, A.; Zhang, H.; Cohen-Or, D.; Chen, B. SmartBoxes for interactive urban reconstruction. *ACM Trans. Graph.* **2010**, *29*, 1–10. [CrossRef]
5. Xiong, B.; Oude Elberink, S.; Vosselman, G. A graph edit dictionary for correcting errors in roof topology graphs reconstructed from point clouds. *ISPRS J. Photogramm. Remote Sens.* **2014**, *93*, 227–242. [CrossRef]
6. Lafarge, F.; Mallet, C. Building large urban environments from unstructured point data. In Proceedings of the 2011 International Conference on Computer Vision, Barcelona, Spain, 6–13 November 2011; pp. 1068–1075.
7. Bulatov, D.; Häufel, G.; Meidow, J.; Pohl, M.; Solbrig, P.; Wernerus, P. Context-based automatic reconstruction and texturing of 3D urban terrain for quick-response tasks. *ISPRS J. Photogramm. Remote Sens.* **2014**, *93*, 157–170. [CrossRef]
8. Zhu, Z.; Stamatopoulos, C.; Fraser, C.S. Accurate and occlusion-robust multi-view stereo. *ISPRS J. Photogramm. Remote Sens.* **2015**, *109*, 47–61. [CrossRef]
9. Toschi, I.; Nocerino, E.; Remondino, F.; Revolti, A.; Soria, G.; Piffer, S. Geospatial Data Processing for 3d City Model Generation, Management and Visualization. *Int. Arch. Photogramm. Remote Sens. Spatial Inf. Sci.* **2017**, *XLII-1/W1*, 527–534. [CrossRef]
10. Rychard, M.; Borkowski, A. 3D building reconstruction from ALS data using unambiguous decomposition into elementary structures. *ISPRS J. Photogramm. Remote Sens.* **2016**, *118*, 1–12. [CrossRef]
11. Aringer, K.; Roschlaub, R. Bavarian 3D Building Model and Update Concept Based on LiDAR, Image Matching and Cadastre Information. In *Innovations in 3D Geo-Information Sciences;* Isikdag, U., Ed.; Springer International Publishing: Cham, Switzerland, 2014; pp. 143–157. [CrossRef]
12. Jarząbek-Rychard, M.; Maas, H.-G. Geometric Refinement of ALS-Data Derived Building Models Using Monoscopic Aerial Images. *Remote Sens.* **2017**, *9*, 282. [CrossRef]
13. Lin, H.; Gao, J.; Zhou, Y.; Lu, G.; Ye, M.; Zhang, C.; Liu, L.; Yang, R. Semantic Decomposition and Reconstruction of Residential Scenes from LiDAR Data. *ACM Trans. Graph.* **2013**, *32*, 1–10. [CrossRef]
14. Verdie, Y.; Lafarge, F.; Alliez, P. LOD Generation for Urban Scenes. *ACM Trans. Graph.* **2015**, *34*, 1–14. [CrossRef]
15. Oesau, S.; Lafarge, F.; Alliez, P. Planar Shape Detection and Regularization in Tandem. *Comput. Graph. Forum* **2016**, *35*, 203–215. [CrossRef]
16. Rottensteiner, F.; Sohn, G.; Gerke, M.; Wegner, J.D.; Breitkopf, U.; Jung, J. Results of the ISPRS benchmark on urban object detection and 3D building reconstruction. *ISPRS J. Photogramm. Remote Sens.* **2014**, *93*, 256–271. [CrossRef]
17. Musialski, P.; Wonka, P.; Aliaga, D.G.; Wimmer, M.; Van Gool, L.; Purgathofer, W. A survey of urban reconstruction. *Comput. Graph. Forum* **2013**, *32*, 146–177. [CrossRef]

18. Haala, N.; Kada, M. An update on automatic 3D building reconstruction. *ISPRS J. Photogramm. Remote Sens.* **2010**, *65*, 570–580. [CrossRef]
19. Wang, R.; Peethambaran, J.; Chen, D. LiDAR Point Clouds to 3D Urban Models: A Review. *IEEE J. Sel. Top. Appl. Earth Obs. Remote Sens.* **2018**, *11*, 606–627. [CrossRef]
20. Yang, B.; Dong, Z. A shape-based segmentation method for mobile laser scanning point clouds. *ISPRS J. Photogramm. Remote Sens.* **2013**, *81*, 19–30. [CrossRef]
21. Vo, A.-V.; Truong-Hong, L.; Laefer, D.F.; Bertolotto, M. Octree-based region growing for point cloud segmentation. *ISPRS J. Photogramm. Remote Sens.* **2015**, *104*, 88–100. [CrossRef]
22. Zhou, G.; Cao, S.; Zhou, J. Planar Segmentation Using Range Images From Terrestrial Laser Scanning. *IEEE Geosci. Remote Sens. Lett.* **2016**, *13*, 257–261. [CrossRef]
23. Kim, C.; Habib, A.; Pyeon, M.; Kwon, G.-R.; Jung, J.; Heo, J. Segmentation of Planar Surfaces from Laser Scanning Data Using the Magnitude of Normal Position Vector for Adaptive Neighborhoods. *Sensors* **2016**, *16*, 140. [CrossRef]
24. Vosselman, G.; Gorte, B.G.; Sithole, G.; Rabbani, T. Recognising structure in laser scanner point clouds. *Int. Arch. Photogramm. Remote Sens. Spat. Inf. Sci.* **2004**, *46*, 33–38.
25. Schnabel, R.; Wahl, R.; Klein, R. Efficient RANSAC for point-cloud shape detection. *Comput. Graph. Forum* **2007**, *26*, 214–226. [CrossRef]
26. Yan, J.; Shan, J.; Jiang, W. A global optimization approach to roof segmentation from airborne lidar point clouds. *ISPRS J. Photogramm. Remote Sens.* **2014**, *94*, 183–193. [CrossRef]
27. Dong, Z.; Yang, B.; Hu, P.; Scherer, S. An efficient global energy optimization approach for robust 3D plane segmentation of point clouds. *ISPRS J. Photogramm. Remote Sens.* **2018**, *137*, 112–133. [CrossRef]
28. Pham, T.T.; Eich, M.; Reid, I.; Wyeth, G. Geometrically consistent plane extraction for dense indoor 3D maps segmentation. In Proceedings of the 2016 IEEE/RSJ International Conference on Intelligent Robots and Systems (IROS), Daejeon, Korea, 9–14 October 2016; pp. 4199–4204.
29. Chen, D.; Zhang, L.; Mathiopoulos, P.T.; Huang, X. A Methodology for Automated Segmentation and Reconstruction of Urban 3-D Buildings from ALS Point Clouds. *IEEE J. Sel. Top. Appl. Earth Obs. Remote Sens.* **2014**, *7*, 4199–4217. [CrossRef]
30. Chen, J.; Chen, B. Architectural Modeling from Sparsely Scanned Range Data. *Int. J. Comput. Vis.* **2008**, *78*, 223–236. [CrossRef]
31. Sampath, A.; Shan, J. Segmentation and Reconstruction of Polyhedral Building Roofs From Aerial Lidar Point Clouds. *IEEE Trans. Geosci. Remote Sens.* **2010**, *48*, 1554–1567. [CrossRef]
32. Zhou, Q.; Neumann, U. 2.5D building modeling by discovering global regularities. In Proceedings of the 2012 IEEE Conference on Computer Vision and Pattern Recognition, Providence, RI, USA, 16–21 June 2012; pp. 326–333.
33. Poullis, C. A Framework for Automatic Modeling from Point Cloud Data. *IEEE Trans. Pattern Anal. Mach. Intell.* **2013**, *35*, 2563–2575. [CrossRef] [PubMed]
34. Dehbi, Y.; Henn, A.; Gröger, G.; Stroh, V.; Plümer, L. Robust and fast reconstruction of complex roofs with active sampling from 3D point clouds. *Trans. GIS* **2020**, *25*, 112–133. [CrossRef]
35. Karantzalos, K.; Paragios, N. Large-Scale Building Reconstruction Through Information Fusion and 3-D Priors. *IEEE Trans. Geosci. Remote Sens.* **2010**, *48*, 2283–2296. [CrossRef]
36. Huang, X. Building reconstruction from airborne laser scanning data. *Geo-Spat. Inf. Sci.* **2013**, *16*, 35–44. [CrossRef]
37. Henn, A.; Gröger, G.; Stroh, V.; Plümer, L. Model driven reconstruction of roofs from sparse LIDAR point clouds. *ISPRS J. Photogramm. Remote Sens.* **2013**, *76*, 17–29. [CrossRef]
38. Tseng, Y.-H.; Wang, S. Semi-automated Building Extraction Based on CSG Model-Image Fitting. *Photogramm. Eng. Remote Sens.* **2003**, *69*, 171–180. [CrossRef]
39. Fayez, T.-K.; Tania, L.; Pierre, G. Extended RANSAC algorithm for automatic detection of building roof planes from LiDAR data. *Photogramm. J. Finl.* **2008**, *21*, 97–109. Available online: https://halshs.archives-ouvertes.fr/halshs-00278397 (accessed on 2 May 2021).
40. Kada, M.; McKinley, L. 3D building reconstruction from LiDAR based on a cell decomposition approach. *Int. Arch. Photogramm. Remote Sens. Spat. Inf. Sci.* **2009**, *38*, W4.
41. Lafarge, F.; Descombes, X.; Zerubia, J.; Pierrot-Deseilligny, M. Automatic building extraction from DEMs using an object approach and application to the 3D-city modeling. *ISPRS J. Photogramm. Remote Sens.* **2008**, *63*, 365–381. [CrossRef]
42. Poullis, C.; You, S. Automatic reconstruction of cities from remote sensor data. In Proceedings of the 2009 IEEE Conference on Computer Vision and Pattern Recognition, Miami, FL, USA, 20–25 June 2009; pp. 2775–2782.
43. Brenner, C. Modelling 3d Objects Using Weak Csg Primitives. *Int. Arch. Photogramm. Remote Sens. Spat. Inf. Sci.* **2004**, *35*, 1085–1090.
44. Vosselman, G.; Dijkman, S. 3D building model reconstruction from point clouds and ground plans. *Int. Arch. Photogramm. Remote Sens. Spat. Inf. Sci.* **2001**, *34*, 37–44.
45. Kada, M.; Wichmann, A. Feature-Driven 3D Building Modeling using Planar Halfspaces. *ISPRS Ann. Photogramm. Remote Sens. Spat. Inf. Sci.* **2013**, *II-3/W3*, 37–42. [CrossRef]
46. Wang, H.; Zhang, W.; Chen, Y.; Chen, M.; Yan, K. Semantic Decomposition and Reconstruction of Compound Buildings with Symmetric Roofs from LiDAR Data and Aerial Imagery. *Remote Sens.* **2015**, *7*, 13945–13974. [CrossRef]

47. Suveg, I.; Vosselman, G. Reconstruction of 3D building models from aerial images and maps. *ISPRS J. Photogramm. Remote Sens.* **2004**, *58*, 202–224. [CrossRef]
48. Hu, P.; Yang, B.; Dong, Z.; Yuan, P.; Huang, R.; Fan, H.; Sun, X. Towards Reconstructing 3D Buildings from ALS Data Based on Gestalt Laws. *Remote Sens.* **2018**, *10*, 1127. [CrossRef]
49. Bauchet, J.P.; Lafarge, F. City Reconstruction from Airborne Lidar: A Computational Geometry Approach. *ISPRS Ann. Photogramm. Remote Sens. Spat. Inf. Sci.* **2019**, *IV-4/W8*, 19–26. [CrossRef]
50. Perera, G.S.N.; Maas, H.-G. Cycle graph analysis for 3D roof structure modelling: Concepts and performance. *ISPRS J. Photogramm. Remote Sens.* **2014**, *93*, 213–226. [CrossRef]
51. Verma, V.; Kumar, R.; Hsu, S. 3D Building Detection and Modeling from Aerial LIDAR Data. In Proceedings of the IEEE Computer Society Conference on Computer Vision and Pattern Recognition, New York, NY, USA, 17–22 June 2006; pp. 2213–2220.
52. Elberink, S.O.; Vosselman, G. Quality analysis on 3D building models reconstructed from airborne laser scanning data. *ISPRS J. Photogramm. Remote Sens.* **2011**, *66*, 157–165. [CrossRef]
53. Xiong, B.; Jancosek, M.; Oude Elberink, S.; Vosselman, G. Flexible building primitives for 3D building modeling. *ISPRS J. Photogramm. Remote Sens.* **2015**, *101*, 275–290. [CrossRef]
54. Xu, B.; Jiang, W.; Li, L. HRTT: A Hierarchical Roof Topology Structure for Robust Building Roof Reconstruction from Point Clouds. *Remote Sens.* **2017**, *9*, 354. [CrossRef]
55. Bassier, M.; Vergauwen, M.; Van Genechten, B. Automated classification of heritage buildings for as-built BIM using machine learning techniques. *ISPRS Ann. Photogramm. Remote Sens. Spat. Inf. Sci.* **2017**, *4*, 25–30. [CrossRef]
56. Ma, L.; Sacks, R.; Kattel, U.; Bloch, T. 3D object classification using geometric features and pairwise relationships. *Comput. Aided Civ. Infrastruct. Eng.* **2018**, *33*, 152–164. [CrossRef]
57. Wichmann, A.; Agoub, A.; Kada, M. RoofN3D: Deep Learning Training Data for 3D Building Reconstruction. *Int. Arch. Photogramm. Remote Sens. Spat. Inf. Sci.* **2018**, *42*, 1191–1198. [CrossRef]
58. Axelsson, M.; Soderman, U.; Berg, A.; Lithen, T. Roof Type Classification Using Deep Convolutional Neural Networks on Low Resolution Photogrammetric Point Clouds From Aerial Imagery. In Proceedings of the 2018 IEEE International Conference on Acoustics, Speech and Signal Processing (ICASSP), Calgary, AB, Canada, 15–20 April 2018; pp. 1293–1297.
59. Zhang, L.; Zhang, L. Deep Learning-Based Classification and Reconstruction of Residential Scenes From Large-Scale Point Clouds. *IEEE Trans. Geosci. Remote Sens.* **2018**, *56*, 1887–1897. [CrossRef]
60. Yu, D.; Ji, S.; Liu, J.; Wei, S. Automatic 3D building reconstruction from multi-view aerial images with deep learning. *ISPRS J. Photogramm. Remote Sens.* **2021**, *171*, 155–170. [CrossRef]
61. Axelsson, P. DEM Generation from Laser Scanner Data Using adaptive TIN Models. *Int. Arch. Photogramm. Remote Sens.* **2006**, *60*, 71–80. [CrossRef]
62. Zhang, W.; Qi, J.; Wan, P.; Wang, H.; Xie, D.; Wang, X.; Yan, G. An Easy-to-Use Airborne LiDAR Data Filtering Method Based on Cloth Simulation. *Remote Sens.* **2016**, *8*, 501. [CrossRef]
63. Yang, B.; Huang, R.; Dong, Z.; Zang, Y.; Li, J. Two-step adaptive extraction method for ground points and breaklines from lidar point clouds. *ISPRS J. Photogramm. Remote Sens.* **2016**, *119*, 373–389. [CrossRef]
64. Huang, R.; Yang, B.; Liang, F.; Dai, W.; Li, J.; Tian, M.; Xu, W. A top-down strategy for buildings extraction from complex urban scenes using airborne LiDAR point clouds. *Infrared Phys. Technol.* **2018**, *92*, 203–218. [CrossRef]
65. Boykov, Y.; Kolmogorov, V. An Experimental Comparison of Min-Cut/Max-Flow Algorithms for Energy Minimization in Vision. *IEEE Trans. Pattern Anal. Mach. Intell.* **2004**, *26*, 1124–1137. [CrossRef]
66. Rusu, R.B.; Cousins, S. 3D is here: Point Cloud Library (PCL). In Proceedings of the 2011 IEEE International Conference on Robotics and Automation, Shanghai, China, 9–13 May 2011; pp. 1–4.
67. Delong, A.; Osokin, A.; Isack, H.N.; Boykov, Y. Fast Approximate Energy Minimization with Label Costs. *Int. J. Comput. Vis.* **2012**, *96*, 1–27. [CrossRef]
68. Isack, H.; Boykov, Y. Energy-Based Geometric Multi-model Fitting. *Int. J. Comput. Vis.* **2012**, *97*, 123–147. [CrossRef]
69. Pu, S.; Vosselman, G. Knowledge based reconstruction of building models from terrestrial laser scanning data. *ISPRS J. Photogramm. Remote Sens.* **2009**, *64*, 575–584. [CrossRef]
70. Vosselman, G.; Coenen, M.; Rottensteiner, F. Contextual segment-based classification of airborne laser scanner data. *ISPRS J. Photogramm. Remote Sens.* **2017**, *128*, 354–371. [CrossRef]
71. Desolneux, A.; Moisan, L.; Morel, J.-M. Gestalt theory and computer vision. In *Seeing, Thinking and Knowing*; Springer: Berlin/Heidelberg, Germany, 2004; pp. 71–101.
72. Nan, L.; Sharf, A.; Xie, K.; Wong, T.-T.; Deussen, O.; Cohen-Or, D.; Chen, B. Conjoining Gestalt rules for abstraction of architectural drawings. *ACM Trans. Graph.* **2011**, *30*, 1–10. [CrossRef]
73. Lun, Z.; Zou, C.; Huang, H.; Kalogerakis, E.; Tan, P.; Cani, M.-P.; Zhang, H. Learning to group discrete graphical patterns. *ACM Trans. Graph. (TOG)* **2017**, *36*, 1–11. [CrossRef]
74. Yang, B.; Huang, R.; Li, J.; Tian, M.; Dai, W.; Zhong, R. Automated Reconstruction of Building LoDs from Airborne LiDAR Point Clouds Using an Improved Morphological Scale Space. *Remote Sens.* **2017**, *9*, 14. [CrossRef]
75. People, C. The Computational Geometry Algorithms Library (CGAL). Available online: https://www.cgal.org/ (accessed on 4 May 2021).

76. Laefer, D.F.; Abuwarda, S.; Vo, A.-V.; Truong-Hong, L.; Gharibi, H. 2015 Dublin LiDAR and NYU Research Data. Available online: https://geo.nyu.edu/ (accessed on 8 May 2021).
77. Elberink, S.O.; Vosselman, G. Building Reconstruction by Target Based Graph Matching on Incomplete Laser Data: Analysis and Limitations. *Sensors* **2009**, *9*, 6101–6118. [CrossRef] [PubMed]
78. Rutzinger, M.; Rottensteiner, F.; Pfeifer, N. A Comparison of Evaluation Techniques for Building Extraction From Airborne Laser Scanning. *IEEE J. Sel. Top. Appl. Earth Obs. Remote Sens.* **2009**, *2*, 11–20. [CrossRef]
79. Rau, J.Y. A line-based 3d roof model reconstruction algorithm: Tin-merging and reshaping (tmr). *ISPRS Ann. Photogramm. Remote Sens. Spat. Inf. Sci.* **2012**, *I-3*, 287–292. [CrossRef]
80. Sohn, G.; Jwa, Y.; Jung, J.; Kim, H. An implicit regularization for 3D building rooftop modeling using airborne lidar data. *ISPRS Ann. Photogramm. Remote Sens. Spat. Inf. Sci.* **2012**, *1*, 305–310. [CrossRef]

remote sensing

Article

An Accurate Digital Subsidence Model for Deformation Detection of Coal Mining Areas Using a UAV-Based LiDAR

Junliang Zheng, Wanqiang Yao *, Xiaohu Lin, Bolin Ma and Lingxiao Bai

College of Geomatics, Xi'an University of Science and Technology, Xi'an 710054, China; jlzheng@xust.edu.cn (J.Z.); xhlin214@xust.edu.cn (X.L.); 004844@xust.edu.cn (B.M.); 19210210050@stu.xust.edu.cn (L.B.)
* Correspondence: sxywq@xust.edu.cn

Abstract: Coal mine surface subsidence detection determines the damage degree of coal mining, which is of great importance for the mitigation of hazards and property loss. Therefore, it is very important to detect deformation during coal mining. Currently, there are many methods used to detect deformations in coal mining areas. However, with most of them, the accuracy is difficult to guarantee in mountainous areas, especially for shallow seam mining, which has the characteristics of active, rapid, and high-intensity surface subsidence. In response to these problems, we made a digital subsidence model (DSuM) for deformation detection in coal mining areas based on airborne light detection and ranging (LiDAR). First, the entire point cloud of the study area was obtained by coarse to fine registration. Second, noise points were removed by multi-scale morphological filtering, and the progressive triangulation filtering classification (PTFC) algorithm was used to obtain the ground point cloud. Third, the DEM was generated from the clean ground point cloud, and an accurate DSuM was obtained through multiple periods of DEM difference calculations. Then, data mining was conducted based on the DSuM to obtain parameters such as the maximum surface subsidence value, a subsidence contour map, the subsidence area, and the subsidence boundary angle. Finally, the accuracy of the DSuM was analyzed through a comparison with ground checkpoints (GCPs). The results show that the proposed method can achieve centimeter-level accuracy, which makes the data a good reference for mining safety considerations and subsequent restoration of the ecological environment.

Keywords: airborne LiDAR; coal mine; surface subsidence; deformation detection; digital subsidence model

Citation: Zheng, J.; Yao, W.; Lin, X.; Ma, B.; Bai, L. An Accurate Digital Subsidence Model for Deformation Detection of Coal Mining Areas Using a UAV-Based LiDAR. *Remote Sens.* **2022**, *14*, 421. https://doi.org/10.3390/rs14020421

Academic Editors: Wanshou Jiang, San Jiang and Xiongwu Xiao

Received: 30 November 2021
Accepted: 14 January 2022
Published: 17 January 2022

Publisher's Note: MDPI stays neutral with regard to jurisdictional claims in published maps and institutional affiliations.

1. Introduction

Deformation detection has been defined as the identification of geometric state differences based on multiple periods of data capturing. The detection of the surface subsidence of coal mining areas is a part of deformation detection and has become a hot topic to mitigate hazards and property loss.

Generally, the surface subsidence caused by coal mining is the main source of danger for the destruction of buildings and structures, inevitably causing surface collapse and environmental damage [1–3]. Therefore, scholars have adopted various methods to observe the surface subsidence of coal mining areas. The traditional geodetic method detects surface subsidence with fixed points on the ground [4,5]. Although high-precision data can be obtained, it is still point-to-point acquisition, which is inefficient and expensive. Moreover, this method only measures local subsidence, and full coverage of a mining area cannot be obtained. Recently, Shi et al. [6–9] tried to detect surface deformation with the method of interferometry synthetic aperture radar (InSAR), which can obtain accurate vertical displacement measurements. However, the speed of the surface subsidence of a shallow coal seam is relatively fast, and a long period of SAR satellite observation easily causes incoherence of SAR images [8,9]. UAV oblique photogrammetry can obtain the three-dimensional (3D) coordinate information of ground features [10–14]. Many scholars have tried to use oblique photogrammetry to detect the surface subsidence of mining areas [15–18]. However, the 3D point cloud generated by UAV oblique images includes a large number of vegetation points [19], which leads to limited accuracy without control points. Martínez-Carricondo et al. [20] improved the accuracy of UAV oblique photogrammetry with high-density control points. However, there are landslides, ground fissures, and other hazards in mining areas, which cannot allow high-density control points. Terrestrial laser scanning (TLS) has been used to detect the deformation of landslide, dam, and mining areas, and accurate detection results have been achieved [21–24]. However, TLS adopts station-type scanning, with which there are problems, such as ground obstructions, narrow fields of view, heavy workloads, and special terrain not being scannable. Airborne LiDAR can capture large-scale, dense 3D point clouds [25]. It can be divided into manned and unmanned airborne LiDAR. Yu and AO [26–28] tried to apply manned airborne LiDAR to the surface deformation monitoring of a large mining area, which can achieve comprehensive observations without being restricted by terrain. Moreover, ground point clouds in the presence of vegetation can be obtained through vegetation gaps. However, manned airborne LiDAR data collection requires a lot of manpower and material resources, making it unsuitable for small-scale observations. Making airborne LiDAR unmanned greatly reduces the costs (equipment cost and data acquisition cost) and improves the efficiency, which we term UAV-based LiDAR in this paper. It has been widely used in topographic surveys [29], power line diagnoses [30], dam deformation monitoring [31], vegetation height measurements [32,33], and so on. However, few scholars have applied UAV-based LiDAR to detecting the subsidence of the working face of a mining area. Therefore, we took a working face as our study area, which has the characteristics of active, rapid, and high-intensity surface subsidence, to explore the utility of UAV-based LiDAR in the deformation detection of the working face of a coal mining area.

The purpose of this study was to determine the potential of UAV-based LiDAR in the deformation detection of the working face of a coal mining area. Two field measurement campaigns using UAV-based LiDAR were performed to collect data during 7 November 2020 and 19 May 2021. The main contributions of this paper are as follows: (1) we proposed and made a DSuM with multiple periods of DEM difference calculation for deformation detection based on UAV-based LiDAR. (2) Data mining was conducted based on DSuM to obtain parameters such as the maximum surface subsidence value, a subsidence contour map, the subsidence range, the subsidence area, and the subsidence boundary angle. (3) The accuracy of DSuM was analyzed through comprehensive comparisons with GCPs. The results showed that the proposed method can achieve centimeter-level accuracy.

The remainder of this paper is organized as follows. Following this introduction, Section 2 describes the study area, reference data, and data processing method. Then, Section 3 describes the experimental results. Section 4 shows the discussion of the experimental results. Finally, the conclusions and future research directions are presented in Section 5.

2. Materials and Methods

2.1. Study Area

2.1.1. Physical Geography and Environment

Liangshuijing Coal Mine belongs to Yushen Coal Field, located in northwest Shaanxi, China, as shown in Figure 1. Its geographical location is 38°47′29″–38°53′24″ north, 110°14′22″–110°21′24″ east, and the coal mine covers an area of 68.91 km². Its altitude is from 1100 to 1326 m above sea level. The terrain is generally high in the west and low in the east.

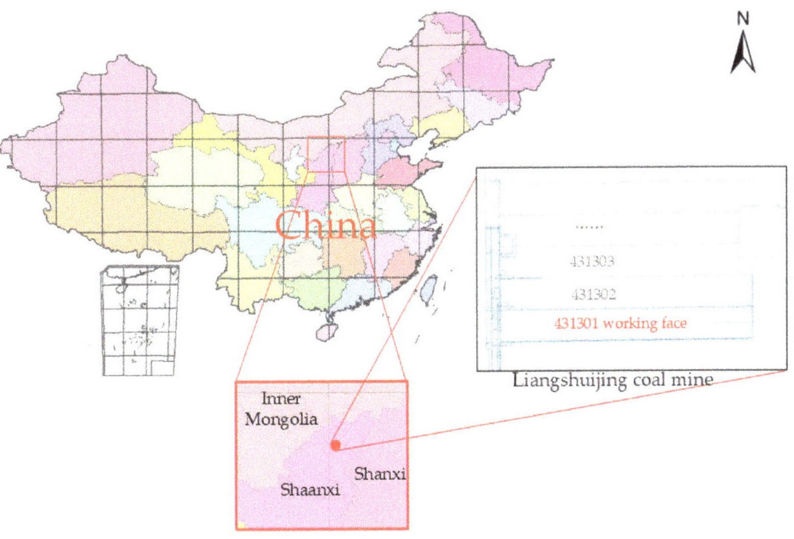

Figure 1. The location of the study area and the working face.

Yushen Coal Field is located in the transition zone of the Muus Desert and the Loess Plateau. The eastern part consists of a loess ridge and a valley; the western part consists of wavy dunes. The study area has a mid-temperate, continental, semi-arid climate. There are regular droughts. There is little rain (mean annual precipitation of 435.7 mm; the annual average evaporation is 1774.1 mm, which is 4–5 times the precipitation), sparse vegetation, and serious soil erosion. The ecological environment is very fragile [34]. Thus, it is of great significance to monitor the deformation of the coal mining areas. In this study, a local part of a working face was selected as the study area. UAV-based LiDAR was used to obtain the surface deformation data, and checkpoints were obtained by leveling.

2.1.2. Mining and Geological Conditions

We took the 431,301 working face of Liangshuijing Coal Mine as the study area. The working face, which mined 3–4 coal seams, is stable and has flat seams. The coal seam thickness is 1.17–1.43 m, the average thickness is 1.3 m, and the average depth of coal seam is 138 m. The working face adopts the longwall, fully mechanized, and full-seam mining method, the roof of which is managed with the all fall method. The ground is covered by loose sandy soil with a strong flow characteristic. The entire study area is 858 m long from east to west and 466 m long from north to south, giving an area of 399,431 m^2. The location of the study area and the working face is shown in Figure 1.

2.2. Reference Data

Our data include airborne laser scanning data and geodetic data. The geodetic data were obtained by laying observation piles on the ground. Then, static and dynamic GNSS [35], total station observation, and precision leveling were used to obtain geodetic data. In the study, the checkpoints include plane (RTK observation) and elevation (leveling observation) data.

2.2.1. LiDAR Data

UAV-based LiDAR was used to collect initial data, and the DSuM was generated by the data for detecting the deformation. The endurance time of UAV in each flight was 35 min, and the effective working time was about 20 min. The laser scanner was a RIEGL miniVUX-1UAV with scanning range of 250 m. The same aerial survey parameters were used to ensure each data acquisition would have the same system error. The parameters included flying height, flying speed, laser scanning speed, pulse emission frequency, and weather conditions. The flying height was 50 m, and the flying speed was 8 m/s. The laser scanning speed was 100 lines/s, and the pulse emission frequency was 100 kHz. Each laser beam had five echoes, and data collection was performed under the same conditions on different dates. Two groups of point clouds were obtained by UAV-based LiDAR.

The diagram of the UAV-based LiDAR data collection and the mining process of the underground working face is shown in Figure 2. The nth LiDAR data collection was carried out to obtain the initial shape of the surface. After a period of time, the coal seams were mined, which caused surface deformation. The m-th LiDAR data collection was carried out to obtain the current surface morphology. With the calculation of two phases of LiDAR data, the surface shape change caused by underground working face mining was obtained, which is called the subsidence basin.

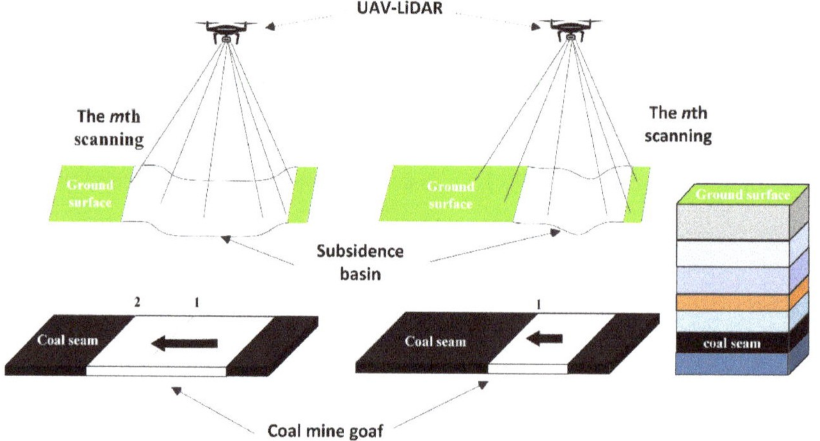

Figure 2. The diagram of ground deformation monitoring by UAV-based LiDAR.

Figure 2 shows the process of mining subsidence. The layered structure on the right of Figure 2 represents the geological structure, which includes the ground surface, sandy soil layer, rock layer, coal seam, and other strata. The main part of Figure 2 shows the relationship between coal seam mining and surface deformation. The surface deformation was different at different mining locations. The black arrow indicates the mining direction of the coal seam. Numbers 1 and 2 indicate the locations of the shearer on different dates. The continuous mining produced a goaf in the coal seam, caused surface deformation, and formed a subsidence basin on the surface.

The data collection dates were 7 November 2020 and 19 May 2021, and the corresponding coal seam mining locations are shown in Figure 3. The original point cloud statistics are shown in Table 1.

Table 1. Statistics of the point cloud in the study area.

Date	Study Area (m^2)	Number of Points	Point Cloud Density (per/m^2)	Number of Ground Points	Ground Point Cloud Density (per/m^2)
7 November 2020	399,431	35,590,816	89	33,227,279	83
19 May 2021	399,431	26,361,111	66	21,776,124	55

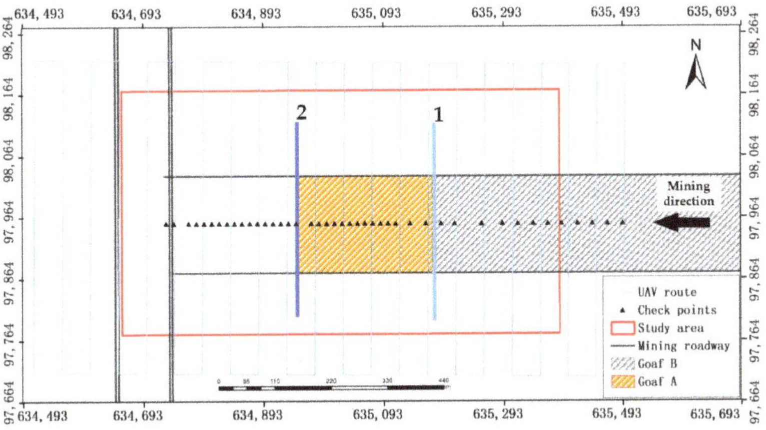

Figure 3. The map of the ground and underground facilities.

Figure 3 shows the 2D relations between the ground facilities and underground facilities, but they had different elevations. The ground facilities included the UAV route, checkpoints, and study area. The UAV route indicated by the light blue line shows the data acquisition range. The black triangles indicate the locations of the ground checkpoints. The underground facilities included a mining roadway and goaf. Mining roadways are indicated by black lines, which were used for transportation and ventilation. Goaf B represents the existing goaf at the time of the first observation. Goaf A represents the area mined during the period between the first and second observations. The light blue line (number 1) and dark blue line (number 2) in Figure 3 represent the positions of the shearer on 7 November 2020 and 19 May 2021, and they have the same meaning as the numbers in Figure 2.

2.2.2. Ground Checkpoints

The fixed observation piles were used as the objects to obtain the partial subsidence value of the ground surface, because there were few fixed markers in the study area. The observation piles were made by precasting concrete, whose structure is shown in Figure 4b, and the locations of the observation piles are shown as black triangles in Figure 3; they are also called ground checkpoints. The processing of GCPs included the design of ground checkpoints' locations, observation pile burying, geodetic surveying, measurement calculations, and subsidence polyline drawing, as shown on the right-hand side of Figure 5. In order to verify the measurement accuracy of the point cloud, DEM, and DSuM, the observation piles were measured on 7 November 2020 and 19 May 2021, the same dates as the UAV-based LiDAR scanning. The results of GCPs were also used to calculate the max subsidence value and boundary angle.

2.3. Data Processing

When the shearer mined different positions on the working face, the two periods of UAV-based LiDAR surveying were carried out to obtain the surface morphology of the working face in different periods, as shown in Figures 2 and 3. The GCPs were mainly used to verify the accuracy of the point cloud, DEM, and DSuM, which were all generated directly or indirectly based on LiDAR. Similarly to previous related studies [7,17,35–37], we used geodetic data as a reference to analyze the accuracy of the point cloud, DEM, and DSuM. The multi-scale morphological algorithm [38] and progressive triangulation filter algorithm [39,40] were chosen for filtering the point cloud, and the Kriging interpolation algorithm [41] was chosen for generating DEM. This paper also analyzes the influence of resolution on DEM accuracy and proposes a new model termed DSuM based on the optimal resolution of DEM to detect the deformation of coal mining areas. The main purposes of coal mining subsidence monitoring are subsidence value acquisition and boundary angle calculation. This section introduces the calculation methods for the subsidence value and boundary angle, and the process of generating DSuM from the UAV-based LiDAR data.

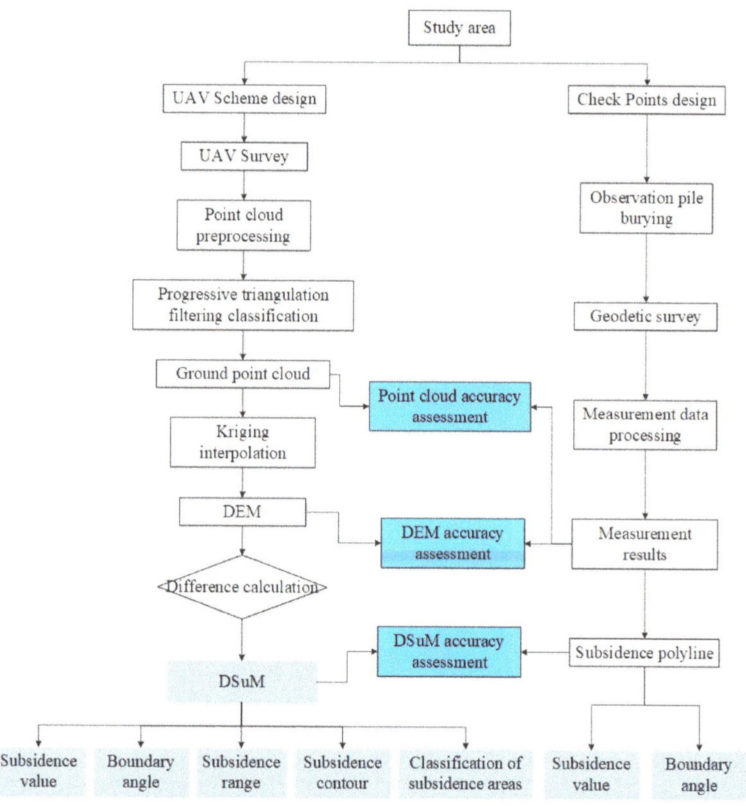

Figure 4. (**a**) Schematic diagram of boundary angle calculation, where $\delta 1$ and $\delta 2$ represent the boundary angle with threshold values of 10 and 100 mm; (**b**) sectional view of the ground observation pile.

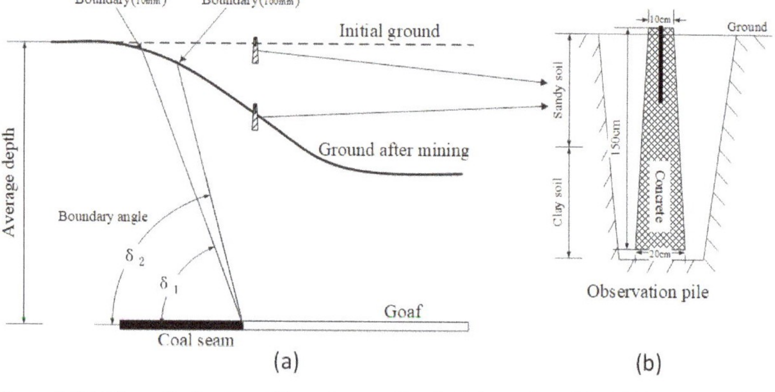

Figure 5. Flow diagram of the proposed method.

2.3.1. Subsidence Value and Boundary Angle

Subsidence data analysis includes data preprocessing, data calculation, subsidence polyline drawing, and boundary angle calculation. The data preprocessing is used to remove error data caused by human factors; data calculation is to obtain the difference between the two periods' data. Then, subsidence polyline diagram is drawn, and boundary angles are obtained. The subsidence of the observation points is calculated by Formula (1):

$$\omega_n = H_n^0 - H_n^m \tag{1}$$

where ω_n represents the subsidence value of point n; H_n^0 and H_n^m represent the elevation of point n during the first and the m-th observations.

The boundary angle is used to determine the range and boundary of the subsidence caused by coal mining. Due to the observation error and the seasonal variations of the soil, an observation point with a subsidence value close to zero could not be accurately determined, so points with subsidence values of 10 and 100 mm were taken as the boundaries of the subsidence basin, as shown in Figure 4. Boundary angles calculated by different boundary thresholds are different. The boundary angle is calculated as follows:

$$\delta_0 = arctan\frac{H_0}{L_1}, \tag{2}$$

where L_1 represents the horizontal distance between the boundary of the underground goaf and the strike or incline boundary of the subsidence basin, H_0 represents the average depth of the coal seam, and δ_0 represents the boundary angle.

2.3.2. UAV-Based LiDAR Data Processing

The data processing of UAV-based LiDAR mainly included data checking, point cloud processing, and interpolation calculation. Data checking was to remove erroneous data; point cloud processing included registration, denoising, and filtering; the purpose of interpolation calculation was to generate the DEM. After data checking, the produced raw data were converted to point clouds. The point cloud was roughly registered by pose data generated by GNSS/INS integration, and then fine registration was carried out based on the iterative closest point (ICP) algorithm. Subsequently, we removed the noise points with the morphological method [42]. Then, the PTFC algorithm was used to obtain the ground point cloud, and the DEM was generated from the ground point cloud. Finally, an accurate DSuM was obtained through multiple periods of DEM difference calculation.

We compared the GCPs and the point cloud data at the same location to verify the point cloud accuracy, which is a direct accuracy verification method. Subsequently, we evaluated the accuracy of the DEM by comparing it with GCPs. The Kriging interpolation algorithm was used to generate the DEM, and the most suitable grid size was determined by trial and error.

There are two methods to evaluate the accuracy of a DEM. The elevation error statistical analysis method compares the DEM with the reference DEM or checkpoints; the logical analysis method is a qualitative method for overall accuracy evaluation, including a visual interpretation method, contour analysis, visual analysis, and other methods [41]. We adopted the method of comparing DEM to checkpoints. The differences in elevation between the checkpoints and the DEM with different resolutions were calculated, and the results were displayed by box plots. Meanwhile, the mean error (*ME*), the mean absolute error (*MAE*), and the root mean square error (*RMSE*), were calculated [19]:

$$ME = \frac{1}{n}\sum_{m=1}^{n}(R_m - Z_m) \tag{3}$$

$$MAE = \frac{1}{n}\sum_{m=1}^{n}(|R_m - Z_m|) \tag{4}$$

$$RMSE = \sqrt{\frac{1}{n} \sum_{m=1}^{n} (R_m - Z_m)^2} \tag{5}$$

where R_m represents the value of the DEM and Z_m represents the value of a checkpoint.

Since directly calculating the difference between two point clouds is difficult, we used the difference between any two DEMs created by the point cloud to represent the ground surface subsidence deformation during the period. The difference between the two DEMs is called the DSuM. Finally, the accuracy of the DSuM was analyzed by GCPs. The analysis of a point cloud, DEM, and DSuM is shown in Figure 5.

2.4. Pipeline of DSuM

The difference between DEMs obtained on any two different dates can be calculated to obtain the ground surface elevation change during this period, in our case, the ground surface subsidence value caused by underground coal mining. In this paper, the difference between the DEMs was calculated, and represented by DSuM, which is a digital model that represents the value of ground subsidence in an ordered array of values. Each pixel value of the DSuM represents the subsidence value of the pixel location caused by coal mining. The schematic diagram of DSuM generation is shown in Figure 6, which is a digital expression of whole ground surface subsidence value. A DSuM can be calculated from any two DEMs obtained on different dates, and can represent the subsidence value of any position in the ground surface during the mining processing. Finally, data mining was conducted based on the DSuM to obtain outputs such as the maximum surface subsidence value, a subsidence contour map, the subsidence range, the subsidence area, and the subsidence boundary angle, as shown in Figure 5.

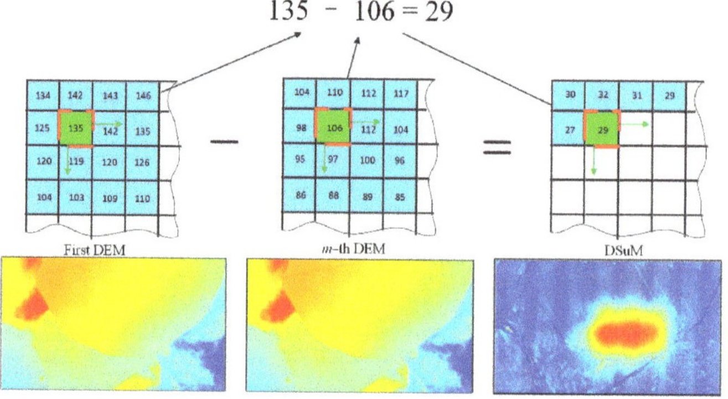

Figure 6. The schematic diagram of DSuM calculation.

After point cloud data collection, the DEMs were generated from clean ground point clouds, and the DSuM was obtained through multiple periods of DEM difference calculation. Then, coal mining subsidence was analyzed based on the DSuM. We adopted point, line, and area analysis from different perspectives. The point analysis used the method of simulating traditional monitoring to extract subsidence values from points, which involves drawing the strike and incline subsidence polyline and obtaining the max subsidence value. The line analysis easily extracted high-density points from the DSuM, which was used to extract points in the strike and incline direction in this paper. It is robust for discrete monitoring points, and it can plot the subsidence curve graph and calculate the strike and incline boundary angle. The area analysis can analyze the whole subsidence area. It was used to calculate subsidence area, maximum subsidence value, and the ratio of subsidence area.

Finally, data mining was carried out based on the DSuM to obtain the maximum subsidence value, subsidence area, subsidence distribution of surface, boundary angle, and other parameters for later analysis.

3. Experimental Results

3.1. GCPs Analysis

The GCPs results were obtained by Formula (1) based on the geodetic data. The results indicate that the maximum ground subsidence value of the working face was 1826 mm, and the minimum subsidence value was 0 mm, during 7 November 2020 and 19 May 2021, respectively. Taking ground observation point A176 as the origin, the subsidence values of all GCPs are shown as blue squares in Figure 7.

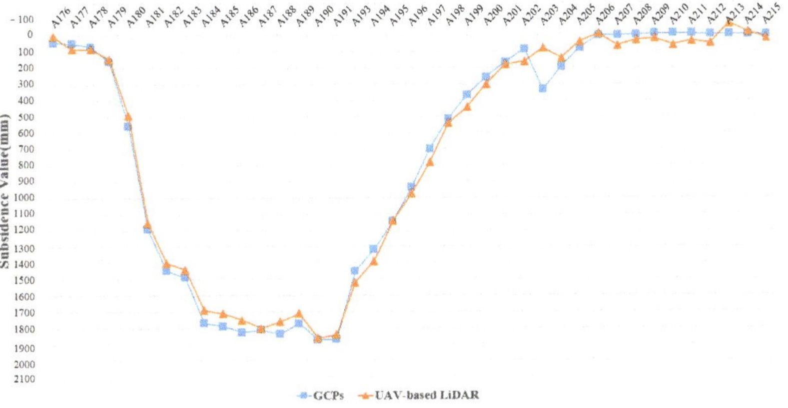

Figure 7. The comparison of subsidence values between the GCPs and DSuM.

The GCPs covered the entire subsidence area, including the non-subsidence area and the maximum subsidence area, and can reflect the surface subsidence in a period of time. The blue polyline in Figure 7 shows that the subsidence boundary with the threshold of 10 mm located between A205 and A206, and the horizontal distances of A205 and A206 from the boundary of goaf were 121 and 136 m, respectively. The average mining depth of the coal seam was 138 m. According to Formula (2), the mining subsidence boundary angle with subsidence threshold of 10 mm was between 45° and 48.7°. The average was 46.9°. Similarly, the boundary angle with a subsidence threshold of 100 mm was 61.2°.

3.2. Accuracy Assessment

The ground point cloud could be obtained after filtering the original point cloud. Since a ground point cloud is composed of a series of discrete points, we took the average elevation values of points near the GCP to calculate the differences between ground point cloud and GCPs, and the distances from the selected points to the GCPs had to be less than a set threshold. The results indicate that the RMSE of elevation was 60.6 mm on 7 November 2020, and 59.9 mm on 19 May 2021, and the results are shown in Table 2.

Table 2. The statistics of the absolute values of point cloud error.

Date	Max (mm)	Min (mm)	Ave (mm)	Med (mm)	RMSE (mm)
7 November 2020	130.0	1.0	50.0	48.5	60.6
19 May 2021	113.0	4.0	51.5	47.5	59.9

Next, we calculated the accuracy of the DEM and analyzed the influence of resolution on DEM accuracy. The differences between DEM and GCPs are shown in Figure 8. As resolution increases, the average error remains basically unchanged, but the error distribution becomes more discrete. The error distribution of the resolution of 0.1 m was most concentrated, and all errors were less than 100 mm.

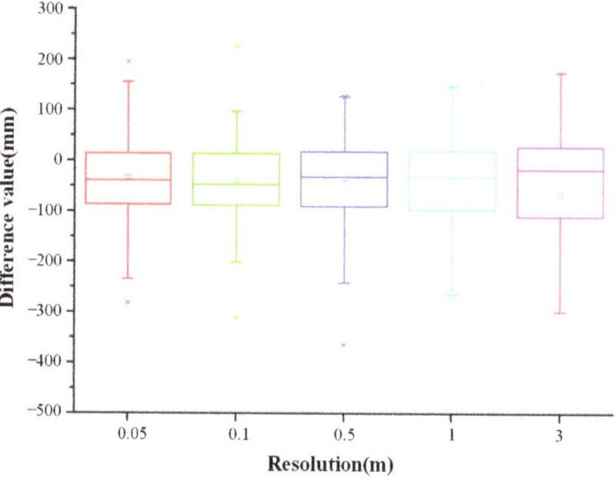

Figure 8. The difference distribution statistics of DEM and the checkpoints.

The difference results of DEM from data at different resolutions and GCPs are shown in Table 3. The results include ME, MAE, and RMSE, and indicate that with different resolutions, the maximum ME is 323 mm and the minimum is –34 mm; the maximum MAE is 537 mm and the minimum is 74 mm; and the maximum RMSE is 97 mm.

Table 3. The statistics of DEM error.

Grid Size (m)	ME (mm)	MAE (mm)	RMSE (mm)
0.05	−34	74	97
0.1	−39	79	106
0.5	−43	77	109
1	−41	76	102
3	−67	106	167
5	−89	157	248
10	−4	165	238
20	323	537	926

The ME did not change, and the optimal state was in the range of 0.05–1 m, but it increased when the grid size was larger than 1 m. The change in MAE was similar to that of the ME. Similarly, the RMSE reached an equilibrium state within 0.05–1 m. When the resolution was greater than 1 m, the RMSE gradually increased as the grid size increased, and it increased sharply when the resolution became greater than 1 m.

According to Figure 8, the data with 0.1 m grid size are the most concentrated. Therefore, we selected 0.1 m as the parameter for conducting the study to ensure research accuracy.

Finally, the accuracy of the DSuM was analyzed. The DSuM of the study area was obtained by the subtraction of DEM in 7 November 2020 from DEM in 19 May 2021, as shown in Figure 9. Different colors represent different subsidence values in the DSuM. Additionally, the main area of subsidence caused by coal mining is mainly distributed in the goaf and its surroundings.

The DSuM contains plane coordinate (X, Y) and subsidence value. The subsidence value of any position can be acquired. The subsidence monitoring accuracy of UAV-based LiDAR can be acquired by comparing the subsidence value of the DSuM with those of the GCPs. The results are shown in Figure 10.

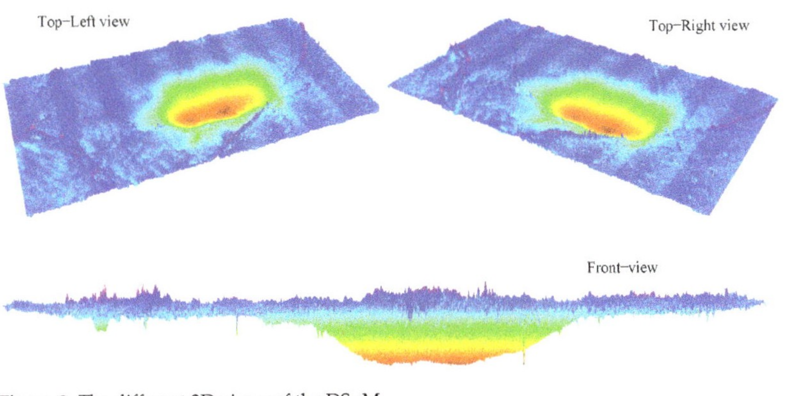

Figure 9. The different 3D views of the DSuM.

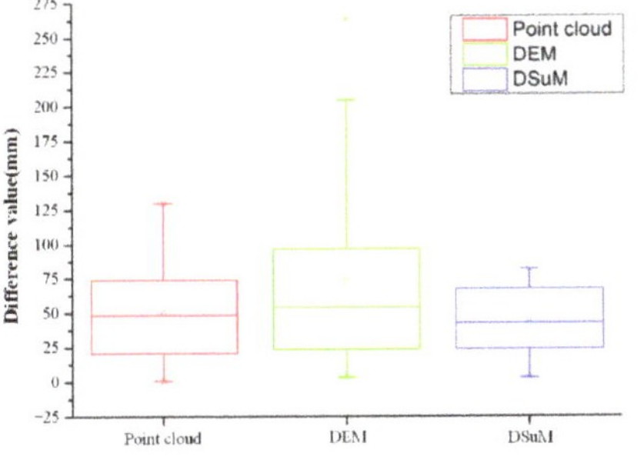

Figure 10. The absolute values of differences.

Figure 7 shows the maximum subsidence value and minimum subsidence value. The subsidence polyline trends obtained by DSuM and GCPs were basically consistent. The maximum subsidence value calculated by DSuM was 1860 mm, and the maximum subsidence value of GCPs was 1872 mm. The inflection points of the subsidence polyline were consistent. After removing the abnormal subsidence value of ground checkpoint (A 203), the maximum and minimum of the subsidence difference's absolute values were 82 and 3 mm, the average difference was 45 mm, and 75% of the difference was within 50 mm. All of the differences were less than 100 mm, and their distribution was relatively uniform. The absolute values of differences among the point cloud, DEM, and DSM are shown in Figure 10.

3.3. Analysis of DSuM

In order to comprehensively verify the accuracy of the DSuM, we conducted point, line, and area analysis based on the DSuM.

3.3.1. Point Analysis

The traditional method of coal mining subsidence monitoring is geodetic surveying by a fixed observation pile, which is a point survey method. For the point analysis of DSuM, we used the method of simulating traditional monitoring to extract subsidence values point by point. The locations of monitoring points were set according to requirements of relevant specifications. The interval between adjacent points was 15 m, the total number was 50, and the whole surface subsidence change area was covered, as shown in Figure 11, by black triangles.

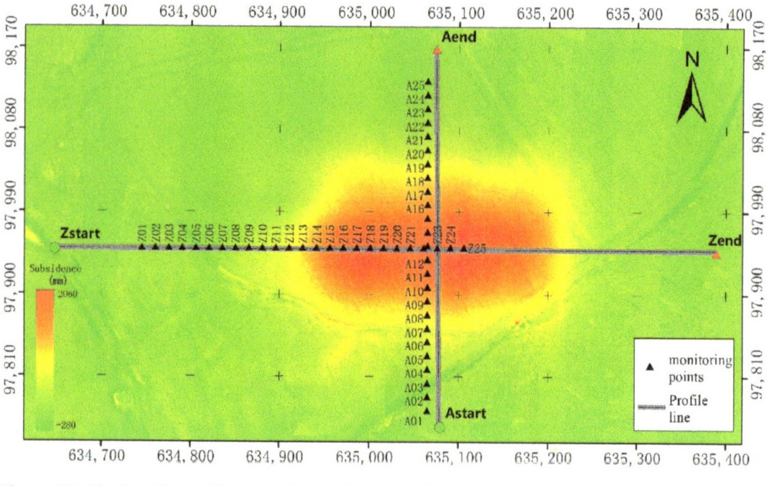

Figure 11. The locations of monitoring points, and the strike and incline observation lines, where Z represents the strike observation line and A represents incline observation line.

The subsidence value of each monitoring point was extracted by DSuM and is shown in the form of a polyline graph in Figure 12.

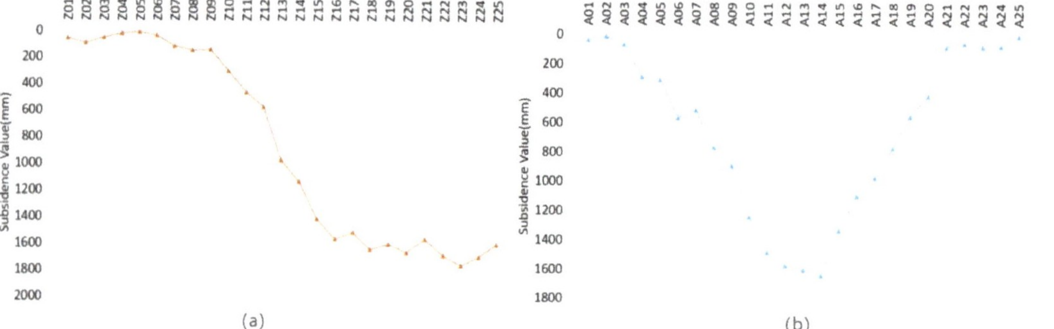

Figure 12. The subsidence values of monitoring points. (**a**) The strike observation polyline; (**b**) the incline observation polyline.

The results indicate that the average maximum value of the subsidence basin is approximately 1650 mm, and the maximum single point subsidence is 1761 mm. The shape of the strike and incline polyline graph is the same as that of the standard mining subsidence polyline graph, whose shape is like a basin or a half-basin. The results can reflect the elevation deformation of the surface caused by coal mining, and meet the requirements of mining subsidence monitoring.

3.3.2. Line Analysis

Line analysis is robust for discrete monitoring points. Therefore, it is necessary to obtain high-density points on the observation line. We extracted high-density points from DSuM in the strike and incline directions and plotted the subsidence curve graph in Figure 13. The locations of strike and cline lines are shown in Figure 11 in gray. The strike observation line started from the point named Zstart and ended at the point named Zend, with a total length of 740 m. The incline observation line started with the point named Astrat and ended with the point named Aend, with a total length of 413 m.

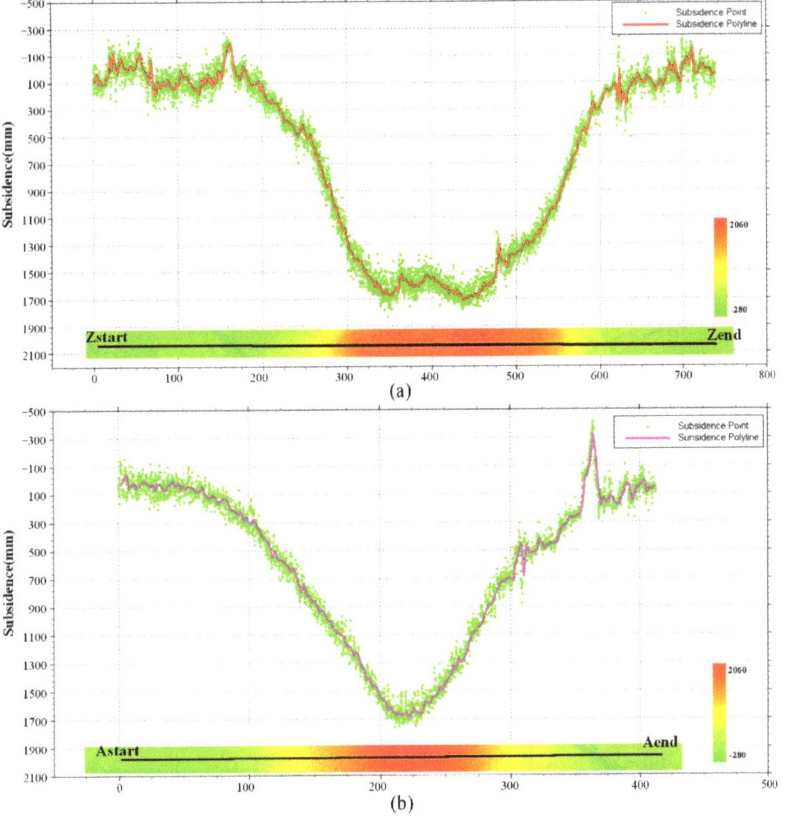

Figure 13. The subsidence of observation curves. (**a**) The strike observation curve; (**b**) the incline observation curve.

We extracted monitoring points from DSuM in an interval of 0.1 m, and drew the points on graphs, as shown in Figure 13a,b—the green points. Finally, we extracted and drew the strike and incline curves, as shown in Figure 13 as the red curves.

The strike curve is shaped like a basin, conforming to the characteristics of mining subsidence, and it shows that a super-full mining state in the strike direction was reached. The incline observation line is in a valley state and is symmetrically distributed due to the small width of the working face. According to the strike and incline curves, we found that the maximum subsidence value is about 1700 mm. Since the average accuracy of UAV-based LiDAR data cannot reach 10 mm, 100 mm was used as the subsidence boundary threshold to extract and calculate the boundary angle by Formula (2). In the strike direction, the horizontal distance between the subsidence boundary and the goaf boundary is 68 m; in the incline direction, the horizontal distance is 60 m, and the average coal mining depth is 138 m. Therefore, when the boundary subsidence threshold is 100 mm, the strike and cline boundary angle are 63.8° and 66.5°.

3.3.3. Area Analysis

In order to accurately express the subsidence values of different areas and the subsidence area with different subsidence values, it was necessary to perform area statistical analysis based on the DSuM. Therefore, we classified and counted the DSuM at the interval of 100 mm. The results are shown in Figure 14.

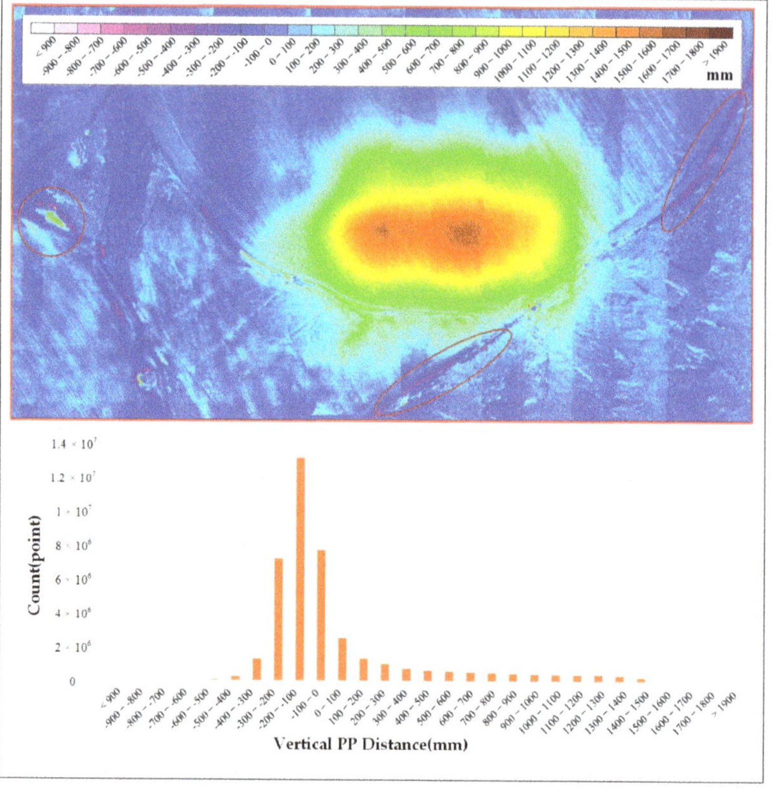

Figure 14. The subsidence classification and statistics map.

The upper half of Figure 14 is the subsidence classification map of the study area. It shows the locations of different subsidence values, and the different colors represent different subsidence values. The lower part of Figure 14 is a histogram which represents the number of pixels for each subsidence value. According to the classification map, there are some areas with negative subsidence values, especially the red circled area, which are caused by man—excavation, road construction, etc.

Figure 14 shows that the subsidence area is distributed in an oval shape. The main subsidence area is located at the center of the study area. Due to the influence of natural rainfall, there is a large area with subsidence values less than 100 m. In particular, there are very few areas where the subsidence value is less than -100 mm, as shown in the red circle in Figure 14. In addition, according to the histogram in Figure 14, we found that the larger the subsidence value, the smaller the number of the corresponding grids. Next, we counted the number of grids with different subsidence values, and calculated the areas of different subsidence values. The results are shown in Table 4.

Table 4 indicates that areas with a subsidence value of less than 100 mm occupied more than 55% of the study area, in which most of the subsidence values are caused by external factors. In order to determine the subsidence area caused by coal mining, we took 100 mm as the minimum subsidence threshold caused by coal mining. The area of subsidence value larger than 100 mm is about 180,075 m^2. The goaf area is 49,789 m^2. The ratio is 3.6:1. According to Table 4, the area with a subsidence value of 100–300 mm is 102,585 m^2, which means the majority of subsidence was caused by coal mining. The maximum subsidence value is about 1700 mm. We found that the larger the subsidence value, the smaller the corresponding area, and the area ratios of different subsidence values are shown in Table 4.

Table 4. Statistics of the areas with different subsidence values.

Subsidence Value (mm)	Number of Pixels	Resolution (m)	Area (m²)	Area Ratio (%)	Subsidence Area Ratio (%)
<100	22,062,688	0.1	220,627	55.1	-
100–300	10,258,459	0.1	102,585	25.6	57.0
300–600	3,033,611	0.1	30,336	7.6	16.8
600–900	1,710,181	0.1	17,102	4.3	9.5
900–1200	1,341,367	0.1	13,414	3.3	7.4
1200–1500	1,116,807	0.1	11,168	2.8	6.2
1500–1700	516,812	0.1	5168	1.3	2.9
>1700	30,187	0.1	302	0.1	0.2

The isoline map is a classical graphical representation, such as a contour map, which can be used to represent elevation of terrain. In this study, we took the isoline map to express the subsidence value, named the subsidence isoline map, which can clearly show the surface subsidence values in different locations. The subsidence isoline map shows different forms when setting different interval values, as shown in Figure 15.

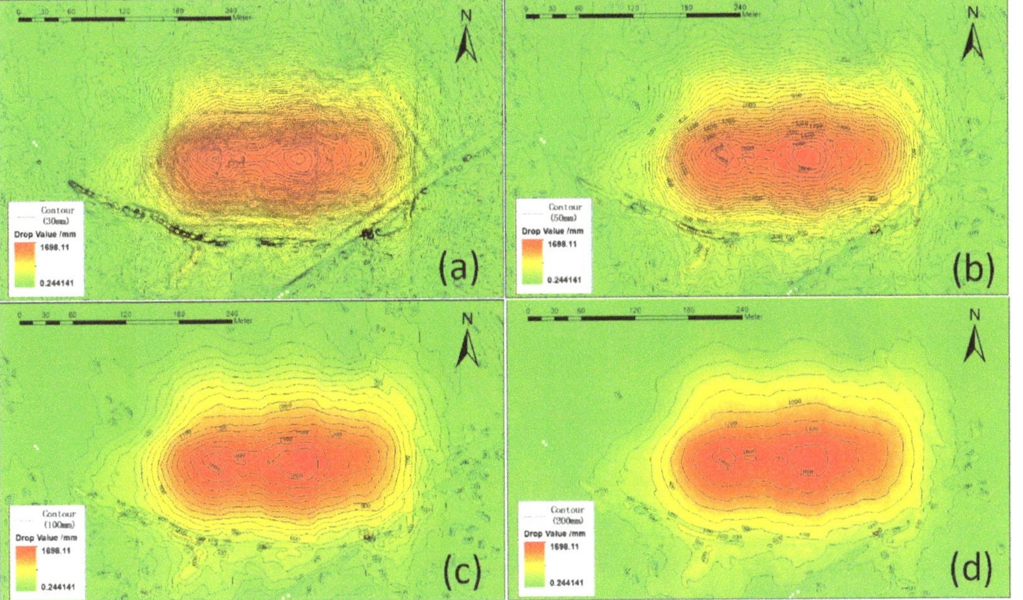

Figure 15. The subsidence isoline map with different interval values: ((**a**): 30 mm, (**b**): 50 mm, (**c**): 100 mm, and (**d**): 200 mm).

It can be seen in Figure 15 that the detailed characteristics of the subsidence isoline map with different interval values are different. The smaller isoline interval values correspond to more detailed subsidence characteristics, but also cause the noise to be more pronounced. Through a comprehensive comparison, we found that the isoline map with a subsidence interval value of 100 mm can retain the details of subsidence characteristics of coal mining and reduce the impact of noise.

After analysis of the subsidence isoline map with an interval value of 100 mm, we took the subsidence value of 100 mm as the minimum subsidence threshold and extracted the subsidence boundary. The result is shown in Figure 16.

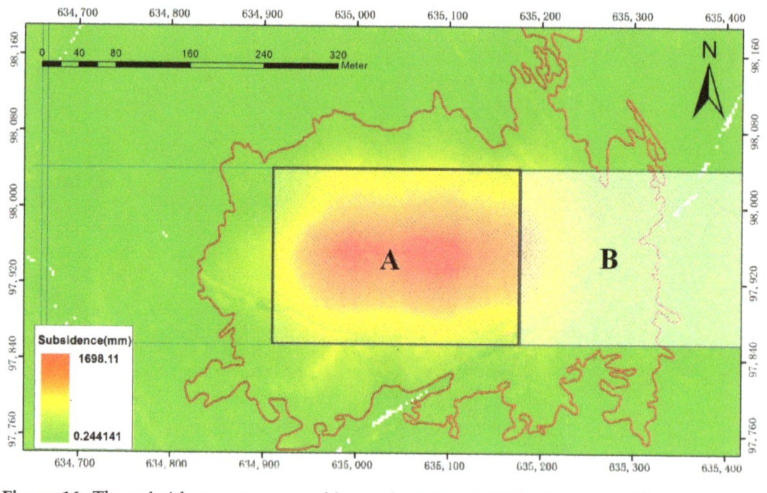

Figure 16. The subsidence area caused by coal mining. (**A**) Goaf generated between 7 November 2020 and 19 May 2021; (**B**) goaf generated before 2020.11.

Figure 16 indicates that the area of subsidence is larger than the goaf area, and the subsidence area is offset to the right relative to the goaf. The subsidence area was 140,271 m^2 during 7 November 2020 and 19 May 2021, which is smaller than the 180,075 m^2 calculated in Table 4, and the ratio of subsidence area to goaf was 2.8:1.

A subsidence value larger than 100 mm is considered to be caused by coal mining. We divided the subsidence values into six grades, and the corresponding area of each grade was calculated, and then expressed in a specific color, as shown in Figure 17.

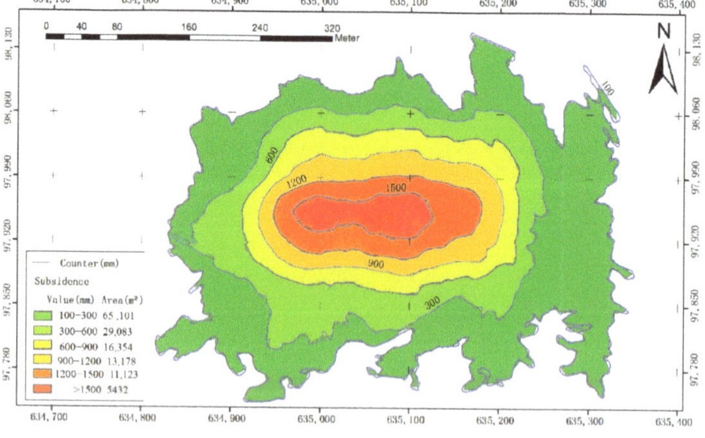

Figure 17. The different subsidence grades.

As shown in Figure 17, the areas with different subsidence grades gradually expanded outward with the goaf from the center. The subsidence can be divided into six grades: 100–300 mm, 300–600 mm, 600–900 mm, 900–1200 mm, 1200–1500 mm, and more than 1500 mm. The corresponding subsidence areas are 65,101, 29,083, 16,354, 13,178, 11,123, and 5432 m^2. The ratio among the subsidence grades is 46:21:12:9:8:4.

4. Discussion

The accuracy of the DSuM was determined by point cloud and DEM. Therefore, we analyzed the accuracy of the point cloud and DEM, and the results are shown in Tables 2 and 3. We also analyzed the influence of interpolation grid size (resolution of DEM) on the accuracy of the DEM. The results indicate that the grid size of 0.1 m is suitable for the study area (Table 3, Figure 8). Compared to the results of Yu H. et al. [26], the average error of DEM generated by UAV-based LiDAR increased from 0.32 to 0.039 m. Although UAV oblique photogrammetry can achieve centimeter-level accuracy by arranging high-density ground control points [43], it is not applicable in coal mining subsidence monitoring due to the difficulty of setting ground control points in the coal mining area. The proposed method can achieve centimeter accuracy without ground control points. This paper analyzed and compared the accuracy of point cloud, DEM, and DSuM. The results indicate that the accuracy of the DSuM is better than the accuracy of point clouds and DEMs, as shown in Tables 2 and 3 and Figures 7 and 10. The reason is that the DSuM generated by two periods of UAV-based LiDAR data adopts the same flight parameters, and the data processing process is exactly the same. Therefore, it can eliminate some systematic errors and improve the accuracy.

We calculated the strike and incline boundary angles with the threshold of 100 mm, which are 63.8° and 66.5°, respectively. The boundary angle value is similar to the strike boundary angle of 61.2° calculated by GCPs. The result indicates that line analysis could be applied to the calculation of boundary angle with the subsidence threshold of 100 mm, but it is difficult to detect a boundary of 10 mm. In the area analysis, there was a series of subsidence values less than zero. The reasons for this may be external factors, such as rain wash and crop farming. Therefore, to further determine the subsidence area caused by coal mining, we set the subsidence value of 100 mm as the maximum threshold value caused by non-coal mining to eliminate the influences of other factors. We calculated and drew the subsidence isoline map with four different interval values. The result indicates that the subsidence isoline map with an interval of 100 mm is suitable for subsidence monitoring; see Figure 15. It can perfectly express the characteristics of coal mining subsidence. We calculated the area of subsidence area according to the entire study area and the isoline map (Figures 14 and 16 and Table 4). The results show that the area calculated by the isoline map is smaller than the area calculated according to the entire study area, the reason for which is that the area calculated by the entire study area includes some subsidence not caused by coal mining. The final subsidence area caused by coal mining should be the result calculated by the isoline map with interval value of 100 mm.

5. Conclusions and Future Works

Subsidence detection is important work for coal mining safety. There are various methods used to detect the deformation of mining areas. However, the accuracy is difficult to guarantee in mountainous areas. In this study, UAV-based LiDAR was used to monitor the ground surface subsidence of the working face of a coal mining area, and a new model termed DSuM is proposed to detect the subsidence deformation of a coal mining area. The accuracy of the DSuM was verified by GCPs, and the data mining was performed based on the DSuM to obtain the parameters required for coal mining subsidence monitoring. Subsequently, point, line, and area analyses of the DSuM were conducted. The results indicate that UAV-based LiDAR can be used to monitor continuous changes of the working face in a coal mining area, which provides basic information for subsidence prediction and damage recovery in mining areas.

Some issues are still worth investigating. The accuracy of UAV-based LiDAR can reach the centimeter level, which is not suitable for areas with small subsidence values, especially when the subsidence values are less than 100 mm. On the other hand, the ground surface deformation also includes horizontal movement, but the DSuM cannot express horizontal deformation. Therefore, future work will be focused on further improving the accuracy of the DSuM and creating a model that can represent horizontal deformation.

Author Contributions: Conceptualization, funding acquisition, project administration, and methodology, J.Z. and W.Y.; data curation and investigation, B.M. and L.B.; formal analysis and writing—original draft preparation, J.Z.; writing—review and editing and validation supervision, X.L. All authors have read and agreed to the published version of the manuscript.

Funding: This research was funded by the National Natural Science Foundation of China under project 41977059.

Data Availability Statement: Not applicable.

Acknowledgments: We thank Shaanxi 185 Coalfield Geology Co., Ltd. for providing the checkpoint data and the staff at Liangshuijing Coal Mine for allowing us to conduct our studies at the experimental site.

Conflicts of Interest: The authors declare no conflict of interest.

References

1. Bell, F.G.; Stacey, T.R.; Genske, D.D. Mining subsidence and its effect on the environment: Some differing examples. *Environ. Geol.* **2000**, *40*, 135–152. [CrossRef]
2. Lechner, A.M.; Baumgartl, T.; Matthew, P.; Glenn, V. The impact of underground longwall mining on prime agricultural land: A review and research agenda. *Land Degrad. Dev.* **2016**, *27*, 1650–1663. [CrossRef]
3. Xiao, W.; Fu, Y.; Wang, T.; Lv, X. Effects of land use transitions due to underground coal mining on ecosystem services in high groundwater table areas: A case study in the Yanzhou coalfield. *Land Use Policy* **2018**, *71*, 213–221. [CrossRef]
4. Guo, W.; Barbato, J.; Dai, H.; Peng, S.S.; Agioutantis, Z.; Adhikary, D.; Qu, Q.; Wilkins, A.H.; Poulsen, B.A.; Guo, H.; et al. Surface subsidence damage, mitigation and control. In *Surface Subsidence Engineering: Theory and Practice*; Peng, S., Ed.; CSIRO Publishing: Clayton, Australia, 2020; pp. 105–131.
5. Stoch, T. *Horizontal Displacement in Mining Area Protection*; AGH University of Science and Technology Press: Kraków, Poland, 2019; ISBN 978-83-66364-19-6.
6. Shi, Y.; Li, J.; Lv, J.; Ma, D. Monitoring and prediction of mining subsidence combined with SBAS-InSAR and support vector regression. *Remote Sens. Inf.* **2021**, *36*, 6–12.
7. Chen, L.; Zhao, X.S. Progress of large gradient deformation monitoring technology in mining area combined with InSAR. *Surv. Mapp. Bull.* **2018**, *7*, 18–23.
8. Pawluszek-Filipiak, K.; Borkowski, A. Integration of DInSAR and SBAS techniques to determine mining-related deformations using Sentinel-1 data: The case study of Rydutowy mine in Poland. *Remote Sens.* **2020**, *12*, 242. [CrossRef]
9. Przyłucka, M.; Herrera, G.; Graniczny, M.; Colombo, D.; Béjar-Pizarro, M. Combination of conventional and advanced DInSAR to monitor very fast mining subsidence with TerraSAR-X data: Bytom city (Poland). *Remote Sens.* **2015**, *7*, 5300–5328. [CrossRef]
10. Rossi, G.; Tanteri, L.; Tofani, V.; Vannocci, P.; Casagli, N. Multitemporal UAV surveys for landslide mapping and characterization. *Landslides* **2018**, *15*, 1045–1052. [CrossRef]
11. Lindner, G.; Schraml, K.; Mansberger, R.; Hübl, J. UAV monitoring and documentation of a large landslide. *Appl. Geomat.* **2015**, *8*, 1–11. [CrossRef]
12. Ge, L.; Li, X.; Ng, H.M. UAV for mining applications: A case study at an open-cut mine and a longwall mine in New South Wales, Australia. In Proceedings of the 2016 IEEE International Geoscience and Remote Sensing Symposium (IGARSS), Beijing, China, 10–15 July 2016.
13. Tong, X.; Liu, X.; Chen, P.; Liu, S.; Luan, K.; Li, L.; Liu, S.; Liu, X.; Xie, H.; Jin, Y.; et al. Integration of UAV-based photogrammetry and terrestrial laser scanning for the three-dimensional mapping and monitoring of open-pit mine areas. *Remote Sens.* **2015**, *7*, 6635–6662. [CrossRef]
14. Lian, X.; Liu, X.; Ge, L.; Hu, H.; Wu, Y. Time-series unmanned aerial vehicle photogrammetry monitoring method without ground control points to measure mining subsidence. *J. Appl. Remote Sens.* **2021**, *15*, 024505. [CrossRef]
15. Ignjatovi Stupar, D.; Roer, J.; Vuli, M. Investigation of unmanned aerial vehicles-based photogrammetry for large mine subsidence monitoring. *Minerals* **2020**, *10*, 196. [CrossRef]
16. Chen, P. Research on Mining Subsidence Monitoring Method of UAV Tilt Photogrammetry. Master's Thesis, Taiyuan University of Technology, Taiyuan, China, 2018.
17. Wikaa, P.; Gruszczyński, W.; Stoch, T.; Puniach, E.; Wójcik, A. UAV Applications for determination of land deformations caused by underground mining. *Remote Sens.* **2020**, *12*, 1733. [CrossRef]
18. Zhou, D.; Qi, L.; Zhang, D.; Zhou, B.; Guo, L. Unmanned aerial vehicle (UAV) photogrammetry technology for dynamic mining subsidence monitoring and parameter inversion: A case study in China. *IEEE Access* **2020**, *8*, 16372–16386. [CrossRef]
19. Rauhala, A.; Tuomela, A.; Davids, C.; Rossi, P.M. UAV remote sensing surveillance of a mine tailings impoundment in sub-arctic conditions. *Remote Sens.* **2017**, *9*, 1318. [CrossRef]
20. Martínez-Carricondo, P.; Mesas-Carrascosa, F.J.; García-Ferrer, A.; Agüera-Vega, F.; Pérez-Porras, F.J. Assessment of UAV-photogrammetric mapping accuracy based on variation of ground control points. *Int. J. Appl. Earth Obs. Geoinf.* **2018**, *72*, 1–10. [CrossRef]

21. Johan, K.; Christophe, D.; Pascal, A.; Pierre, P.; Marion, J.; Eric, V. Application of a terrestrial laser scanner (TLS) to the study of the Séchilienne landslide (Isère, France). *Remote Sens.* **2010**, *2*, 2785–2802. [CrossRef]
22. Guo, C.; Li, J.; Feng, H.; Zhao, J. Application of 3D laser scanning technology in dam subsidence monitoring in mining area. *Mine Surv.* **2014**, *6*, 70–72. [CrossRef]
23. Bai, W. Monitoring mining subsidence by three-dimensional laser scanning technology. *Metal Mines* **2017**, *1*, 132–135. [CrossRef]
24. Gu, Y.; Zhou, D.; Zhang, D.; Wu, K.; Zhou, B. Study on subsidence monitoring technology using terrestrial 3D laser scanning without a target in a mining area: An example of Wangjiata coal mine, China. *Bull. Eng. Geol. Environ.* **2020**, *79*, 3575–3583. [CrossRef]
25. Brede, B.; Lau, A.; Bartholomeus, H.M.; Kooistra, L. Comparing RIEGL RiCOPTER UAV LiDAR derived canopy height and DBH with terrestrial LiDAR. *Sensors* **2017**, *17*, 2371. [CrossRef] [PubMed]
26. Yu, H.; Lu, X.; Gang, C.; Ge, X. Detection and volume estimation of mining subsidence based on multi-temporal LiDAR data. In Proceedings of the 19th International Conference on Geoinformatics (IEEE 2011), Shanghai, China, 24–26 June 2011; pp. 1–6. [CrossRef]
27. Yu, H.; Lu, X.; Ge, X.; Cheng, G. Digital terrain model extraction from airborne LiDAR data in complex mining area. In Proceedings of the 18th International Conference on Geoinformatics (IEEE 2010), Beijing, China, 18–20 June 2010.
28. Ao, J.; Wu, K.; Wang, Y.; Li, L. Subsidence monitoring using lidar and morton code indexing. *J. Surv. Eng.* **2016**, *142*, 06015002. [CrossRef]
29. Elsner, P.; Dornbusch, U.; Thomas, I.; Dan, A.; Bovington, J.; Horn, D. Coincident beach surveys using UAS, vehicle mounted and airborne laser scanner: Point cloud inter-comparison and effects of surface type heterogeneity on elevation accuracies. *Remote Sens. Environ.* **2018**, *208*, 15–26. [CrossRef]
30. Siranec, M.; Hger, M.; Otcenasova, A. Advanced power line diagnostics using point cloud data—Possible applications and limits. *Remote Sens.* **2021**, *13*, 1880. [CrossRef]
31. Bakuła, K.; Pilarska, M.; Salach, A.; Kurczyński, Z. Detection of levee damage based on UAS data—Optical imagery and LiDAR point clouds. *Int. J. Geo-Inf.* **2020**, *9*, 248. [CrossRef]
32. Chen, Q.; Hang, M.; Li, J.; Wang, X. Study on extraction method of vegetation restoration height in mining area based on lidar. *Coal Sci. Technol.* **2020**, *48*, 113–119.
33. Streutker, D.R.; Glenn, N.F. LiDAR measurement of sagebrush steppe vegetation heights. *Remote Sens. Environ.* **2006**, *102*, 135–145. [CrossRef]
34. Song, S. Quantitative Evaluation of Ecological Environment Damage Caused by Coal Mining in Yushenfu Mining Area. Master's Thesis, Xi'an University of Science and Technology, Xi'an, China, 2009.
35. Hsieh, Y.C.; Chan, Y.C.; Hu, J.C. Digital elevation model differencing and error estimation from multiple sources: A case study from the Meiyuan Shan landslide in Taiwan. *Remote Sens.* **2016**, *8*, 199. [CrossRef]
36. Chen, Q.; Wang, H.; Zhang, H.; Sun, M.; Liu, X. A point cloud filtering approach to generating DTMs for steep mountainous areas and adjacent residential areas. *Remote Sens.* **2016**, *8*, 71. [CrossRef]
37. Salach, A.; Bakua, K.; Pilarska, M.; Ostrowski, W.; Górski, K.; Kurczyński, Z. Accuracy assessment of point clouds from LiDAR and dense image matching acquired using the UAV platform for DTM creation. *Int. J. Geo-Inf.* **2018**, *7*, 342. [CrossRef]
38. Yang, B.; Huang, R.; Dong, Z.; Li, J. Two-step adaptive extraction method for ground points and breaklines from lidar point clouds. *ISPRS J. Photogramm. Remote Sens.* **2016**, *119*, 373–389. [CrossRef]
39. Axelsson, P. DEM generation from laser scanner data using adaptive TIN models. *Int. Arch. Photogramm. Remote Sens.* **2000**, *33*, 110–117.
40. Jiao, X.H. Research on Airborne LiDAR Point Cloud Filtering Algorithm and DEM Interpolation Method. Master's Thesis, Taiyuan University of Technology, Taiyuan, China, 2018.
41. Kang, S.; Ji, L.; Jiao, Q.; Zhang, J. Comparative study on interpolation methods based on ground lidar point cloud data. *Geod. Geodyn.* **2020**, *4*, 400–404. [CrossRef]
42. Dougherty, E.R. *An Introduction to Morphological Image Processing*; SPIE Optical Engineering Press: Bellingham, WA, USA, 1992.
43. Maciuk, K.; Lewińska, P. High-rate monitoring of satellite clocks using two methods of averaging time. *Remote Sens.* **2019**, *11*, 2754. [CrossRef]

MDPI
St. Alban-Anlage 66
4052 Basel
Switzerland
Tel. +41 61 683 77 34
Fax +41 61 302 89 18
www.mdpi.com

Remote Sensing Editorial Office
E-mail: remotesensing@mdpi.com
www.mdpi.com/journal/remotesensing